FROM WOOL TO CLOTH

The Triumph of the Suffolk Clothier

Nicholas R. Amor

RefineCatch Limited

First published in 2016 by
RefineCatch Limited

Broad Street, Bungay, Suffolk, NR35 1EF
www.refinecatch.com

ISBN: 978-0-9955085-0-7

A catalogue record for this book is available
from the British Library

Typeset by RefineCatch Limited, Bungay, Suffolk
Printed and bound by
PRINTER TO COME

Contents

List of Illustrations

Dedication

To Julia Smith
The best mother in the world

Acknowledgements

Many debts must be acknowledged. First, Mark Bailey provided insightful comments on each chapter and, in doing so, made this an immeasurably better volume than it would otherwise have been. Robert Palmer and his team at the University of Houston were valued but unwitting research assistants because, in addition to those plea rolls that I have personally transcribed from their website, they have created a library of indices of others. Jean Rogers and Jane Gosling of Lavenham Guildhall helped me to understand the practical skills of medieval cloth making. Sue Andrews and Margaret Woods shared with me their unrivalled knowledge of medieval Hadleigh. Liz Wigmore provided me with a copy of her (sadly) unpublished work on Long Melford, while Vivienne Aldous shared with me research on Monks Eleigh carried out by herself and John Weller. Vivienne Aldous, Abby Antrobus and Jo Sear found articles for me and Jo shared data from her own doctoral work. Charles Farrell tracked down elusive alnage accounts at TNA. Joan Lyell of the Colchester and Ipswich Museum Service showed me their collection of alnage matrices and seals. David Addy put his enthusiasm for digital maps to good use by providing me with the hundredal maps for Essex, Norfolk and Suffolk. Philip Pantelis sowed the seeds of chapters 3 to 6 by referring me to the drapers' window in the church of Our Lady at Semur en Auxois, while my son Benjamin accompanied me on a visit and later used his IT expertise to help me with the maps. Christopher Dyer and David Dymond both provided helpful comments on earlier drafts, while my 'mother-in-law' Pat Smith and my friends David and Rona Gottesmann were intelligent lay readers. I was privileged to have had the opportunity to discuss this project with both John Munro and Richard Britnell shortly before their untimely deaths in 2013. I have incorporated helpful comments made by those attending the presentation of my paper at the 2016 Economic History Conference. The staff of the British Library, the National Archives, and the Bury St Edmunds and Ipswich branches of the Suffolk Records Office have, as always, been co-operative and helpful. It need hardly be added that any remaining errors are entirely mine.

Preface

On 29 July 1981 my father and I visited Lavenham and lunched at the Swan Hotel. It was the day of the ill-starred wedding of Charles and Diana. Five years earlier he had persuaded me, for all the right reasons, to abandon my plans to study medieval history at university and to become a lawyer instead. Now my interest was rekindled by this fascinating small town. In time, I would come to have the same interest in Hadleigh, Long Melford, Sudbury and the other cloth towns and villages of south Suffolk. I am delighted that we now live so close to them in one of the most beautiful parts of the county and that, so my neighbours tell me, a tributary of the river Brett flows (sometimes) in front of our home.

In the early 1990s I was fortunate to meet Mark Bailey and David Dymond who taught, and continue to teach me the skills of the medievalist. Another ten years after that I began my work on the Suffolk alnage accounts of the late 1460s, subsequently published in the Proceedings of the Suffolk Institute of Archaeology and History. In 2009, as I was nearing completion of my doctorate, my supervisor Carole Rawcliffe suggested that I look at Robert Palmer's new AALT website where I would find scans of the plea rolls of the court of common pleas. It soon became clear that anyone who was rash enough to devote the necessary time and attention could open up a new window on the cloth industry by identifying hundreds of card-makers, spinners, weavers, fullers, dyers and clothiers. With my PhD behind me, I set to work and this is the result of my study.

I dedicate this volume to my partner Julia Smith. She has devoted two decades of her life to looking after our chronically ill daughter Rosalind. She has woken early, gone to bed late and been both mother and carer. This is a tribute to your love, intelligence, courage and indomitable spirit.

Nick Amor
Stanningfield, Suffolk
1 May 2016

Abbreviations

Ag. Hist. Rev.	Agricultural History Review
Annals	The Annals of Ipswiche . . . by Nathaniel Bacon serving as Recorder and Town Clerk . . . 1654
BL	British Library, London
CCA	Canterbury Cathedral Archives
CCR	Calendar of Close Rolls
CFR	Calendar of Fine Rolls
CPR	Calendar of Patent Rolls
CUL	Cambridge University Library
DNB	Oxford Dictionary of National Biography
Econ. Hist. Rev.	Economic History Review
NCC	Norwich Consistory Court
Nonae	Nonarum Inquisitiones 1341
NRO	Norwich Record Office
OED	Oxford English Dictionary
PSIAH	Proceedings of the Suffolk Institute of Archaeology and History
Rot. Parl.	Rotuli Parliamentorum ut et Petiones et Placita in Parliamento
SROB	Suffolk Record Office (Bury St Edmunds)
SROI	Suffolk Records Office (Ipswich)
Statutes	The Statutes of the Realm (1101–1713)
TNA	The National Archives, Kew
TNA CP 40A	The plea rolls for Hilary term every five years from 1390 to 1510
TNA CP 40B	TNA CP 40/883, 885A, 888, 895, 907, 909, 911, 919, 931, 943 and 951

Glossary

Alien	A foreign national
Alizarin	An orange-red dye extracted from madder root (genus *Rubia*)
Alnage	The fee or duty charged for determining whether cloth conforms to particular standards of shape and quality, as required under English law
Alnager	An officer appointed to examine cloth and certify its quality
Alum	An astringent mineral salt, typically occurring as colourless or whitish crystals, that was used as a mordant for dyeing
Andiron	A utensil, consisting of an iron bar sustained horizontally at one end by an upright pillar or support usually ornamented or artistically shaped, at the other by a short foot; in a kitchen fire-place the upright support carried a rack in front for the spit to turn in
Approver	A royal official who managed the alnage on behalf of the Crown
Assize of cloth	Edict, ordinance or enactment prescribing the size and/or weight of cloth
Back stool	A stool with a back
Balance	An apparatus for weighing, consisting of a beam poised so as to move freely on a central pivot, with a scale pan at each end
Bolster	A long stuffed pillow or cushion used to support the sleeper's head in a bed
Broadcloth	A woollen that was generally between 22 and 33 yards long and 7 quarters of a yard wide
Burling	Removing knots and other imperfections from cloth
Carding	Brushing small tufts of short-staple wool between two cards ready for spinning
Cards	Wooden boards (measuring about 30cm × 15cm) with leather on one side impregnated with hundreds of fine, short, sharp wires used to prepare short-staple wool for spinning
Chafour	A vessel containing burning charcoal, for keeping food hot
Chapman	A travelling salesman
Chattel	Property; goods; money (not land)
Christopher	A figure of St Christopher, often made of silver
Clothier	Someone who made, or organized the making of cloth, often by the putting out system

Combing	Jigging wool with combs, so as to separate the fibres, bring them into parallel order, and separate the shorter from the longer
Common Bench	A short-form name for the court of common pleas
Couching	Crushing woad balls into powder, piling them into a heap, sprinkling them with water and leaving them in woad houses to ferment
Covenant	A mutual agreement between two or more persons to do or refrain from doing certain acts; a compact, contract, bargain
Coverlet	The uppermost covering of a bed; a counterpane, quilt
Demesne	A landed estate possessed or occupied by the lord of the manor himself, and not held of him by any tenant
Detinue	A legal claim to recover a chattel
Distaff	A cleft staff about 3 feet long, on which wool was wound
Dozen	A cloth of varying width, but at least 12 yards long
Dry drapery	The production of worsteds
Ell	A measure of length equivalent in England to 45 inches
Enclosure	Surrounding or marking off land with a fence or boundary; the action of thus converting pieces of common land into private property
Engrossment	Buying up land in large quantities, often in adjacent parcels
Ewe	A female, adult sheep
Ewer	A pitcher with a wide spout, used to bring water for washing
Farm	Lease of the proceeds or profits of land, office, customs, taxes etc for a fixed payment
Finished	Cloth that was dyed and sheared
Fold	A pen or other enclosure for confining sheep on arable land
Foreigner	Someone who did not have the freedom of the town in which they lived or traded
Gigs	Teasels
Gridiron	A cooking utensil formed of parallel bars of iron or other metal in a frame, usually supported on short legs, and used for broiling flesh or fish over a fire
Handle	A rectangular wooden frame with a long haft, fitted tightly with teasels to draw across the cloth and so raise the nap
Hanse	The political and commercial league of Germanic towns whose merchants traded in England
Headport	A port that was the centre of customs administration for itself and neighbouring ports
Heddle	The small cords through which the warp was passed in a loom after going through the reed, and by means of which the warp

	threads were separated into two sets so as to allow the passage of the shuttle bearing the weft
Hilary	The first court term of the calendar year, running from January to March and so named because the feast day of St Hilary of Poitiers, 14 January, falls during this term
Hogg	A young sheep from the time it is weaned until its first shearing
Husbandman	Someone who held and cultivated a small agricultural estate
Hythe	Small landing place for boats
Indigo	A substance obtained in the form of a blue powder from woad and used as a blue dye
Intercessory	Pleading for others, particularly the deceased, often through prayer
Jigging	Transferring long-staple wool between two combs ready for spinning
Kersey	A coarse narrow cloth woven from short-staple wool and deemed equivalent to one third of a broadcloth
Kirtle	A man's tunic or coat, reaching to the knees or lower, sometimes forming the only body-garment, but more usually worn with a shirt beneath and a cloak or mantle above
Lashing	Transferring long-staple wool from one comb to another ready for spinning
Latten	A mixed metal of yellow colour, either identical with, or closely resembling, brass; often hammered into thin sheets
Laver	See Ewer
Leet court	A local court that managed tithings, and controlled and punished infringement of regulations and other petty crime
Litmus	A blue dye prepared from certain lichens, esp *Roccella tinctoria*
Luteolin	A flavonoid found in weld which produced a bright yellow dye
Madder	A dyestuff or pigment prepared from the root of the plant *Rubia tinctorum* from which the reddish-purple dye alizarin was extracted
Mark	A sum of money equal to 6s. 8d. (£0.333)
Mazer	A bowl, drinking cup, or goblet, usually without a foot, made from a burr or knot of a maple tree and frequently mounted with silver or silver-gilt bands at the lip and base
Medley	Cloth woven from yarn of two or more colours
Messuage	A portion of land occupied, or intended to be occupied, as the site for a dwelling house
Mordant	A substance which combined with a dye that had low affinity for wool, enabling it to become firmly fixed in the wool
Musterdevillers	Woollen cloth, originally from Montivilliers in Normandy, usually of a mixed grey colour

Nap	The rough layer of projecting threads or fibres on the surface of a woollen fabric
Noils	Short pieces and knots of wool combed out of a long staple
Obit	A requiem mass held to pray for the soul of a deceased person (at the request and usually the expense of that person or his or her family) on the anniversary of his or her death, or at some other appointed time
Orchil	A red or violet dye prepared from certain lichens, esp *Roccella tinctoria*
Particular	Detailed accounts prepared by a local official
Perch	A wooden bar, or frame of two parallel bars, over which pieces of cloth were pulled for examination and burling
Pewter	A grey alloy of tin, with about 20 per cent lead (and sometimes other elements) and a small quantity of copper, used chiefly for plates, tankards and other utensils
Piece	A standard length of cloth
Plate	Gold or silver vessels and utensils
Potash	An alkaline substance consisting largely of impure potassium carbonate and obtained by leaching the ashes of burnt plant material and evaporating the solution in large iron pots
Putting out/ Domestic system	A system of production whereby the clothier purchased materials, put them out to craftsmen, paid for their work, sold the finished cloth and pocketed the profit
Ray	A striped cloth
Reed	Part of a loom consisting of a set of evenly spaced slender pieces of reed or cane fastened between two parallel horizontal bars and used for separating, or determining the spacing between, the warp threads, and for beating the weft into place
Rolag	A spongy, soft, air-filled roll of carded wool ready for spinning
Roving	A long thin roll of combed wool drawn out and slightly twisted ready for spinning
Russet	A coarse cloth of a reddish-brown or subdued colour, used for clothing esp by country people and the poor
Scribbling	Separating the wool fibres and removing foreign matter ready for carding
Sealed	Confirmation by the affixing of a lead seal that the cloth conformed to particular standards of shape and quality, as required under English law
Shearing	Raising, cutting, and smoothing short fibres of cloth known as the nap
Shed	The opening made between the threads of the warp by the motion of the heddles for the shuttle to pass through

Shuttle	An instrument used in weaving for passing the thread of the weft to and fro from one edge of the cloth to the other between the threads of the warp
Sliver	A continuous ribbon or band of loose, untwisted, parallelized fibres of wool ready for drawing, roving, or slubbing
Slubbing	Drawing and twisting of wool slivers in preparation for spinning
Spindle	A simple instrument employed in spinning by hand, consisting of a slender rounded wooden rod, tapering towards each end, that was made to revolve and twist into thread the fibres drawn out from a rolag or roving
Spruce	Of the country of Prussia in Germany
Staple	Fibre of wool that could be long or short, fine or coarse
Stocks	The wooden mallets that hammered cloth in a fulling mill
Strait	A coarse narrow cloth woven from short-staple wool and deemed equivalent to one quarter of a broadcloth
Subsidy	Tax
Tabby weave	Simple pattern produced by causing the weft threads to pass over and under threads of the warp in regular succession
Teasel	Plant of the genus *Dipsacus* with large prickly heads that were dried and used in cloth-finishing
Teasing	The pulling apart of fibres of wool
Tenter	A wooden framework on which cloth was stretched after being fulled, so that it set or dried evenly and without shrinking
Tenterhook	One of the hooks or bent nails set in a close row along the upper and lower bar of a tenter, by which the edges of the cloth were firmly held
Tester	A canopy over a bed, supported on the posts of the bedstead or suspended from the ceiling; or alternatively the wooden framework supporting the canopy and curtains
Tithing	A company of ten householders within a community, intended to keep order and procure compensation for damage done
Tithingman	The chief man of a tithing
Tramel	A net; or alternatively a hopper used in a mill
Transom	See Bolster
Treadle	A lever worked by the foot to produce rotary motion in a spinning wheel or reciprocating motion in a loom
Trespass	Any violation or transgression of the law, including any actionable wrong committed against the person or property of another
Twill weave	Complex pattern of parallel diagonal ridges or ribs, produced by causing the weft threads to pass over one and under two or more threads of the warp, instead of over and under in regular succession

Vesse	A worsted fabric
Warp	The threads which were extended lengthwise in the loom, usually twisted harder than the weft, with which these threads were crossed to form the piece
Weft	The threads that crossed from side to side of a piece, at right angles to the warp threads with which they were interlaced
Weld	A dyestuff, known colloquially as 'dyer's rocket', prepared from the leaves of the flowering plant *reseda* from which the yellow dye luteolin was extracted
Wet drapery	The production of woollens
Wether	A male sheep, a ram; esp a castrated ram
Woad	A dyestuff prepared from the leaves of *Isatis tinctoria* powdered and fermented from which the blue dye indigo was extracted
Woollen	A fabric made from short-staple wool in which the weft, and usually the warp was carded
Worsted	A fabric made from combed long-staple wool
Yearling	A sheep that was one year old; in its second year
Yeoman	Someone who held and cultivated a more substantial agricultural estate
Yolk	The natural oils in a fleece

Sources: Many of these definitions are based on entries in the *Oxford English Dictionary*, Oxford University Press, online edn, 2016 [http://wwwoedcom, accessed 22 March 2016]

Introduction

Jigsaw puzzles have lost none of their appeal and the satisfaction in finding a missing piece remains as strong as ever. Revealing the history of Suffolk's medieval cloth industry has been like assembling an old jigsaw. Some pieces have been found by pioneering economic historians whose names grace the Dictionary of National Biography; others by eminent post-war academics; and yet more by distinguished local historians. Some have yet to be found, and others will forever be missing. Slowly a picture, still incomplete, has emerged. But why have so many clever people devoted so much time over so many years to this topic?

Late medieval England was a predominantly rural and agricultural society. More than 95 per cent of the population lived in the countryside or small towns, the majority of working people were engaged in farming and less than a quarter of the labour-force worked in industry.[1] Suffolk has rarely been at the forefront of the nation's industrial development. Yet, in the fifteenth century, when Parliament rather dramatically declared cloth-making to be 'the greatest occupation and living of the poor commons of this land', it became England's leading textile county.[2] Its production of sealed woollens, as recorded in the alnage accounts, increased by nearly five fold between the 1350s and the 1390s, and probably doubled again by the late 1460s.[3]

Nearly three-quarters of this production was in the hundreds of Babergh and Cosford (Map I.1). By any measure we know, this 160 square miles of south Suffolk was one of the most industrialised and economically successful districts in the

1 Broadberry et al (2015), pp. 153, 195.

2 *Rot. Parl.*, v, p. 274. To put matters in perspective, Postan calculated that, in 1400, cloth-making provided direct full-time employment for less than 1 per cent of the national population, a computation that probably erred 'on the side of generosity': Postan (1950), p. 232. Of course, many cloth-workers only worked part time in the industry and in textile districts such as Babergh and Cosford, the percentage was significantly higher: Amor (2004), pp. 419–20.

3 Gray (1924), p. 34; Britnell (1986), p. 188; Amor (2004), p. 417. As explained in Appendix A, the alnage accounts of the 1350s did not record narrow cloths, so this almost certainly exaggerated the true increase in actual cloth production between the 1350s and the 1390s.

Map I.1: *Suffolk Hundreds*

Notes: The hundreds of Babergh and Cosford are shaded.
Map format attribution: David Addy.

kingdom.[4] There were no satanic mills, but cloth-workers were kept very busy in their homes. Historians have written of an 'industrial revolution of the fifteenth century', or, more soberly, of 'one of western Europe's prominent false starts along the road to industrial capitalism'.[5] In the mid-fifteenth century, while most of the country languished in an economic downturn that Hatcher christened the 'Great Slump', the area enjoyed a boom.[6] For a time, the towns of Lavenham, Great and Little Waldingfield, Brent Eleigh and Acton comprised a discrete area for the farming of the alnage, the domestic tax on cloth which we will discuss later. This, in itself, marked the importance of cloth-making there and its value as a source of royal revenue. Between Michaelmas 1425 and Easter 1457 the amount payable for that farm rose by a striking 34.2 per cent.[7] This industrial success came to depend on the putting out or domestic system whereby clothiers purchased materials, put them out to craftsmen to do the

4 Cornwall (1988), p. 14.
5 Lipson (1921), p. 88; Bailey (2007), p. 8.
6 Hatcher (1996), pp. 237–72.
7 *CFR 1422–30*, p. 114; *CFR 1452–61*, p. 174.

work in their own homes, paid for their work, sold the finished cloth and pocketed any profit.[8]

The people of Babergh and Cosford enjoyed an exceptionally high rate of growth in assessed wealth after 1334.[9] Few of the parishes in these two hundreds claimed tax relief in 1428 or 1449 on the grounds that they were small or impoverished.[10] By 1524/25 Lavenham, Hadleigh, Long Melford, Sudbury and Nayland were numbered among the wealthiest towns in England.[11] The Spryngs of Lavenham became England's first great industrial dynasty, while many other lesser-known clothiers made their family fortunes. Out of 1,403 taxpayers in Babergh, 102 were assessed with personal wealth of over £40. The top forty-two, with assets of not less than £100, paid more than half the total tax bill for the hundred. At a more humble level, over half of all husbandmen enjoyed personal wealth of at least £5, while even at the bottom of the social ladder a third of labourers owned goods worth at least £2. Such affluence was unknown anywhere else in the country outside London.[12]

As the high tide of economic development receded almost as quickly as it had come in, the medieval cores of these cloth-making towns escaped redevelopment in later centuries and so, in a sense, were frozen in time.[13] The great cloth churches and glorious medieval houses still stand as a testament to past glories. But while concentrated in Babergh and Cosford, Suffolk cloth-making was not limited to that district, but could be found in the great monastic centre of Bury St Edmunds and in other towns and villages lying along the river valleys of the county. In this light, the fascination of historians for the subject is understandable and their work is considered next.

Historians of Suffolk's medieval cloth industry

George Unwin became the first professor of economic history in the British Empire when he accepted the position at the University of Manchester in 1910.[14] By then he had already authored the sections on the old and new draperies in the second volume of the *Victoria County History, Suffolk* published in 1907. In the first of these sections he drew attention to references by Jocelyn of Brakelond to early

8 Braudel (1982), p. 299, 316–8.

9 Darby et al (1979), p. 258.

10 Dymond and Virgoe (1986), pp. 80, 82, 98.

11 A. Dyer (2000), pp. 765–7.

12 Cornwall (1988), pp. 24–5, 71–3.

13 This volume does not address the period of crisis and decline in the sixteenth century, but those wanting to learn more about it are directed to Betterton and Dymond (1989), pp. 45–74; and more recently Lee (2015), pp. 340–6.

14 See http://www.oxforddnb.com/view/article/62701

cloth-making in Bury. He cited textile-related entries in the Ipswich Domesday, published in 1870 in a volume of the Rolls Series, and also in the Ipswich General Court Register that he would have found in the British Library. He noted the number of Hadleigh cloth-workers listed in the 1381 poll tax return transcribed by Edgar Powell in 1896. With his deep interest in guilds, however, it is unsurprising that he found the Bury weavers' ordinances of 1477 'a much more interesting document', and one in which he saw signs of 'the increasing influence of capital on the industry'. He concluded his commentary on the medieval period with a brief account of the Springs of Lavenham.[15]

Louis Salzmann was a contemporary of Unwin albeit a much longer lived one. Both were active at a time when 'English economic history was still relatively unexplored'. He served as general editor of the Victoria County History between 1934 and 1949.[16] In 1913 he published his classic work *English Industries of the Middle Ages* that, naturally enough, included a chapter on cloth-making, together with a brief analysis of the alnage accounts of the 1390s and, in particular, those of Suffolk for 1395. Salzmann also drew attention to Suffolk's 'considerable trade in a cheap coarse variety of cloth known as "Vesses or set cloths" '.[17]

Ephraim Lipson's foremost claim to fame was as founding editor of the Economic History Review.[18] Earlier in his career, in 1921, he had written *The History of the Woollen and Worsted Industries* as a general account of these manufactures from the thirteenth to the early twentieth century. He commented, by reference to the 1395 Suffolk alnage accounts, that 'capitalism had already established a footing in the cloth trade'. He recognized, perhaps on the strength of Hubert Heaton's work, that, by 1470, 'in the list of textile counties Suffolk ranked first'.[19] However, a work of such broad scope of space and time left little room for a detailed examination of the sources.

Heaton was Yorkshire born, but spent most of his professional career in Australia and North America where he held the position of professor of economic history at the university of Minnesota.[20] Before he went on his travels, his doctoral thesis was published in 1920 as *The Yorkshire Woollen and Worsted Industries, from the Earliest Times up to the Industrial Revolution*. In this volume he analysed alnage accounts from all around the country for the late 1460s and early 1470s, including that of Suffolk for 1468/69. He concluded that '62 per cent of the total [English] production came from five counties', but that of these only Suffolk produced over 5,000 cloths in any one year. So, albeit on rather slender evidence, he was the first person

15 Unwin (1907), pp. 254–56.
16 http://www.oxforddnb.com/view/article/31652
17 Salzmann (1913), pp. 157, 168.
18 http://www.oxforddnb.com/view/article/34550
19 Lipson (1921), pp. 43, 223.
20 http://adb.anu.edu.au/biography/heaton-herbert-6626

to recognize Suffolk as England's leading cloth-making county.[21] He also considered, in some detail, the consequences for Yorkshire's cloth industry of Edward III's invitation to Flemish weavers to come and work in England, concluding that the Flemish element 'was small, and exerted little influence'.[22]

Barbara McClenaghan echoed both his conclusions in her short book, published in 1924, on *The Spryngs of Lavenham and the Suffolk Cloth Trade in the XV and XVI Centuries*. She stated that 'at one time Suffolk was making the greatest number of woollen cloths of any county in the whole of England', but that 'the rapid increase of the industry in the 15th century, had, at least in Suffolk, no connection with Flemish labour'.[23] These are both themes to which we will return. McClenaghan commendably drew on a more eclectic mix of sources than any previous student of the Suffolk cloth industry, including aliens' subsidies, the Babergh military survey of 1522, chancery rolls, the Howard household books, inquisitions post mortem, Parliament rolls, statutes and wills. For her secondary material she relied, of course, on Unwin, Salzmann, Lipson and Heaton. The one glaring omission from McClenaghan's sources was the alnage accounts. In this, she may have been influenced by her director of studies Eileen Power.[24] Power's volume *Medieval People* was published in the same year as *The Springs of Lavenham* and included a chapter on the Essex clothier Thomas Paycocke of Coggeshall, so she would have been familiar with the sources for the study of the cloth industry. In her Ford lectures, published in 1941 as *The Wool Trade in English Medieval History*, she dismissed the alnage accounts as being unreliable.[25] Furthermore, she helped establish the *Economic History Review* to which another of her protégés, Eleanora Carus-Wilson, contributed, in 1929, *The Aulnage Accounts: A Criticism*.[26] Carus-Wilson stingingly described them as 'second-hand compilations of doubtful veracity, often abbreviated, distorted and repeated again and again'.[27] She was to become one of the twentieth century's foremost authorities on textile history, although she only ever referred to Suffolk in passing.

If McClenaghan steered clear of the alnage accounts, Gladys Thornton both embraced and used them as the basis for the section on the 'Suffolk cloth trade in the Middle Ages' in her 1928 book *A History of Clare Suffolk*. Although she was aware of the attack that Carus-Wilson was about to launch, she bravely and correctly

---

21 Heaton (1920), pp. 84–8. Heaton's sample was geographically broad, but included only one account from any given county.

22 Heaton (1920), p. 20.

23 McClenaghan (1924), pp. 2, 6.

24 http://www.oxforddnb.com/view/article/35581

25 Power (1941), p. 37.

26 http://www.oxforddnb.com/view/article/30905; Carus-Wilson (1929), passim.

27 Carus-Wilson (1929), p. 123. Her criticism was directed at West-Country alnage accounts.

concluded that there was 'no reason to doubt' the accounts prepared by the local alnager and approver William Whelpdale in the late 1460s. She thought that those of his successor John Flegge of Needham Market were also reliable until 1473, but that thereafter 'the entries were copied almost exactly from those of the previous year'. Thornton continued with a detailed analysis of the alnage accounts and, for the first time, sought to list the Suffolk cloth towns and villages in order of importance.[28] She also analysed the careers and wills of Clare's leading medieval clothiers, John Horrold and his son Thomas the Younger and noted their trading relationship with London mercers.[29]

With the publication of Thornton's history of Clare, a period of intense interest in Suffolk's medieval cloth industry came to an end. In the 1950s John Pilgrim studied the early modern history of the East Anglian cloth industry, authored the section on 'The cloth industry in East Anglia' in *The Wool Textile Industry in Great Britain* and included a few paragraphs on the medieval period. In 1961 Joan Thirsk's essay *Industries in the Countryside* was first published and helped kick start the debate on proto-industrialisation.[30] In this she sought explanations for the growth of the medieval cloth industry, including that of Suffolk. She tentatively proposed, as 'common factors' in the early development of rural handicraft industries:

> A populous community of small farmers, often freeholders (as in Suffolk) or customary tenants with a tenure almost as good as freehold (as in the Yorkshire Dales), pursuing a pastoral economy. This may rest upon dairying in which case the farms are usually early enclosed, and manorial organisation and cooperative farming, in consequence, are weak or non existent.[31]

Notwithstanding these valuable contributions from Pilgrim and Thirsk, interest in Suffolk's medieval cloth-making remained largely dormant for fifty years until rekindled by David Dymond. As a lecturer for the Board of Extra-Mural Studies at the University of Cambridge, and in his many other roles, Dymond has done more than anyone else to encourage the study of local history in Suffolk. In the early 1970s he tutored an extra-mural class in Lavenham that ultimately led to the

---

28 Thornton (1928), pp. 144–54. Thornton lived most of her life in Clare and the author was privileged, in 2014, to hear her daughter Jennifer Ward present her volume on *Elizabeth de Burgh, Lady of Clare* to the annual general meeting of the Suffolk Records Society held in the town's church.

29 Thornton (1928), pp. 181–3.

30 Clarkson explicitly describes Thirsk's essay as one that 'foreshadowed' the 'features of proto-industrialisation': Clarkson (1985), p. 13.

31 Thirsk (1984), p. 231.

publication, in 1982, of the first edition of *Lavenham Industrial Town* (then entitled *Lavenham, 700 Years of Textile Making*). He and his co-author Alec Betterton, with the assistance of Peter Northeast, made particularly good use of 'a remarkable number of wills [...] to plot the town's rising prosperity through the fifteenth century [on the strength of] the making and trading of its famous woollen cloth'.[32] They transcribed the wills of three successive Thomas Springs in full, as well as drawing on many others. They also analysed the alnage accounts for Lavenham of the 1460s. Dymond subsequently contributed to the writing of the popular booklet *Suffolk Threads Heritage Trail*.

Richard Britnell was professor of history at the university of Durham. He initially took an interest in Suffolk cloth-making from outside the county through his study of Colchester, *Growth and Decline in Colchester, 1300–1525*, possibly the best account of a late medieval English town ever written. In particular he considered the degree of concentration of industrial entrepreneurship in Colchester compared with that in Lavenham, concluding that 'the leading Lavenham merchants were bigger operators than any in Colchester'. He analysed in considerable detail the alnage accounts prepared by William Whelpdale in the 1460s which, 'if not a perfect mirror of reality, were at least the fruit of an attempt to make them so'. Their most striking feature was 'the prominence of Suffolk, where the annual number of sealings had almost doubled from the estimated number in the 1390s'. Britnell was less positive about the fortunes of cloth-making in Sudbury and Clare after the Black Death, than those of Lavenham and Hadleigh.[33] Thirteen years after the publication of his Colchester volume, Britnell returned to the subject of Suffolk cloth-making with a short essay in *The Ricardian* on *The Woollen Textile Industry of Suffolk*. He summarized our then-current state of knowledge and concluded with a series of tables based on data drawn from Colchester borough court rolls and alnage accounts.

Mark Bailey, professor of later medieval history at the University of East Anglia, has made several important contributions to the study of the Suffolk cloth industry. In his 1989 work, *A Marginal Society? East Anglian Breckland in the Later Middle Ages*, he found a limited number of references to cloth-making. 'In relative terms, the region compares unfavourably with the major centres of East Anglian production. It neither dominated the west Suffolk market nor did it possess the high degree of labour specialization prevalent in the leading "industrial" villages of south Suffolk.'[34] Eighteen years later, in *Medieval Suffolk: An Economic and Social History 1200–1500*, he became perhaps the first scholar to consider cloth-making

32 Betterton and Dymond (1989), pp. 4–5.

33 Britnell (1986), pp. 184–92.

34 Bailey (1989), p. 180.

in its wider context of 'the most industrialised and urbanized county in England'. He cited Thirsk as developing a 'multi-faceted explanation' for the growth of the cloth industry in south Suffolk, but emphasized the importance of an 'existing pool of experience' in the old urban centres of Bury St Edmunds, Clare and Sudbury, and an intelligent response to market demands.[35] In his third contribution, *Technology and the Growth of Textile Manufacture in Medieval Suffolk*, Bailey focused on fulling and, in particular, fulling mills. While identifying many more such mills in Suffolk than once thought, he downplayed the importance of new technology as being 'far less influential' than the evolution of the organizational structure of cloth-making.[36]

The author closely analysed the Suffolk alnage accounts and cloth-makers' wills for his paper *Merchant Adventurer or Jack of All Trades: The Suffolk Clothier in the 1460s* that was published in the Proceedings of the Suffolk Institute of Archaeology and History. He found that a majority of them were still operating on a relatively modest scale. In his later work on Ipswich he discussed the importance of cloth-finishing in the town in closing years of the fifteenth century.

Having concluded this brief historiographical survey, we now turn to the sources used in writing this volume. As will already be apparent, much valuable work has been done on alnage, borough, manorial and testamentary records, but this work will rely on other sources that significantly expand our understanding of the medieval cloth industry.

Sources

The plea rolls of the court of common pleas (alias common bench) are a key source. The origins of the court can be traced back to the seventeenth clause of Magna Carta which provided that 'common pleas shall not follow our court but shall be held in some certain place'. It generally sat in Westminster Hall, London during four terms each year – Hilary, Easter, Trinity and Michaelmas. Sergeants at law honed their skills in legalistic pleadings before the justices, but rarely dirtied their hands with the facts of the case, or the evidence to support it, since the trial itself would have taken place at the county assizes. Together with the court of king's bench it evolved, in the thirteenth century, out of the earlier *Curia Regis,* and by the fourteenth and fifteenth centuries was 'much the busier of the two great central courts of common law', often being overloaded with work.[37]

35 Bailey (2007), pp. 8, 298–300.

36 Bailey (2009), p. 18.

37 Hastings (1971), pp. 16, 190.

The plea rolls are probably our best source of information about private litigation between lords, burgesses and freemen.[38] With the decline of serfdom, very few were excluded from the court's jurisdiction. 'The tremendous bulk and number of surviving plea rolls', four from nearly every year, with thousands of entries in each roll, are a database of epic proportions.[39] Even though the total number of pleas issued each term declined during the course of the fifteenth century, the rolls remain particularly valuable for that period from which few manorial or borough court records 'survive in continuous sequence [...] leaving a more fragile velum trail'.[40] Between 70 and 80 per cent of pleas were debt actions for sums ranging from 40s to many thousands of pounds.[41] Others asserted title to land; initiated actions of detinue, account, and covenant; and alleged trespass, including infringement of the Statute of Labourers and other legislation. They allocated each plea to a county, thereby telling us where the cause of action arose.[42] Following the enactment of the Statute of Additions 1413, they invariably gave the name, place of residence and occupation of the defendant; as well as the name, they often gave the place of residence and occupation of the claimant.[43] Occasionally, they provided a little more detail.

In order to become embroiled in a debt dispute for a sum of not less than 40s it is likely, not certain, that the litigant was someone of substance and in business on his own account. The same does not apply in cases of trespass or infringement of the Statute of Labourers. Indeed, in such cases it is likely, again not certain, that the defendant was someone with limited political and economic power or independence. So, when considering cases involving the practitioners of any particular craft, the lower the proportion of them litigating debts, the lower their wealth, status and/or independence would probably have been.

As a source, these plea rolls have their limitations. As Hastings stated, they 'often conceal rather than reveal what happened in the action', and tell us almost nothing about the trial itself. Indeed, medieval litigation being such 'a long and tedious business', as well as a very expensive one, almost certainly very few of the actions were ever tried. Issuing a writ was often just a way of drawing your

38 The figures, maps and tables in this volume draw on those rolls that the author has studied himself, and those indexed and published on the AALT website up to and including 2 April 2015. Selective use has been made of rolls indexed and published on the AALT website on 25 March 2016.

39 Hastings (1971), p. 7.

40 Stevens (2012), pp. 227–8; Goddard (2010), pp. 82–3.

41 Analysing a much smaller sample of common pleas disputes involving Londoners, Stevens found that 69 per cent of 'lawsuits which reached the stage of pleading' were writs of debt: Stevens (2012), p. 240.

42 Hastings (1971), p. 159.

43 *Statutes*, 2, p. 171.

opponent into arbitration.[44] Most of the entries are monotonous writs of mesne process, by which an action began, and nearly all the rest are pleadings or final process. Nevertheless, they offer us the exciting prospect of tracing personal details for millions of those living in late medieval England, a sizeable proportion of the total population. They could potentially put the study of medieval economic history on firmer quantitative footings.

According to Stevens, 'for Londoners common pleas was an essential tool for resolving mercantile disputes, and heavily relied upon by both the city's merchants and tradesmen'.[45] Was this equally true for merchants and tradesmen elsewhere in the realm who did not enjoy close and easy access to the court at Westminster? In a recent paper Stevens included a table of entries for the Michaelmas terms of 1400, 1450 and 1501 and a calculation of the number of entries per 1,000 persons in each county.[46] This serves as a useful tool in identifying any geographical bias in the court rolls. Inevitably, some counties had a greater count than others, but the only two that were grossly over represented were London and Middlesex. This may be because Londoners were at the epicenter of national commerce and so had more to litigate about. Alternatively, it may be because Londoners, and those who came to London on business had easier access to the court. The possibility of a London bias needs to be kept in mind. In a volume about the Suffolk cloth industry, it can be safely assumed that there was no bias towards that county.

The author is certainly not the first to use these plea rolls as a source. Bridbury referred to them, in his *Medieval English Clothmaking: An Economic Survey*, in order to identify the counties of residence of persons sued by Salisbury citizens. In this way he could form some idea of their business contacts and 'draw some tentative conclusions about the cloth trade for which Salisbury was a focal point'.[47] In her study of *Local Markets and Regional Trade in Medieval Exeter*, Kowaleski used the plea rolls to help examine Exeter's role as a regional centre of commerce; determine the occupation, status and residence of outsiders involved in trade with the city's merchants; and show how different commodities affected the size and shape of its hinterland.[48] Since the late 1990s Keene and his team at the Centre for Metropolitan History, Institute for Historical Research, including Galloway and Stevens, have used the plea rolls for their projects. One explored the economic hinterland of the Capital, while another assessed the importance to Londoners of the court of common pleas by analysing all those actions involving them that reached the stage

44 Hastings (1971), p. 157; Rawcliffe (1991), pp. 99–100.

45 Stevens (2012), p. 244.

46 Stevens (2012), pp. 232–4. Norwich was over represented in 1450 and 1501, but this is almost certainly because its economy was particularly buoyant as it rose to be the nation's second city.

47 Bridbury (1982), pp. 70–1.

48 Kowaleski (1995), p. 280.

of pleadings. The results of this latter project were discussed by Stevens in his 2010 paper *Londoners and the court of common pleas in the fifteenth century* and are now available on-line.[49] In her 2001 doctoral thesis *The Drapers and the Drapery Trade of Late Medieval London, c.1330–c.1500,* Quinton studied the London drapers' debts enrolled in the court for the Michaelmas terms of 1384, 1403, 1424 and 1480. She rather downplayed their significance in stating that 'the best that most Common Pleas cases can offer, even in the fifteenth century, is a skeleton of geographical patterns of domestic supply and distribution, and more anecdotal information from other sources is needed to add some flesh to the bones'.[50] It might, however, fairly be observed that fifteenth-century economic and social history enjoys a surfeit of flesh, but suffers a paucity of bone.

To understand the alnage accounts, one needs some historical background. The office of alnager dates from the time of King Richard I. His primary responsibility was to police the cloth industry within the region for which he was responsible by checking the size and quality of cloth. If approved, he applied his seal for all the world to see and, without that seal, the cloth could not legally be sold.[51] Colchester and Ipswich Museums have a small collection of matrices and seals, one of which is illustrated (Plate I.1). As we shall see in chapter 4 and Appendix A, the Government frequently enacted legislation to proscribe cloth sizes, and later either repealed or amended that legislation, or granted licences to exempt certain cloths. So, the alnager needed to keep up to date with the law. He also had to be on his guard because less scrupulous cloth-makers were known to tack or plait cloths together before sealing, so that they appeared to be of the required length.[52] From time to time, his enforcement duty was backed up by a power to forfeit substandard or non-conforming products. For instance, in 1434–35 he forfeited the cloth or fined various cloth-makers or merchants in Suffolk.[53] Three of the offenders, George Hasell, Hans Fox and Hans Pew, were German, probably from Cologne and members of the Hanseatic League. Overseas shippers may have been seen as easy targets as the preamble to an earlier statute, that had temporarily curtailed the alnager's power of forfeiture, complained of his harassment of foreign merchants.[54] In addition to policing the industry, as from 1353, he was empowered to collect a sealing fee of ½d. for a whole cloth and ¼d. for a half cloth; and to levy a subsidy on

49 Galloway (2003), pp. 121–5; http://www.british-history.ac.uk/no-series/common-pleas/1399-1500

50 Quinton (2001), pp. 97, 141.

51 The alnager's seal was first mentioned in a statute of 1328. As from 1390, the cloth-maker was also required to apply his seal: Egan (2004), p. 2.

52 *Statutes*, 2, p. 164.

53 TNA E 101/342/21.

54 *Statutes*, 1, p. 330.

certain cloths 'which are to be sold' for the first time. The subsidy payable on a whole cloth was 4d., and on a half cloth 2d. Narrow cloths were probably exempt until 1394, but thereafter were subject to a fee of ⅛d. and a subsidy of 1d. (Appendix A).[55]

The alnager's accounts recorded these payments and the more detailed particular accounts from the 1390s gave the names of the Suffolk payers and, in addition, in the 1460s they included the places at which the cloth was presented for sealing. The author has made extensive use of both sets and has made passing reference to others, the authenticity of which is more open to question. These particular accounts are summaries of day-books written by the alnager at the time of sealing.[56] They appear to offer compelling quantative evidence for cloth production in the various counties of the realm, but in reality they do not provide a complete record of production. Throughout the Middle Ages certain cloths were exempt from the alnager's jurisdiction including worsted and those made by anyone for the clothing of himself and his household.[57] Others, and in particular narrow cloths, were not recorded until 1394. Cloth was listed under the town of presentation to the alnager rather than the town of manufacture, so it can be difficult to know exactly how much was produced where.[58] The 1460s entries for Bury St Edmunds and Ipswich both refer to 'various strangers coming to the market' from outside the town, and those for Bury are even more specific in mentioning the 'Monday market'.[59] Some entries, however, do state where the strangers came from. Furthermore, in the industrial belt of Babergh and Cosford the scale of production made it worthwhile for the alnager to visit most of the small towns, so cloth-makers did not need to travel to have their produce sealed.

In addition to these exemptions, Bridbury highlighted the prospect of significant evasion of the alnage which was 'an excise, and excise is notoriously a tax which is very much easier to evade than a duty which is imposed where trade must concentrate, as it concentrates normally for purposes of loading and unloading at a port'.[60] He thought that evasion would be much easier in the countryside than in the town, so that alnage accounts might overstate the importance of urban cloth-making. This may be true for local trade in low quality cloth focused on retail sales

---

55 *Statutes*, 1, pp. 330–1; *Statutes*, 2, p. 88. The subsidy was specific, not *ad valorem*, and only represented a tiny fraction (less than 1 per cent) of the value of the cloth. The same legislation imposed a higher charge on premium quality cloths called scarlets, but these were not manufactured in Suffolk.

56 Bridbury (1982), p. 63.

57 Statutes, 1, p. 331; Sutton (1989), pp. 207–9.

58 Bridbury pondered whether Salisbury's cloth production was under recorded because rays made in the town were produced to the alnager at Westminster fair and so recorded there: Bridbury (1982), pp. 73–5; Hare (1999), p. 11.

59 TNA E 101/343/4.

60 Bridbury (1982), pp. 52–3.

to husbandmen unfamiliar with an alnager's seal. It is far less likely to be the case in a textile industry largely geared, as was Suffolk's, towards sales to hard-nosed Hanseatic shippers and members of the City's leading merchant companies – drapers, grocers and mercers. If no alnager's seal appeared on the cloth that they were offered, the supplier would almost certainly have been firmly rebuffed.

A more serious defect in the alnage accounts arises from Government fiscal policy. For long periods the collection of the alnage was taken away from approvers, who were civil servants reporting direct to the Exchequer, and, in order to raise immediate cash, farmed out to the highest bidder. Farmed accounts are much less likely to be an accurate reflection of reality. Henry Kyng was appointed as farmer of the alnage for Norfolk and Norwich for six years from 29 September 1445 for a fixed sum, but subject to the proviso that he should match any increased offer for the farm made during that period.[61] Commercial pressure did not always make such farmers honest men and few were averse to creative accountancy. In this respect Carus-Wilson's criticism of the alnage accounts was justified. Unfortunately, the baby went out with the bath water. For a generation or more textile historians steered clear of this valuable source material. In 1971 Ponting, in his book *The Woollen Industry of South-West England,* opined that the alnage accounts could not be trusted.[62] The following year Geraint Jones edited an otherwise excellent series of papers in *The Wool Textile Industry in Great Britain* none of which made use of them. Slowly, as we have seen, the alnage accounts crept back into favour as scholars learned to discriminate between those that were fabrications and those that sought to be accurate.[63] They have in recent years been used to good effect in the study of the cloth industry in Berkshire-Oxfordshire, Devon, Essex, London, Suffolk and Wiltshire.[64] Perhaps the last word should go to Heaton who wrote, as long ago as 1920, that 'the returns are useful in affording a rough general comparison of the production for sale in various areas'.[65] He might have added 'across various periods'.

Lipson once described cloth-making as 'the favourite child of the legislature'.[66] The indexes of the first two volumes of the *Statutes of the Realm* are filled with

61 *CFR 1445–52*, p. 6.

62 Ponting (1971), p. 19.

63 As early as 1962 A.R. Bridbury sought to rehabilitate the alnage accounts in *Economic Growth: England in the Later Middle Ages.* His view was 'thus what the early aulnage returns have to say about home consumption is very likely to be somewhere near the truth': Bridbury (1975), p. 35.

64 Even when they were not fictitious, the alnage accounts are not always easy sources to work with – see the criticism of/comment on Gray's figures by Lloyd (1983), p. 449; and Kowaleski (1995), p. 23.

65 Heaton (1920), p. 84.

66 Lipson (1921), p. 5.

references to wool and cloth, reflecting a 'bewildering maze of legislation'.[67] These measures illustrate the importance of wool textiles to medieval society, as well as contemporary concerns, and so are of great interest to the economic historian. Then, as now, law-makers could not resist the temptation to interfere with industry, nor forbear from changing their minds on a regular basis. As we have already seen, the alnage rules changed repeatedly. Until 1394 narrow cloths were exempt from sealing and free of subsidy, then the exemption was lost, only to be temporarily reinstated five years later, in respect of those cloths worth no more than 13s. 4d., for the 'ease and relief of the poor common people of the realm'. The exemption was removed once again in 1433 and this change in the law may explain why the German shippers, George Hasell and his compatriots, were caught out in the following year.[68] Henry IV was an unrepentant tinkerer. His 1405 statute required cloths of ray to be 28 yards long by 6 quarters wide to avoid 'deceit and great damage of the common people of the realm'.[69] The Act was annulled and repealed two years later for being 'very grievous and prejudicial', but re-enacted in 1411 with directions that it should be 'firmly holden and kept'.[70] No doubt it depended on who had the king's ear at any particular time – cloth-maker or cloth buyer. Both central and local government were anxious to protect textile jobs, introducing protectionist measures that were often flouted by powerful interest groups and were rarely effective. At the same time, they wanted to prevent fraud in manufacture not only to safeguard the domestic consumer, but also to maintain the good standing of English cloth overseas. As we shall see in chapters 4 and 5, most stages of cloth production were regulated in one way or another. These two policies were not always compatible. In their defence, the king and his councillors had been taught to fight and pray, not to run an increasingly sophisticated industrial economy.

The Ipswich borough archive at the town's branch of the Suffolk Record Office and the archive of the manor of Hadleigh Hall at Canterbury Cathedral add significantly to our understanding of the medieval Suffolk cloth industry. For instance, among the Ipswich archives the leet court rolls recorded incidents of petty crime, telling us about dyers who caused pollution; while the composite court books of the late 1400s provide a long and continuous catalogue of debt litigation, including disputes involving dyers and shearmen. From these we can learn much about cloth finishing in Ipswich. Among the Hadleigh archives, the court rolls provide names and details of offences of local dyers from as early as the late-thirteenth century; while the reeve's accounts, sergeant's accounts and farmer's view trace the history of

---

67 Heaton (1920), p. 130.

68 *Statutes*, 2, p. 119. Statutes, 2, p. 285.

69 A 'ray' was a kind of striped woollen cloth, generally made of wool of different colours.

70 *Statutes*, 2, pp. 154, 160, 168.

the local fulling mill from the same period until the late fifteenth century. Together they tell us that commercial cloth-making in Hadleigh began long before the Black Death and cast fresh light on the transition of the textile industry from its urban roots into the countryside.

Wills provide indispensable source material for any medieval historian, often offering the only insight into the everyday lives of cloth-makers.[71] Strictly speaking we should talk of wills and testaments, the former containing directions for the disposal of land and other real property; and the latter concerned with personal property such as goods and chattels, book debts and cash. By the fifteenth century, more often than not, the two were combined and in this volume the term 'will' is used to describe both. Some wills included details of precious possessions, while a few referred to industrial premises and goods – plant, equipment and stock. In Bury St Edmunds testators frequently stated their occupation, so it is relatively easy to identify who did what. The sacrist of the abbey may possibly have had his own agenda in encouraging them to do so.[72] Elsewhere, such occupational labels are rare. In the registers of the archdeacon of Suffolk up to 1520, only nine cloth-makers mentioned an occupation in their will. In seeking to identify textile workers one must either cross-reference to other sources, such as the plea rolls, or rely on the invaluable work of Betterton and Dymond, Northeast and Favey. Moreover, only a minority of people left a will, maybe as few as one in ten and certainly no more than one in three.[73] Even when they did, it may not have survived or even been copied.[74] Indeed, there are few original medieval wills in our archives and we know of more only because they were copied, hopefully accurately, into probate registers. Finally, in using wills, one must remember Swanson's words of caution that they are expressions of aspiration that give 'a rather better insight

71 Historians of the early modern cloth industry have made good use of probate inventories, but these were not obligatory until 1530 and the author has found and considered in chapter 6 only two relating to Suffolk clothiers – namely John Reynham of Nayland and John Hyll of Long Melford.

72 Among his many powers over, and responsibilities for the town, the sacrist executed, on the abbot's behalf, archidiaconal authority over testamentary matters: Gransden (2015), p. 103. Outside Bury St Edmunds, wills could be proven for probate purposes in the court of the archdeacon of Sudbury (in the west of the county) or Suffolk (in the east), the bishop of Norwich or the archbishop of Canterbury. Some Suffolk parishes, such as Hadleigh and Monks Eleigh, were peculiars and subject to the direct jurisdiction of the archbishop of Canterbury whose court alone could accept the probate: Grimwade (1984), pp. vi–vii. Some Ipswich wills were copied in the records of the petty court.

73 Northeast refers to a third of the population making a will, but, in medieval times, the author considers this too optimistic: Northeast (2001), p. xxxvii.

74 Many entries in the plea rolls describe parties as executors when no record survives of a will ever having been proven in any of the ecclesiastical courts.

into the state of mind of the testator than into the state of his or her finances'.[75] Notwithstanding all these caveats, the occupational labels in the plea rolls have enabled the author to track down and use the wills of many testators whose involvement in the textile industry had not hitherto been recognised, including some of the forty-five rich clothiers whose wills were proven at Canterbury (Appendix C).

Late-fourteenth- and fifteenth-century sources predominate so as to give this work a medieval rather than early modern bias. As we shall see, by 1500, Suffolk clothiers had triumphed with an output that far exceeded that of contemporaries in other counties. Nevertheless, the cut-off date is not rigidly fixed at 1500, but has been extended to include, for instance, the Babergh military survey of 1522 and the wills of cloth-workers and clothiers who would have been active before 1500, albeit that they died afterwards.

Why Another Volume?

In the light of the existing small library of work on the history of Suffolk's medieval cloth industry, what does this book add to our understanding?

Firstly, the author has had the benefit of sources that earlier scholars have not used, particularly: the plea rolls of the court of common pleas that are now easily accessible on the AALT website; the Thorney archives that the author read for a PSIAH paper; the Ipswich borough archives and particular customs accounts that the author read for his PhD; the manorial archives of the cathedral priory of Christchurch, Canterbury, namely Hadleigh Hall studied by Margaret Woods and Monks Eleigh studied by Vivienne Aldous; and the Melford Hall archives studied by Liz Wigmore. Each of them has kindly shared their material with the author. These new sources also shed fresh light on the material that was available to earlier scholars. Together they have revealed a much clearer picture of the county's medieval cloth industry.

Secondly, eminent scholars researching other places in other periods have added greatly to our general knowledge of the cloth industry in past times. One naturally thinks first of the late John Munro, but there have been many others whose names grace the bibliography. Their work has been of great value in re-interpreting Suffolk's industry. As will become clear in the next chapter, Caroline Fenwick's work on the 1381 poll tax returns is also important. Finally, information technology has provided new resources and study tools. Sources can be more easily found than ever before using search engines, and more quickly analysed using computer programs. For instance, the Proceedings of the Suffolk Institute of Archaeology and

75 Swanson (1989), p. 156.

History, dating back to 1848, are now on-line and searchable. On a grander scale The National Archives website has an exemplary search function, as do several other archives such as that of Canterbury Cathedral. Suffolk Record Office is making great strides to catch up.

The Structure of this Volume

Chapter 1 puts the Suffolk industry in context by analysing the early development of the county's cloth-making, and comparing it to the industry's fortunes elsewhere in England at the end of the Middle Ages. To this purpose all cloth-workers and clothiers, throughout the country, cited in ten plea rolls, between 1480 and 1500, are listed in Appendix B. Chapter 2 seeks to explain the various factors that influenced the development of the cloth industry in late medieval England in general, and Suffolk in particular, under three main headings – physical geography; economy and technology; and society. Chapter 3 looks at wool and the wool trade and in particular: wool growing in different parts of Suffolk, the role of woolmen and the procurement from throughout England of the wool used to produce the county's cloth. Chapters 4 and 5 describe the various manufacturing processes by which wool became finished cloth, principally carding/combing, spinning, weaving, fulling, dyeing and shearing; and also the people whose effort and skill made this possible. The distinction drawn between cloth production and finishing is a useful, but rather blurred one. Fullers engaged in finishing, while dyers often worked with wool or yarn. Nevertheless, at a time when so much cloth was leaving the country unfinished the distinction was significant for both cloth-makers and legislators. The purpose of this volume is not to provide a manual of how to make cloth, but to explain the processes in sufficient detail to understand their historical significance. For this reason, and for ease of reference, two charts are included at the end of this chapter showing the various stages in the making of woollens and worsteds (Figures I.1 and I.2). Chapter 6 looks at the figure of the clothier who emerged, in Suffolk, in the second half of the fifteenth century, and his impact on the industry and the society in which he lived. The book closes with general conclusions in chapter 7.

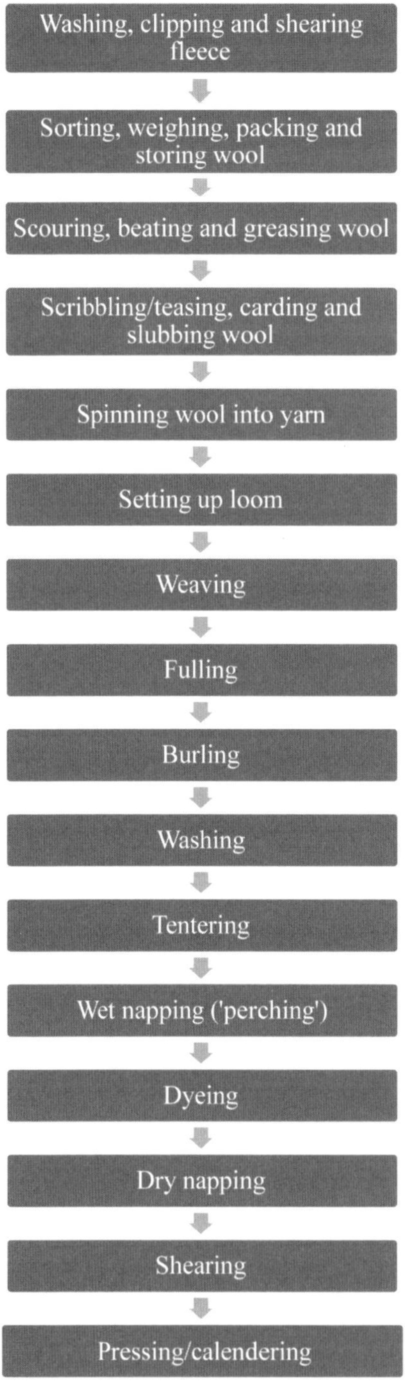

Figure I.1: *Production of Woollens*

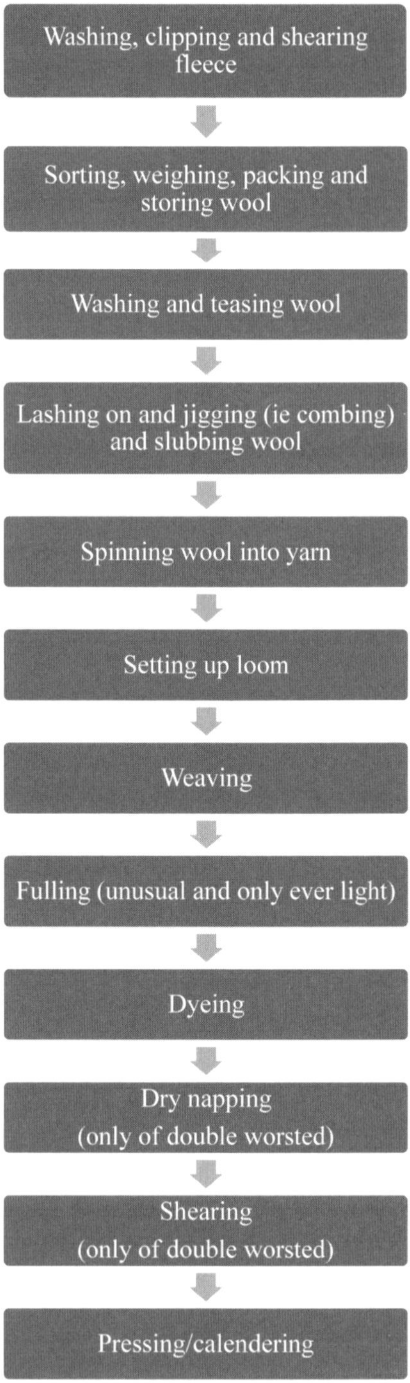

Figure I.2: *Production of Worsteds*

– Part One –

England

– 1 –

The Wider Context

At the close of the Middle Ages people were making cloth in every county in England. They were all helping to meet 'basic and universal human needs' for clothing and bedding.[1] In Lancashire Cecilia and Katherine Faryngton of Leyland span wool into yarn; in Cumbria Thomas Barton of Carlisle wove yarn into cloth; and in Westmoreland Henry Garnet of Kendal dyed cloth to suit his customer's tastes.[2] Cornish production spread across at least twelve different towns and villages, as did that of Staffordshire. Nevertheless, only in a limited number of places was cloth targeted at distant markets (Map 1.1). The location of these textile regions remained broadly similar during the three-hundred years between 1200 and 1500, but such geographical continuity hides significant changes in the nature of the cloth and the scale of production.[3] In this chapter we examine the nature of those changes and, in the next, possible reasons for them. We study thirty particular cloth-making centres and their hinterlands, their fluctuating fortunes, the groups and individuals who were prominent within them and the types of cloth they produced. This chapter concludes with two sections, one on the translocation of cloth-making from town to country and the other on the vexed question of whether the English textile industry grew or contracted. What follows is a sketch, rather than a comprehensive review of cloth-making in the late Middle Ages.

The evidence presented in the various figures and tables in this chapter is derived from two sources. While neither is perfect, they allow quantitative analysis that has hitherto not been possible. They are serendipitously spaced a hundred years apart, allowing comparison to be made between cloth-making at the end of the fourteenth and at the end of the fifteenth century.

---

1 Miller and Hatcher (1995), p. 93.

2 TNA CP 40/931, m. 317v (AALT IMG 1382); CP 40/931, m. 278v (AALT IMG 1307); CP 40/943, m. 29r (AALT IMG 61). The plea rolls disclose little evidence of the cloth-making that Margaret Beaufort encouraged so ruthlessly on her Lakeland estates: Fryde (1999), p. 262.

3 Bridbury was broadly correct in saying that 'clothmaking continued, for the most part, to favour the regions of its earliest choice': Bridbury (1982), p. 59.

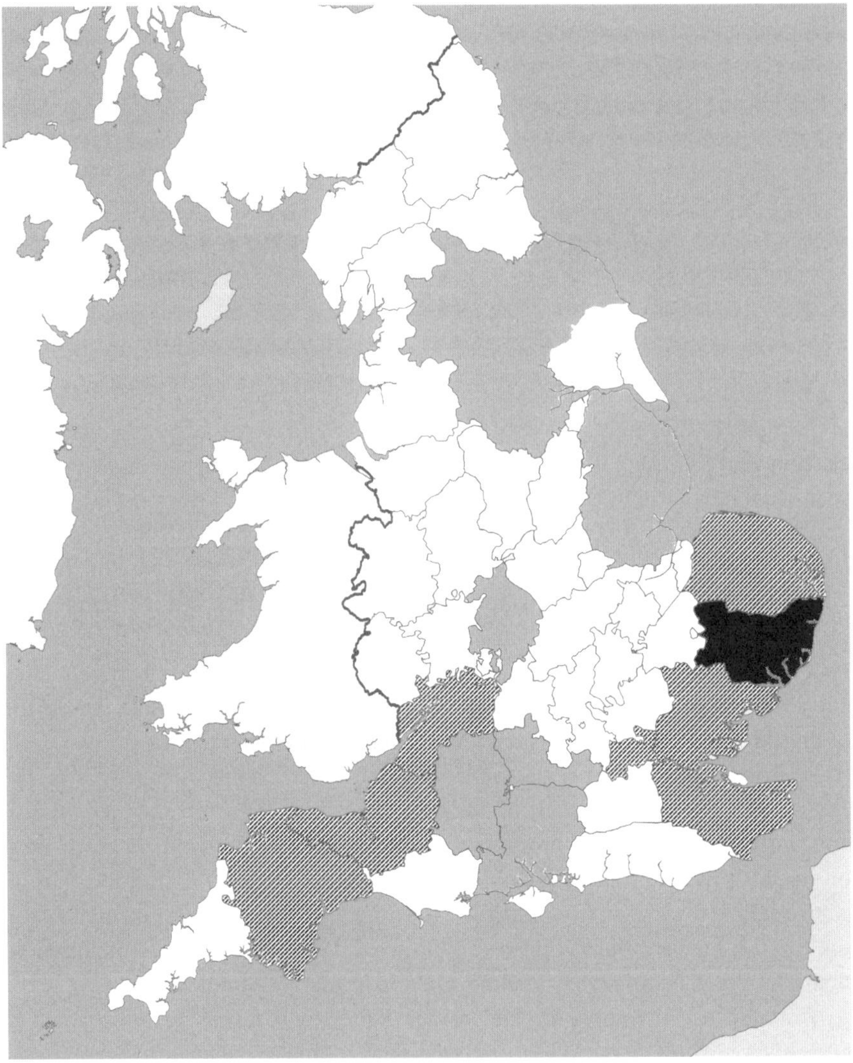

Map 1.1: *English Textile Counties 1480–1500*

Notes: This map shows the old county boundaries. Black shading means not less than two hundred cloth-workers and clothiers recorded. Grey shading means not less than one hundred cloth-workers and clothiers recorded. Light grey means not less than fifty cloth-workers and clothiers recorded.

Source: TNA CP 40B (Appendix B)

Map format attribution: MRSC and Dr Greg.

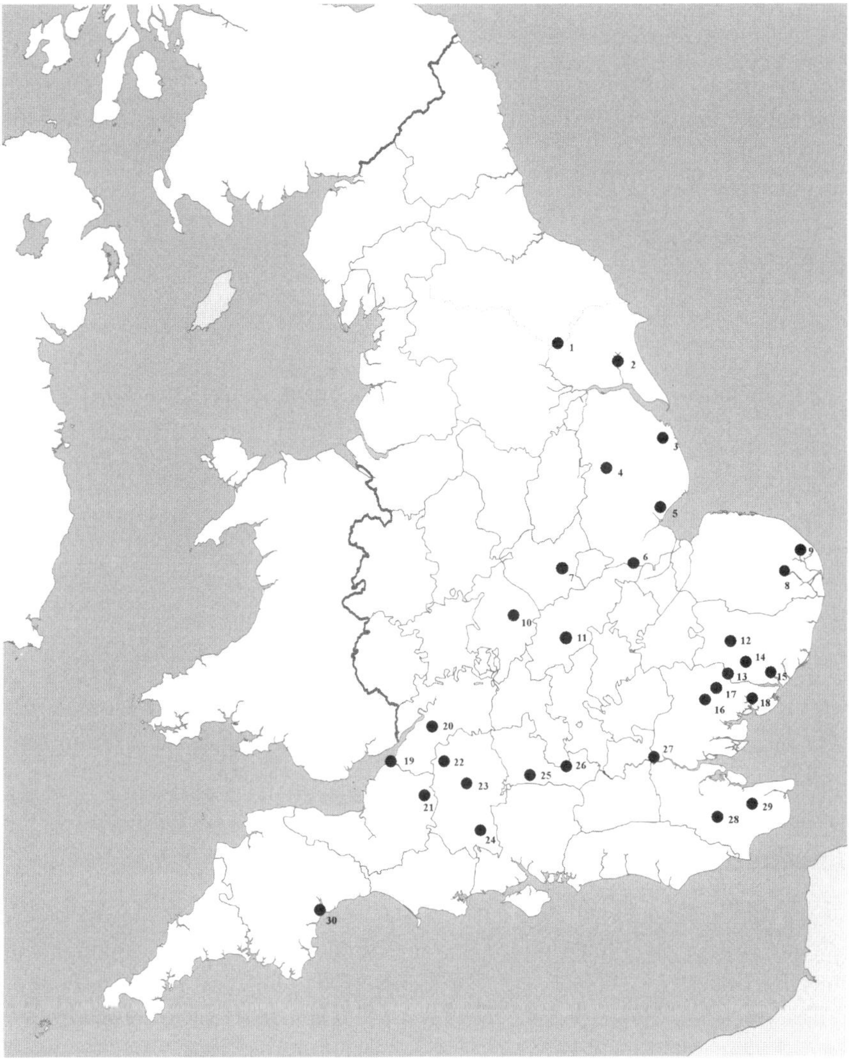

Map 1.2: *English Medieval Textile Centres*

Notes: This map shows the approximate location of the following places mentioned in this chapter: 1. York 2. Beverley 3. Louth 4. Lincoln 5. Boston 6. Stamford 7. Leicester 8. Norwich 9. North Walsham 10. Coventry 11. Northampton 12. Bury St Edmunds 13. Sudbury 14. Hadleigh 15. Ipswich 16. Braintree & Bocking 17. Halstead 18. Colchester 19. Bristol 20. Stroud 21. Frome 22. Castle Combe 23. Devizes 24. Salisbury 25. Newbury 26. Reading 27. London 28. Cranbrook 29. Canterbury 30. Exeter.

Map format attribution: MRSC and Dr Greg.

The first source is the poll tax returns of the late-fourteenth century and, in particular, those of 1381. The terms of the Parliamentary grant of this poll tax were that all men and women aged fifteen plus should pay on average 12d., and that each community should ensure that the rich paid more, but not more than 20s., and the poor less, but not less than 4d.[4] When a taxpayer paid for himself and their spouse or other member of their family, who was not taxed separately, then, for the purposes of calculating the tax paid per head, the total tax liability is divided by the number of members of that family. The 1381 poll tax returns, where they survive at all, are often fragmentary or torn, so much data is missing. The reassessment for Canterbury lists weavers, who were named therein, but does not state the tax they paid. The extant return for Salisbury is 'crumpled', illegible and incomplete. That for Suffolk includes only a few hundred and, for many of those, is incomplete. That for Tunstead Hundred includes only seven of the twenty-four settlements, namely Bacton, North Walsham, Paston, Ridlington and Crostwight, Smallburgh, Witton and Worstead. That for York is badly damaged.

The second source is a database of cloth-workers and clothiers extracted from the plea rolls, including eight rolls for the period 1480 to 1500, and set out for each city, town and village in Appendix B. The numbers given are not a direct proxy for the amount of cloth produced in each place. In areas where the textile industry was longer established, such as the West-Country, production may have been concentrated in fewer hands. Nevertheless, in the absence of reliable alnage accounts, the plea rolls are the only source that enables any comparison to be made for this period.

Lowland Eastern Plain

In the 1200s commercial cloth-making was concentrated in the towns of the lowland eastern plain that drained into the Humber and the Wash, among them Beverley, Leicester, Lincoln, Louth, Northampton, Stamford and York.[5] The fame of producers in this region spread across Europe. 'It is here that we can detect Europe's second most important cloth-producing region at this time'.[6] These towns produced cloth that was exported through the east-coast ports and was much in demand in the Mediterranean. Ipswich's late thirteenth-century customs of the cloth market refer to 'clothys of colour of Beverlie or of Lincoln'.[7] The quality of cloth from these centres at this time has been a subject of debate. Carus-Wilson regarded them as

4 Fenwick (1998), 1, p. xvi.

5 With the exception of Louth, each of these towns paid a fine of at least £5 for exemption from the size of cloth in 1202: Miller and Hatch (1995), p. 101.

6 Carus-Wilson (1987), p. 633.

7 Twiss (1873), p. 197.

'fine quality cloths', while Chorley conceded that England was capable of 'producing cloth of the very highest quality'. Nevertheless, the latter contended that, as the century wore on, English exports increasingly comprised inexpensive fabrics.[8] Even in the early-fourteenth century some cloth exports by alien merchants were of better quality, but more were light, coarse and unfinished.[9] Perhaps most were, like Northampton's, of modest pretensions, and a few, like Lincoln 'scarlets', fit for royalty. Indeed, Munro has described the English cloth export trade of this period as having 'a surprising bi-polar tendency'.[10] In any event, by the late-thirteenth century these seven eastern towns were suffering an industrial crisis and their trade was in decline so that, for a short time, England became a net importer, rather than exporter of cloth.[11] Their subsequent textile fortunes varied, but none of them ever again showed quite the same vigour.

Beverley had once been famous for its blue cloths, and continued to supply high-quality fabric in the early 1300s, but by the fifteenth century had 'drifted into an economic backwater'.[12] In the late 1400s it was still the leading textile town of the East Riding, but in this respect was up against very little competition. Leland reported in the 1530s that '*Ther was good cloth made at Beverle: but that is nowe much decayid*'.[13]

Leicester was an important centre where, as early as the thirteenth century, a putting out system was implemented. Merchant entrepreneurs such as Henry Houhil and Richard of Shilton subjugated the city's manufacturing craftsmen to their control. Ultimately, this led to industrial conflict between merchants and artisans from which the textile industry may never have fully recovered. By 1322 fullers had virtually deserted Leicester and weavers too were very unhappy.[14] Six of them were working there in the middle years of the fifteenth century, but by the end only a rump of cloth finishers remained (Appendix B).

Although the textile industry in Lincolnshire declined after 1300, it never disappeared and, indeed, remained widespread across the county (Figure 1.1). The town of Boston appears to have had the most active cloth industry in the county at the end of the fifteenth century, for all its troubles as a late medieval port. In Lincoln itself cloth-making may have flourished again by 1348, while fourteen fullers were

8 Carus-Wilson (1987), pp. 632–3; Chorley (1988), pp. 3–6; Munro (1999), pp. 108–9, 116–9.

9 Childs (1996), pp. 124–34, 141–3. For this early period, we know more about cloth exports by alien merchants than those by English ones because the latter were not subject to customs duties until 1347: Carus-Wilson and Coleman (1963), p. 194.

10 Munro (1999), p. 118.

11 Miller (1965), p. 70; Miller and Hatcher (1995), pp. 108–10.

12 Heaton (1920), p. 4; Kermode (1999), p. 270.

13 Leland (1964), pp. 1, 47

14 Miller (1965), p. 70; Carus-Wilson (1967), pp. 228–33.

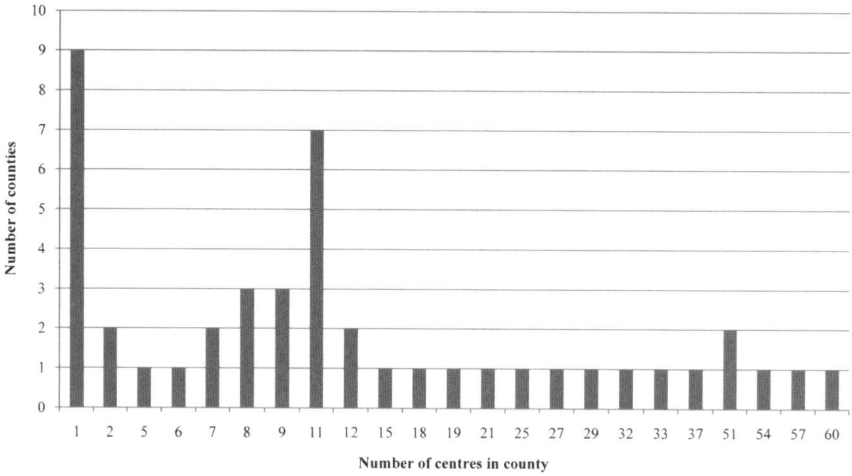

Figure 1.1: *Number of Cloth-making Centres in Different Counties 1480–1500*

Notes: A centre was a town or village where a cloth-worker or clothier was recorded in the plea rolls. There was one centre in Cheshire, Cumbria, Denbighshire, Durham, Isle of Wight, London (disregarding Middlesex and Southwark), Pembrokeshire, Rutland and Westmoreland; two in Huntingdonshire and Lancashire; five in Leicestershire; six in Middlesex; seven in Herefordshire and Northamptonshire; eight in Cambridgeshire, Shropshire and Worcestershire; nine in Derbyshire, Nottinghamshire and Oxfordshire; eleven in Bedfordshire, Berkshire, Buckinghamshire, Dorset, Hertfordshire, Surrey and the East Riding of Yorkshire; twelve in Cornwall and Staffordshire; fifteen in Warwickshire; eighteen in the North Riding of Yorkshire; twenty-one in Sussex; twenty-five in Gloucestershire; twenty-seven in Wiltshire; twenty-nine in the West Riding of Yorkshire; thirty-two in Lincolnshire; thirty-three in Hampshire; thirty-seven in Essex; fifty-one in Kent and Somerset; fifty-four in Devon; fifty-seven in Suffolk; and sixty in Norfolk.

Source: TNA CP 40B (Appendix B)

taxed in Louth in 1379.[15] Nevertheless, by the late 1400s manufacturing in both towns had sunk to a similarly low ebb as that of Leicester (Appendix B). In Stamford cloth production may have survived longer. In 1379 only one fuller was taxed, as well as a dyer and two shearmen, none of them thriving. However, Rogers discovered that 'between 1465 and 1492, thirteen weavers were admitted to the freedom, and weavers, dyers and fullers formed one of the largest craft groups'.[16] Four cloth-workers were cited in the plea rolls (Appendix B).

Further west, the textile industry of Northampton was in ruder health. In 1334 the authorities had lamented that not a single weaver was left in town, but later in

15 Miller (1965), p. 79; Bridbury (1982), pp. 32–3; Munro (1999), p. 121; Fenwick (2001), 2, p. 57.
16 Fenwick (2001), 2, pp. 28–9; Rogers (2012), p. 127.

the century fullers began to re-establish their presence.[17] John Brixworthe was a long-term resident at that time.[18] Between 1440 and 1465 at least sixteen different fullers were cited in the plea rolls. In 1455 a gang of local thugs, who included a dyer, four weavers and a fuller, was accused of ambushing and assaulting a John Makesey outside Northampton.[19] By the end of the century the town's cloth production may well have been in decline, but at least ten dyers continued to thrive (Appendix B).

The story of cloth-making in York has oft been told and generally as one of woe, with certain rural settlements in the West Riding as the villains of the piece. Historians have argued that the over-taxed and over regulated city craftsmen were undercut by artisans raised in the fresh air of Central Pennine valleys. As Heaton wrote:

> The progress of the industry in the West Riding began to be specially marked in the first three-quarters of the fifteenth century, and that period synchronized with a serious decline in the output from the York looms.[20]

At least sixty-nine weavers, spread across nineteen York parishes, paid poll tax in 1381.[21] Only a handful paid more than 12d. per head, and over a third paid less than that sum. Indeed, as a group, York weavers were less affluent, or at least more lightly taxed than those in other centres, giving credence to their repeated pleas of poverty (Figure 1.2). Nevertheless, according to Swanson, the city enjoyed a considerable revival in its industrial fortunes for well over a hundred years from the 1330s to the 1470s. In the second half of the fourteenth century 17 per cent of citizens were cloth-workers and in the 1450s 111 broadlooms were in operation. The alnage accounts for 1468 and 1472 recorded that York still produced a third of all the county's cloth.[22] Even in the final twenty years of the fifteenth-century the plea rolls disclose more weavers working in York than the whole of the West Riding, and nearly twice as many fullers (Appendix B). As yet the cloth-workers of Bradford, Halifax, Leeds and Wakefield had made little mark on cases in the common bench. Nevertheless, the monopoly rights across the county that York weavers had once enjoyed no longer protected them from competition in the west.[23]

More towns and villages in the West Riding were textile centres than in Gloucestershire or Wiltshire (Figure 1.1). Nationally, they numbered among those

17 Munro (1999), p. 121.

18 Fenwick (2001), 2, p. 226; TNA CP 40/555, m. 157r (AALT IMG 332).

19 CP 40/776, m. 290v (AALT IMG 1685).

20 Heaton (1920), pp. 44, 59.

21 Fenwick (2005), 3, pp. 140–56.

22 Miller (1965), p. 79; Swanson (1989), pp. 28–9; Wild (1972), p. 199.

23 Miller and Hatcher (1995), p. 106.

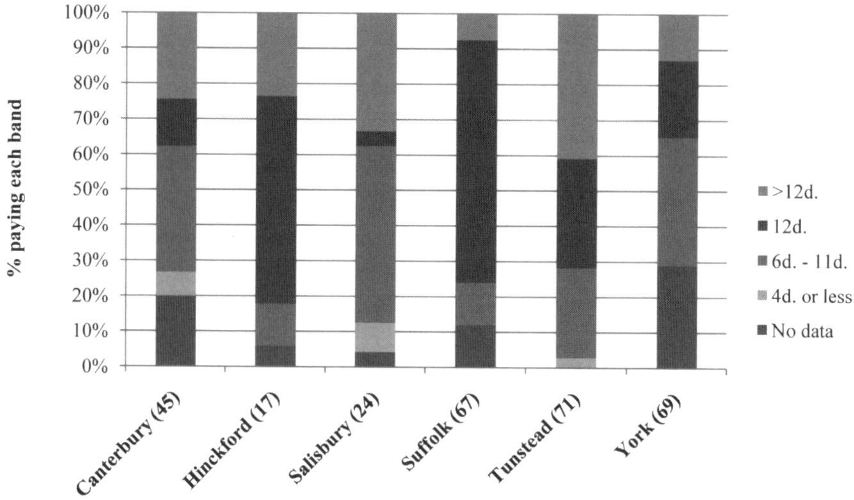

Figure 1.2: *Poll Tax Payable by Weavers in 1381*

Notes: Figure 1.2 records and compares the percentage of known weavers in each city, hundred or county who paid each of several different bands of poll tax, or for whom the poll tax charge is either illegible or missing. The figures in parentheses indicate the total number of recorded weavers. Apprentices and servants of weavers have been disregarded. Figures 1.5 and 4.1 are in a similar format.

Sources: Fenwick (1998), 1, pp. 206–22, 418–33; Fenwick (2001), 2, pp. 196–9, 505–36; Fenwick (2005), 3, pp. 123–5, 140–56.

settlements that enjoyed the highest rates of growth in lay wealth between 1334 and 1524/25.[24] The small number of recorded cloth-makers in each of these centres may be explained by the dispersed pattern of cloth-making there, which meant that few of those involved were men of substance. Less than a third of common pleas enrolments involving West Riding weavers were writs of debt, suggesting that relatively few of their transactions were worth more than the 40s. threshold for common bench claims. For centuries to come, their lives would demand hard work both in the fields and at the loom. 'Industrial labour on the part of the clothier and his family was the general lot and was accompanied in many cases by comparative poverty.'[25]

The eastern towns were not the limit of cloth-making in thirteenth-century England. Many more centres sought to protect their own industries by buying exemptions from the assize of cloth in 1202, notably Coventry that is considered next.[26]

24 Darby et al (1979), pp. 258, 260.

25 Heaton (1920), p. 92.

26 Miller and Hatcher (1995), pp. 100–1.

Not so Gloomy in Coventry

In the Middle Ages Coventry was the regional capital of the Midlands and consistently figured among the most populous and wealthy cities in England.[27] It enjoyed 'continuous development throughout the Middle Ages' and was among those districts that enjoyed the highest rate of growth in lay wealth between 1334 and 1524/25.[28] Its medieval image is, however, a gloomy one because the leading historian of the city made possibly the most pessimistic of all diagnoses of late medieval and early modern urban fortunes. The title of Phythian-Adams's well-known volume about Coventry, 'Desolation of a City', speaks for itself. Without doubt, his careful study of the 1522/23 assessment detected weaknesses in Coventry's economy and social fabric, but the fortunes of the cloth industry at the close of the fifteenth century were far from desolate (Appendix B). Coventry was home to a good mix of cloth producers and finishers, with only three places in England having more resident cloth workers and clothiers. They faced little apparent competition from rural Warwickshire, making up two-thirds of all those recorded in the county. Card-making which, as we shall see in chapter 4, involved making the cards that were used to prepare wool for spinning, appears to have been a particular local speciality. Indeed, together with the saddlers, they were responsible for Coventry's Corpus Christ pageant and had their own chapel.[29] Card-makers have left little mark on history. Nationwide only seventeen were cited, of whom four carried on their craft in Coventry (Appendix B). A hundred years earlier, in 1379, at least four card-makers and three, probably four carders had paid tax there.[30] In 1450 seven had contributed armour for the defence of the city.[31] Until 1478 dyers took delivery of hundreds of cartloads of woad and other dyestuffs from Southampton, but thereafter looked more towards London for their raw materials.[32] So, it is difficult to credit that, in 1500, 'the running down of the textile industry was surely a major factor in the long-term decline of medieval Coventry'; or that 'the first stage of Coventry's decline may thus be explained in large measure by the combination of lowered returns from a stagnant textile industry' and the impact of the civil war; or that the '*fin de siècle* recovery' while remarkable was 'illusory'.[33] Even an analysis of the 1522/23 military assessment reveals signs of resilience in the city's cloth-making. More households were devoted to textiles than to any other category of trade, 60 per

---

27 A. Dyer (2000), pp. 755, 758, 761, 765.

28 Darby et al (1979), pp. 258, 261–2.

29 Phythian-Adams (1979), p. 44.

30 Fenwick (2001), 2, pp. 679–88. This tax list gives relatively few occupational descriptions, so those of card-makers and carders are intriguing.

31 Stephens (1969), pp. 151–7.

32 Deliveries to Coventry were eight-fold those to Northampton: Hare (2015), p. 109.

33 Phythian-Adams (1979), pp. 42–4.

cent of those households had servants, and the mean number of servants was higher than for any other group except merchants.[34] As Cornwall remarked, 'association with the cloth trade was a necessary step towards getting to the top in Coventry'.[35]

The Rise of the West

By the time the English textile industry began to recover in the second quarter of the fourteenth century, its focus had switched to the West Country – Gloucestershire (including Bristol), Somerset, Wiltshire and Devon and, to a lesser extent, Dorset and Hampshire. More than half of England's broadcloths were produced there in the late 1350s.[36] In the following century and a half they were among those regions that enjoyed the highest rate of growth in lay wealth in England and that growth was almost certainly associated with commodities such as wool and cloth.[37]

By 1400 Salisbury had almost certainly become the nation's leading cloth city.[38] The extant 1381 poll tax return records just twenty-four weavers because of its fragmentary condition. A higher proportion of them paid in excess of 12d. per head than anywhere else save Tunstead Hundred in Norfolk, providing evidence of their prosperity (Figure 1.2). Some weavers were doing remarkably well. Seven of them had resident servants. The wealthiest of all, Henry Bonde, paid 5s. tax for himself and his wife, and employed five male servants, probably journeymen weavers, who together contributed a further 3s.[39] Forty years later, 112 master weavers and 208 journeymen attended a protest meeting against Henry IV's legislation regulating cloths of ray. They were joined by fifty-three master fullers and twenty-seven journeymen. All told, a quarter of the city's adult male population may have been present.[40] During the first three-quarters of the fifteenth-century, cloth-making continued to provide a livelihood for over a third of independent traders. The alnage accounts suggest that Salisbury dominated Wiltshire textile production until 1421, but thereafter 'fell into relative decline'.[41] Soon relative decline become absolute. The city must have been badly shaken by the Great Slump of the mid-fifteenth century. Urban decay contributed to Salisbury becoming, in

34 Phythian-Adams (1979), pp. 208–10. The category of textiles did not include Coventry's famous cappers.

35 Cornwall (1988), p. 23.

36 Gray (1924), p. 34.

37 Schofield (1965), pp. 504, 507; Darby et al. (1979), pp. 259–60.

38 Gray (1924), pp. 21, 22, 34; Miller (1965), p. 79.

39 Fenwick (2005), 3, pp. 123–5. The 1379 return is much clearer and complete, but does not state occupations: Fenwick (2005), 3, pp. 110–9.

40 Bridbury (1982), pp. 68–70; Hare (2011), pp. 180–1.

41 Hare (1999), p. 11.

1449, one of the flash points in Cade's rebellion.[42] Thirty-seven weavers were recorded in six terms of the common bench between 1396 and 1440, four in four terms between 1450 and 1470 and nine in ten terms between 1480 and 1500 (Appendix B). Furthermore, the volume of woad, madder and alum arriving by cart from Southampton declined sharply in the final thirty years of the century, and the number of recorded dyers fell to two (Appendix B).[43]

Salisbury ultimately lost its earlier dominance due to the rise of production in small towns and villages of west Wiltshire along the boundary with Somerset. Castle Combe is the best-known example.[44] Situated ten miles north-east of Bath, the village lies in the valley of the river By-Brook, and spreads around the decayed castle, down to Nether Combe and up to Over Combe. It has been described as 'the prettiest village in England', but that would not have been the case in the fifteenth century when it grew explosively as an industrial centre. In 1377 only 138 residents had paid poll tax and there had been just one fulling mill.[45] In 1415 Castle Combe earned its own heading in the alnage accounts, with a list of fourteen different cloth-makers producing seventy-one cloths.[46] By the 1450s the one fulling mill had become three together with a gig mill, probably for use in wet napping, and the only known example from the Middle Ages. Fifty new houses had been built. Some seventy master weavers, fullers and dyers worked furiously in Nether Combe, down in the valley by the stream, each craft supervised by two wardens. Their products were premium quality scarlets, well known in London as 'Castle Combes'. The local industry was stimulated by bulk orders from Sir John Fastolf, who was absentee lord of the manor for nearly fifty years, and, as a warrior, one of the stars of the Hundred Years War.[47] He bought £100 worth of cloth each year from the village in order to dress his troops in a livery of red and white.[48] Fortunes could be made whatever your background or social status. William Haynes was descended from a family who were 'villein by blood', but went on to become one of Wiltshire's most successful clothiers. At his death in 1435 his goods and chattels were worth between £200 and £2,000, and he left a sizeable bequest for the building of a new bell tower. Nevertheless, Sir John still insisted on William's widow paying a hefty entry fine of £140 to retain possession of his moveable goods,

42 Lee (2015), p. 329.

43 Hare (2015), pp. 60, 155–8.

44 Beresford and St Joseph (1979), pp. 267–9; Carus-Wilson (1959), pp. 197–205; Hare (2011), pp. 185–7.

45 Fenwick (2005), 3, p. 12; Beresford and St Joseph (1979), p. 268.

46 Hare (2011), p. 183.

47 http://www.oxforddnb.com/view/article/9199?docPos=1

48 Carus-Wilson (1959), pp. 198–9.

cottage and lands.[49] Walter Power was an Irish immigrant who began his career as a landless artificer and went on to build his own business, employing servants and apprentices and furnishing his house with precious metal.[50]

Castle Combe expanded so rapidly, and provided such opportunities for making fortunes and headway up the social ladder, that it conjures up visions of a gold-rush town. According to Carus-Wilson, 'the throng of craftsmen with their servants and apprentices presented a novel social problem', judging by some of the local ordinances passed in its manor court that prohibited gambling and late hours in taverns.[51] Other ordinances, restraining servants and apprentices from poaching in manorial fishponds, have a more homely tone. In 1455 a member of the local gentry sought compensation for trespass from seven weavers, five fullers and three dyers.[52] Castle Combe's boom years were probably over by 1470. Chevage payments by landless artificers to the lord of the manor were in decline, from seventy in 1443 to thirty-two in 1481, probably a sign of a falling population.[53] The evidence of the plea rolls reinforces this sense of decline. In the final twenty years of the century no cloth-workers at all were cited during ten terms of the common bench, just two clothiers both in 1498 (Appendix B).

The success of a local textile centre often turned on the nature of cloth that it produced. In the closing years of the fifteenth century the strongest demand was no-longer for rays or scarlets, but for heavily fulled and unfinished broadcloth to feed the London and continental markets. In Wiltshire the clothiers of Devizes were the most geared up to meet this demand. 'Cloths of Devizes' became as well known in the City as 'Castle Combes' had been thirty years before. Large undyed broadcloths, known as barstards, were carried by road to Southampton and London.[54] The textile industry was long established in the town. Eleven weavers had been taxed in 1379, two paying as much as 2s., and 140 cloths sealed in 1414/15.[55] In the final twenty years of the fifteenth century maybe a third of the county's clothiers were resident there and they enjoyed perhaps half its London business (Appendix B). Among them, Nicholas Mere was the second generation of clothiers in his family, a Devizes man of the same name having been cited in the plea rolls in 1460. Nicholas committed to supply the London grocer Edward Woode

49 The higher valuation was made by Sir John's Council, which had its own reasons for exaggerating Haynes' wealth, and the lower by a local jury, while the truth no doubt lay somewhere in between: Carus-Wilson (1959), p. 200.

50 Carus-Wilson (1959), p. 202.

51 Carus-Wilson (1987), p. 688.

52 TNA CP 40/776, mm. 403r, 313v (AALT IMG 796, 1730).

53 Hare (2011), p. 186. For a full description of chevage: Bailey (2014), pp. 42–6.

54 The Southampton brokerage books noted the arrival of barstards in the city: Hare (2015), p. 27; Hare (2015), p. 154.

55 Fenwick (2005), 3, pp. 54–5; Hare (2011), p. 183.

with £200 worth of merchandise, almost certainly cloth, but was unable to honour his promise.[56] Like some contemporaries in Suffolk, whom we shall meet later, he had overreached himself.

Somerset had long been a power house of English cloth-making. Between the 1350s and the 1390s the number of the county's sealed cloths increased ten fold to over 12,000 per annum.[57] No county would ever again approach this volume of production during the Middle Ages. In order to maintain output, the putting out system may have evolved rather earlier than in most regions of the country. Five Somerset clothiers were cited in the plea rolls for the Easter 1418 term, as against none from eastern England.[58] Furthermore, even though they lived a hundred miles from London, as from 1430 much of the clothiers' production was targeted at the City (Figure 1.3). Textile manufacture was widespread (Figure 1.1), but it was most concentrated along the banks of the river Frome, dividing the counties of Wiltshire and Somerset.[59] And in the town of Frome, in the Mendip Hills, John Stokton was particularly prominent. He was a frequent visitor to London in the

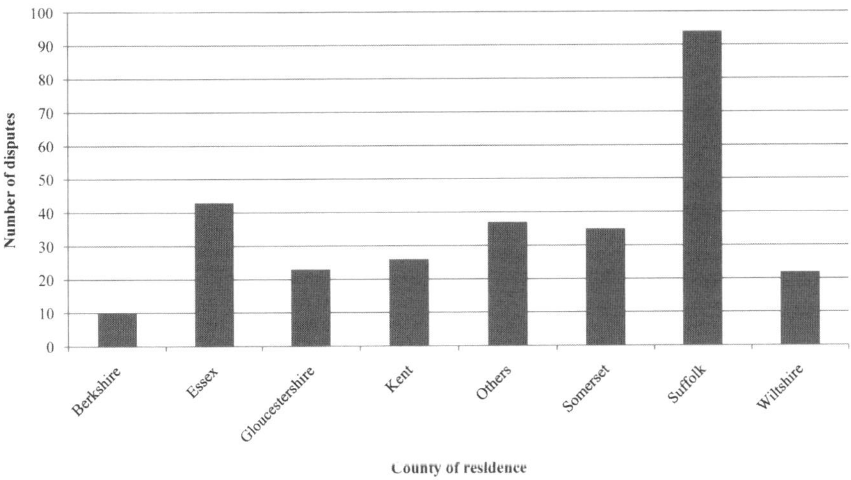

Figure 1.3: *London Trade of Clothiers 1480–1500*

Notes: This Table records all claims arising from transactions entered into in London.

Source: TNA CP 40B

56 TNA CP 40/796, m. 265v (AALT IMG 1371); TNA CP 40/911, m. 14v (AALT IMG 27d).

57 Gray (1924), p. 34. As explained in Appendix A, the alnage accounts of the 1350s did not record narrow cloths, so this almost certainly exaggerated the true increase in actual cloth production between the two periods.

58 TNA CP 40/629, mm. 399r, 329v (AALT IMG 808, 1542).

59 Ponting (1971), p. 14. In his study of alnage accounts Hare found fifty-three centres of cloth-making in Somerset, two more than the author: TNA E 101/343/28.

1460s and appears to have enjoyed close ties with mercers.[60] He may even have been related to another John Stokton who was warden of the London livery company of mercers in 1462 and again in 1468.[61] Within a quarter of a century the main focus of clothiers' operations seems to have moved from Frome to the village of Mells three miles away; while other centres, both in the town and country, had come to the fore, namely Axbridge, Bridgwater, Shepton Mallet and Taunton (Appendix B). In 1500 Somerset remained the second or third most important textile county, after Suffolk and possibly Kent.

Gloucestershire has long been regarded as the poor relation to Somerset in terms of cloth production. It is a view supported by the evidence – the alnage accounts of the 1350s, 1390s and 1460s; and also the plea rolls (Appendix B and Figures 1.1 and 1.3). Nevertheless, even poor relations have interesting tales to tell. The history of the wool textile industry in Bristol can be traced back at least as far as the mid-fourteenth century when it was still part of the county. In 1346 the city authorities enacted ordinances relating to weavers and fullers that 'probably marked an attempt to impose order on a situation created by a period of very rapid development of textile manufacture'.[62] This development is echoed in the alnage accounts that record over 2,000 cloths per annum sealed in the 1350s, nearly double that number in the 1390s, and just over 3,500 in 1468/69.[63] In 1453 the farmers of the alnage paid the Crown £60 per annum, three times that paid for London, and equivalent to alnage and subsidy on 3,200 cloths. By 1485 the payment had fallen to less than half, just £27 13s. 4d.[64] Furthermore, very probably much of the alnage was paid on cloths produced outside, but brought into the city for sealing. The plea rolls suggest that, from mid-century, Bristol's industry was in trouble. Eight weavers and seven fullers were recorded in just one court term in 1430, but only two and five respectively in four terms between 1450 and 1480. It was dyers who were most visible in the closing years of the fifteenth-century (Appendix B). Although Bristol's export trade in cloth remained buoyant at that time, it had few resident clothiers and only limited business with those elsewhere in the county.[65] By this time many of the cloths exported from the port may have come from north Devon.

Two-thirds of Gloucestershire's recorded clothiers gathered together in the valley of another river Frome – in Stroud, Minchinhampton and Rodborough – and in

60 The mercers with whom Stokton dealt included John Don the Elder, William Geffrey and John Marshall: TNA CP 40/814 mm. 24r, 211v, 351v (AALT IMG 50, 1392, 1663).

61 Sutton (2005), p. 557.

62 Miller and Hatcher (1995), pp. 124–5.

63 Gray (1924), p. 34; Heaton (1920), p. 85.

64 *Rot. Parl.*, v, pp. 247; 6, 300.

65 Carus-Wilson and Coleman (1963), p. 143. TNA CP 40B records seven contracts involving Gloucestershire clothiers concluded in Bristol, against twenty-one in London.

their close neighbour Dursley. In the fourteenth century these had been no more than a string of isolated hamlets along 'a mere stretch of water flowing through a deep cleft in the Cotswolds'.[66] No cloth-workers were recorded in Bisley, Minchinhampton or Rodborough in the 1381 poll tax return.[67] In the fifteenth century the Stroud valley enjoyed 'remarkable growth of industrial activity'.[68] Soon men who had once called themselves fullers, such as William Haliday and Robert Kyn, adopted the title clothier. Haliday's enterprises extended beyond cloth manufacture and he took a first step towards gentility by leasing the arable lands of the manor of Minchinhampton together with the manorial buildings and sheepfold. His son Edward is remembered by a brass in Minchinhampton church.[69] Another Robert Kyn, also a clothier and probably son or nephew of the first, was embroiled in several disputes before the common bench in the closing years of the fifteenth century. He pursued two claims against his neighbour and fellow clothier Roger Ferrour for the substantial sums of £21 and £40. Although there is no evidence of his ever taking the road to London, Kyn did well for himself. He died in 1492 with a sufficiently valuable estate to merit proving his will at the Prerogative Court of Canterbury. He left bequests to the church at Minchinhampton and chapels at Rodborough and Stroud, as well as £60 to his son and £40 to his daughter.[70] A Dursley clothier, Robert Rychard, owned a warehouse, wool loft, dye-house and shop, together with stocks of wool, yarn, cloth and dyestuffs worth, at the time of his death, £173.[71] Although the Frome flowed into Bristol only twenty-five miles away, these clothiers took three times as much trade to London as to their local metropolis. None of them did business in the Capital with mercers. One of them, John Plymmeswode of Dursley dealt direct with the Hanseatic merchant Matthew Batterswyk.[72] Another Hanseatic merchant admiringly described their cloth as 'well dyed and costly'.[73] In this light the paucity of recorded dyers in the district is surprising (Appendix B).[74] If we accept Ponting's view that Stroud was 'beginning to emerge

66 Carus-Wilson (1959), pp. 190–1.

67 Fenwick (1998), 1, pp. 285, 303–4. Bisley was the parish from which Stroud grew.

68 Carus-Wilson (1959), p. 196.

69 Carus-Wilson (1959), pp. 195–6.

70 TNA CP 40/919, m. 88v (AALT IMG 915); CP 40/931, m. 257 (AALT IMG 529); PROB 11/9/124.

71 C. Dyer (2005), p. 167.

72 TNA CP 40/888, m. 24v (AALT IMG 685).

73 Carus-Wilson (1959), p. 190.

74 Carus-Wilson wrote of the importance of the Stroud Valley as a centre of cloth finishing, but all the successful industrialists she mentioned were fuller/clothiers. No dyer or shearman was recorded in Minchinhampton or Rodborough, and only one of each in Stroud. Conversely, one fuller was recorded in Minchinhampton, four in Rodborough and nine in Stroud (Appendix B).

as a centre for coloured broadcloths' then, even if the clothiers sold more of their products in London, they may have relied heavily on the services of skilled dyers in Bristol.[75]

Further west still, in Devon, cloth-making was widely spread, but more thinly than anywhere else in the country. One hundred and twelve cloth-workers and five clothiers were recorded, operating in fifty-four different centres (Appendix B). On the south coast the textile industry was well established in Exeter at the beginning of the thirteenth-century when the city supplied the bishop of Winchester with green and blue cloth of reasonable quality.[76] Fifteen master weavers, eleven fullers, three dyers and two shearmen were resident in 1377, of whom the dyers were, as a group, by some margin the wealthiest.[77] Its overseas cloth trade enjoyed a growth spurt in the final twenty years of the fifteenth century that bypassed London and was targeted direct at Brittany, Gascony and Spain. The exported fabrics were kerseys – cheap and light, yet finished before they left Devon.[78] Kowaleski believed that 'Much of Exeter's fifteenth-century prosperity was based on its expanding cloth industry, particularly in cloth finishing'.[79] Much can also be attributed to its role as a regional market centre for the textiles produced in the small towns and villages of south Devon and west Somerset. They were linked together by traffic along the coast from centres such as Sidmouth and down roads along the valleys of the rivers Exe, Yeo, Culm and Lowman from Crediton, Uffculme and Tiverton.[80] Leland remarked on cloth making in Crediton – '*The toun usith clothing, and most therby lyvith*'.[81] Nearly half of all recorded Devon cloth-producers and most of its cloth-finishers lived in or within twenty miles of Exeter (Table 1.1). In terms of population, between 1377 and 1524/25 Exeter rose from twenty-third to sixth position among English towns, and in terms of taxable wealth, between 1334 and 1524/25 from twenth-eighth to sixth.[82] Thanks to textiles, even Phythian-Adams accepted that the city was 'the only unambiguous example of urban success since 1377.'[83] In 1524/25 Crediton, Tiverton and Ottery St Mary – another centre of cloth finishing within Exeter's orbit – all numbered among the hundred most wealthy

75 Ponting (1971), p. 19.
76 Miller and Hatch (1995), p. 103.
77 Kowaleski (1995), p. 128.
78 Carus-Wilson and Coleman (1963), p. 145; Childs (1978), pp. 62, 88; Kowaleski (1995), p. 237; Oldland (2007), p. 109.
79 Kowaleski (1995), pp. 93–5.
80 'River transport was rare in Devon': Kowaleski (1995), p. 40.
81 Leland (1964), 1, p. 239.
82 A. Dyer (2000), pp. 756, 758, 761–7.
83 Phythian-Adams (1979), p. 18.

TABLE 1.1: CLOTH WORKING IN THE DEVON HINTERLAND OF EXETER 1480–1500

	Cloth Production		Cloth Finishing	
Distance from Exeter	*Centres*	*Workers*	*Centres*	*Workers*
Exeter	1	6	1	15
Less than 10 miles	5	7	0	0
More than 10, but less than 20 miles	16	23	4	13
Total	22	36	5	28
All Devon	51	77	10	34
Percentage of all Devon	43.1	46.8	50	82.4

Notes: For the purposes of this Table and also Table 1.2, workers in cloth production were spinsters, weavers and fullers; and workers in cloth finishing were dyers and shearmen. No cardmaker was recorded in Devon.

Source: TNA CP 40B (Appendix B)

towns in England.[84] As a whole the county of Devon enjoyed one of the highest rates of growth in lay wealth in late medieval England.[85]

London and the Home Counties

London, Berkshire and Kent were the main centres of cloth-making in south-east England.[86] In the 1200s master weavers in London produced a variety of cheap, light and coarse cloths known generically as burrels. As the century wore on they lost their independence to an elite of merchant burellers, supported by the City government, who took control of the industry and implemented an early form of the putting out system. They thought that a burrel could be woven in 'two days or three' while the weavers insisted that it took at least four. Ultimately, this and other disputes led to a conflict between weavers and burellers that triggered a decline of cloth production in the Capital.[87] Between 1374 and 1486 only 5 per cent of testators gave their craft as textiles, less than half the number of leather or metal workers, and significantly smaller than the proportion in many other English towns and cities.[88] As London increasingly became the focus of cloth exports, the late-fifteenth century witnessed an inexorable rise in the volume of unfinished cloth flowing into

84 A. Dyer (2000), p. 766.

85 Schofield (1965), pp. 506, 508; Darby et al (1979), pp. 258–9, 261.

86 Since nearly all Essex cloth production was in the north-east of that county, it has been treated as part of East Anglia.

87 Miller and Hatcher (1995), pp. 112–3, Munro (1999), pp. 111–4.

88 Barron (2004), p. 66.

the Capital from the provinces. Much of this was shipped overseas without any further work. Enough remained to make London, including Southwark on the south bank of the Thames, the most important centre of cloth finishing in England.[89] Over a quarter of all shearmen cited in the plea rolls lived there, and over a third of all recorded debts owed to or by shearmen were contracted there (Appendix B).

Textile manufacture flourished in parts of Berkshire in the late Middle Ages. Its manufacturing centres included Abingdon (now in Oxfordshire) and Wokingham, but it was Newbury and Reading that fared best.[90] Newbury was the wealthiest and most populous Berkshire town in the 1300s, but by 1524/25 Reading had overtaken it, in terms of both population and wealth, to become, according to Leland, '*the best toun of al Berkshire*'.[91] In the fifteenth century the two appear to have followed different cloth-making trajectories. A comparison of the respective sets of data (Appendix B) suggests that Reading had significantly more cloth finishers. Indeed, dyers had been active there since at least the thirteenth century.[92] Three paid poll tax in 1381, as well as two shearmen.[93] Leland comments that the waters of the river Kennet, flowing through the town, were '*very commodious for diers*'.[94] The town had fewer clothiers and, among them, only William Lynacre, who was mayor in 1464, was recorded as venturing as far as London.[95]

In Newbury, the industry was dominated by clothiers, almost certainly to the detriment of master dyers and shearmen.[96] In the course of the fifteenth century they successfully wrestled back business from Abingdon and by the end of the third quarter were responsible for between 15 and 20 per cent of all cloth sealed in Berkshire and Oxfordshire. Much of their production was kerseys, many of which were headed for the same ultimate destinations as those of Exeter.[97] The clothiers were committed, perhaps over committed, to trade with Southampton merchants. For many years, men such as John Doget, John Underwood and later John Bacon, made the forty-mile trip south.[98] In the 1490s their destination appears to have switched to London, but they continued to trade with Southamptoners. Nicholas

89 Oldland (2007), pp. 107, 117.

90 Darby et al (1979), pp. 257–8, 262.

91 Leland (1964), 1, p. 109; A. Dyer (2000), pp. 755–6, 759, 761–2, 765.

92 Ditchfield and Page (1923), pp. 342–64 http://www.british-history.ac.uk/vch/berks/vol3/pp342-364

93 Fenwick (1998), 1, p. 40.

94 Leland (1964), 1, p. 111.

95 TNA CP 40/888, m. 247v (AALT IMG 1098); Ditchfield and Page (1923), pp. 342–64.

96 Only three testators gave their occupation as dyer and just two as shearman in Newbury wills during the period 1403–1580: Yates (2007), p. 107.

97 Yates (2007), pp. 83–5.

98 TNA CP 40/883, m. 247v (AALT IMG 449d). The Southampton brokerage books noted the number of carts arriving with loads of kersey from Newbury: Hare (2015), p. 154.

Frere was particularly prominent. He sued Anthony White and John Brigge in the common bench and, on his death in 1495, owned book debts due from others such as the Southampton based Italian merchant William Ambrose.[99] His estate also included four packs and twenty-four pieces of white kerseys in his warehouse and more in his shop.[100] Attracted by the lure of the Capital and/or frightened away by piratical attacks on their galleys, Ambrose's compatriots deserted Southampton for London in the final quarter of the fifteenth century. Their departure triggered a dramatic slump in cloth exports from the port and probably also had a detrimental effect on Newbury's fortunes.[101] The town's share of total cloths sealed in Berkshire and Oxfordshire fell to 6.8 per cent in the period 1476–82.[102] Road traffic from Southampton, bringing woad and other dyestuffs for the textile industry, was at its busiest in the mid- and late 1440s, but generally much reduced in the second half of the fifteenth-century.[103] By 1522 the town was suffering from 'the decline of the local broadcloth and kerseys'. Furthermore, according to Yates, what wealth remained 'was concentrated into the hands of a few, particularly the capitalist clothiers in the town'.[104] It may have been a combination of two factors, unfaithful Italians and concentrated wealth, which dragged down the general wellbeing of Newbury below that of Reading. Happily, two successive John Winchcombes, father and son, would soon revive its fortunes.

Cranbrook is situated in the centre of the Kent Weald, twelve miles south of Maidstone. By the end of the fifteenth century it was, together with neighbouring villages, second only to Babergh in Suffolk as the most industrialised district in the country (Table 1.2). More clothiers lived in, or within six miles of Cranbrook than in all but three English counties (Appendix B). Zell's analysis of the residence of Wealden clothiers for the longer period of 1480 to 1549 found that over 80 per cent lived within this same radius.[105] The importance of Cranbrook might have come as a surprise to earlier historians. Heaton identified only 1,027 cloths sealed in the whole of Kent in his survey of the alnage accounts for 1469/70.[106] In Carus-Wilson's survey of late-fifteenth-century cloth-making, Kent received one curt mention, Cranbrook none at all.[107] Jenkins' volume on *the Wool Textile Industry in Great*

---

99 TNA CP 40/931, m. 46r (AALT IMG 92); CP 40/943, m. 14v (AALT IMG 860); PROB 2/95.

100 TNA PROB 2/95 (source courtesy Margaret Yates).

101 Carus-Wilson and Coleman (1963), p. 149; Platt (1973), pp. 139–40; Oldland (2007), p. 108. Yates referred to 'a short set back in the 1490s': Yates (2007), p. 236. But Hare argued persuasively that the late-fifteenth century enrolled customs accounts understate the true level of cloth exports from Southampton: Hare (2015), pp. 27–28.

102 Yates (2007), p. 84.

103 1478 was an exception and the busiest year in the fifteenth-century: Hare (2015), pp. 107–8.

104 Yates (1999), p. 624. See also Cornwall (1988), p. 34.

105 Zell (1991), p. 155.

106 Heaton (1920), p. 85.

107 Carus-Wilson (1987), p. 679.

TABLE 1.2: CLOTH WORKING IN KENT IN AND AROUND CRANBROOK 1480–1500

	Cloth Production		*Cloth Finishing*		*Clothiers*	
Distance from Cranbrook	*Centres*	*Workers*	*Centres*	*Workers*	*Centres*	*Clothiers*
Cranbrook	1	9	1	2	1	20
Less than 6 miles	9	20	1	1	10	39
More than 6, but less than 10 miles	6	12	1	1	3	10
Total	16	41	3	4	14	69
All Kent	37	73	13	30	23	79
Percentage of all Kent	43.2	56.2	23.1	13.3	60.9	87.3

Notes: Workers in cloth production were spinsters, weavers and fullers; and workers in cloth finishing were dyers and shearmen. No card-maker was recorded in Kent.
Source: TNA CP 40B (Appendix B)

Britain included no regional study of Kent. Yet the plea rolls reveal conclusively the importance of textile manufacture in this part of that county.

There were early signs of industrial activity in Cranbrook – fulling mills in the thirteenth century and, according to local tradition, a Flemish weaver in the early fourteenth.[108] By 1377 it was among the hundred most populous towns in England. Blackbourne Hundred, under which it was then assessed, was the most populous and heavily taxed in Kent; and, within that hundred, Cranbrook was the most populous and heavily taxed settlement.[109] On the back of textile production the Cranbrook region enjoyed economic growth, manufacturing 750 sealed cloths in 1471/72, and would continue to do so in the sixteenth century.[110] In 1524/25 the mean tax paid in Cranbrook Hundred was 5.7s. per taxpayer, nearly three times the average of 2s. paid in many Wealden parishes.[111]

Cranbrook clothiers became renowned for their production of unfinished broadcloths, so it is no surprise to find that two of their London creditors were prominent mercers – John Matthew and Robert Southwood, both wardens of the

108 Zell (1991), p. 153; Miller and Hatcher (1995), p. 118.

109 A. Dyer (2000), p. 759. The Hundred, as listed in the return, included the other cloth villages of Benenden, Biddenden, Frittenden, Hawkhurst, Rolveden, Sandhurst, Staplehurst and Tenterden: Fenwick (1998), 1, pp. 390–1. In 1524/25 the separate Hundred of Cranbrook was, per 1,000 acres, by a considerable margin the most populous and wealthy in the Weald: Zell (1991), p. 142.

110 Schofield (1965), pp. 504, 508–9; Darby et al (1979), pp. 257–8; Mate (2006), p. 12.

111 Mate (2006), p. 177.

company in the 1480s. Sutton comments that 'Kent's prominence among the common pleas debtors [of mercers] in 1484 [...] may reflect the growing Wealden cloth production'.[112] Nevertheless, bearing in mind their proximity to London, a surprisingly few recorded debts, only four, were incurred there by Cranbrook clothiers and only twenty-six by all Kent clothiers (Figure 1.3).

By 1510 six of the town's clothiers had accumulated sufficient land and wealth to have their wills proven in the Prerogative Court of Canterbury.[113] One of them, Stephen Karkeregge, was perhaps typical. He owned a workhouse next to his home in which he made his cloth. Although he left no bequests to spinsters, weavers, fullers or even servants, and probably operated on a relatively modest scale, he was sufficiently successful to fund an annual obit for thirty years after his death and to distribute 6s. 8d. to the poor of Cranbrook at each of those services. His son John, and probably his grandson of the same name followed him into the family business. Stephen enjoyed close relationships with various other local cloth-making families. He may have operated in partnership with the clothier Robert Coweper. Together they owed Sir James Browemer the sum of 22 marks 3s. 10d. from whom, one may speculate, they had bought wool. Stephen served as executor for Peter Courthorp and, in turn, his executors included Richard Brykenden, both members of known cloth-making families.[114]

Cloth-making had been well established in Canterbury by the early thirteenth century where the product, as in London, may have been burrels.[115] In 1381 the city's artisans were more concerned with cloth production than finishing. More than forty each of spinsters, weavers and fullers paid the poll tax, but only six dyers and three shearmen did so. Some of the weavers prospered, three paying the extraordinarily high sum of 9s. Below this elite, however, a long tail of artisans contributed very little (Figure 1.2). Four dyers called themselves 'woader', which means that they were trading in, and/or using woad in their craft. They, too, appear to have been quite prosperous, with John Torold contributing 4s. Another two were described as 'fuller and dyer'.[116] Together the Canterbury cloth-makers must have made a major contribution to annual Kent production of between 1,000 and 1,350 sealed cloths in the second half of the fourteenth century.[117] However, at some time

---

112 TNA CP 40/888, mm. 218r, 328r (AALT IMG 405, 612); Zell (1991), pp. 159, 220; Sutton (2005), pp. 222, 557.

113 John Brekenden, William Cosschemay, Thomas Henley, Simond Huche Stephen Karkeregge and Ralph Lynch.

114 TNA PROB 11/12/116; TNA CP 40/885A, m. 29r (AALT IMG 51); TNA CP 40/883, m. 73v (AALT IMG 141d); Zell (1991), p. 153 n. 3.

115 Miller and Hatcher (1995), p. 102.

116 Fenwick (1998), 1, pp. 418–33. Fenwick's list includes the 1381 re-assessment.

117 Gray (1924), p. 34.

early in the 1400s cloth production took flight to the countryside and cloth finishing concentrated in the towns. In 1471/2 just eighty-nine cloths were presented for sealing in the city.[118] But all was not lost. Those towns on the edge of or outside the Weald, in particular Canterbury, Dartford and Tunbridge Wells, each had a sizeable community of dyers (Appendix B). In contrast to the clothiers of the Cranbrook district, Kent's cloth finishers did much of their business in London. Three quarters of their recorded disputes arose from debts or obligations incurred in the capital. The dyer, Richard Butler, may have commuted back and forth from Tunbridge Wells, since he was said to be resident both there and in London. Three of his colleagues, John Hert, John Casteldoun and Edmund Clerk, traded with London grocers from whom they probably bought their dyes.[119]

East Anglia

Just as the West Country had dominated English cloth-making in 1400, so East Anglia did so by 1500. Each of the industries of Norfolk, Essex and Suffolk possessed distinctive and idiosyncratic features, but those of Norfolk were especially unusual.

The textile industry of Norfolk stood out from the rest of the country in several important respects. Its staple products were not woollens, but worsteds made from longer, straighter pasture wools. The difference between the two is explained in chapter 4. In a county of worsteds, the weaver was king. A few fullers, and rather more dyers and shearmen, had their role to play, but they were heavily outnumbered. Weavers also assumed the role of entrepreneur that elsewhere was filled by the clothier, which explains why only one clothier was recorded in Norfolk in the plea rolls (Appendix B). The industry was spread across more centres, sixty, than in any other county, but at the end of the fifteenth-century 40 per cent of all recorded cloth-workers and finishers lived in Norwich (Appendix B). The city succeeded in securing and retaining cloth production when so many other large towns failed. In doing so it rose to become the second city of the kingdom both in terms of population and taxable wealth.[120] With the possible exception of the period 1440 to 1460, the prosperity of Norwich 'was effectively sustained throughout the fifteenth century' (Table 1.4).[121]

A 'sudden explosion of worsted making' in Norwich in the final quarter of the fourteenth century, was followed by another period of particular prosperity for weavers in final third of the fifteenth. In the 1400s about a third of all freedom

118 Mate (2006), p. 12.

119 TNA CP 40/943, m. 67r (AALT IMG 135); CP 40/895, m. 196r (AALT IMG 359); CP 40/919, m. 198v (AALT IMG 1130); CP 40/931, m. 16r (AALT IMG 33).

120 A. Dyer (2000), pp. 761, 765.

121 Dunn (2004), p. 219.

registrations were awarded to those involved in textiles.[122] The number of weavers recorded in Norwich in the plea rolls was significantly higher than anywhere else in the realm, indeed as high or higher than in any county other than Suffolk (Appendix B). The key to Norwich's success was the double worsted, a cloth that was finer, lighter, more closely woven than fine broadcloth and worth more than three times as much as single worsted. So Norwich produced a premium quality cloth that not only required the superior skills of the urban weaver, but also could bear the higher costs of urban manufacture and regulation.[123] Some, such as William Neweman, John Davy and John Bronde the Elder, styled themselves distinctly as 'thick wool weaver'.[124] At a time when silk was in great demand among society's elite for clothing and home furnishings, double worsted had a similar texture that made it a good second best for well heeled customers, such as the Pastons, both domestically and overseas. While exports of single worsted were in sharp decline, by the 1430s those of double worsted increased through Great Yarmouth and, even more so, London.[125] But were there other reasons for the success of textile manufacture in Norwich and its decline in old heartlands such as Tunstead Hundred?

North Walsham is a picturesque market town, about fourteen miles north-east of Norwich, less than five miles from the north Norfolk coast and the largest settlement in the hundred of Tunstead (Map 1.3). It is an unlikely place for the desperate last stand of the rebels of the Peasants Revolt, but such is the case. It was there, in late June 1381, that the warrior bishop of Norwich, Henry Despenser, 'like a wild boar gnashing its teeth', easily overcame a motley band led by Geoffrey Lister, a local dyer, who was summarily executed after the skirmish. Evidence of rebel activity is to be found in all but two of the larger settlements in the hundred.[126] But those who followed Lister into battle were not all oppressed serfs choking on the yoke of feudalism.[127] For Tunstead was not only a hotbed of rebellion, but also the location of probably the most intensive rural cloth-making in England in the late-fourteenth century (Table 1.4) – 'the heart of the medieval rural worsted industry'.[128] Indeed, the village of Worstead, adjacent to North Walsham, gave its name to the cloth that was first made before 1300, would be the staple product of the county in the

122 Dunn (2004), pp. 214, 216, 220, 234; Oldland (2011), pp. 182, 184.

123 Bailey (2007), p. 299. As discussed in chapter 4, the urban weaver was more likely than his rural counterpart to have undertaken a formal apprenticeship training and a period as a journeyman.

124 TNA CP 40/758, m. 43v (AALT IMG 706); CP 40/814, mm. 275v, 473v (AALT IMG 1516, 1914).

125 Carus-Wilson and Coleman (1963), pp. 199–200; Oldland (2011), pp. 181–5, 190.

126 Cornford and Reid (1993), pp. 86–7: Barker (2014), pp. 348–51.

127 Hilton (1973), pp. 172–4; Fryde (1999), p. 51.

128 Whittle (2000), p. 252.

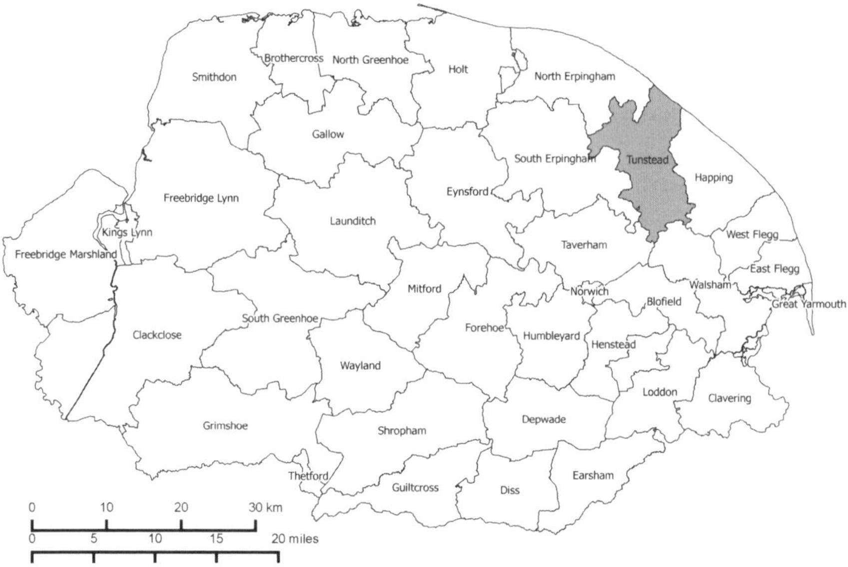

Map 1.3: *Norfolk Hundreds*
Map format attribution: David Addy

fifteenth century and would become 'famous throughout the world'.[129] Worstead had a sizeable community of weavers by 1327 and, together with North Walsham and Tunstead village, 'was conspicuously wealthier than neighbouring places'.[130]

Fifty years later the list of Tunstead Hundred taxpayers in 1379 included the names of 119 weavers, including twenty-nine in North Walsham and no less than forty-five in Worstead. They comprised over 5 per cent of the total taxpaying population of the hundred, nearly 10 per cent of those in North Walsham and nearly 15 per cent of those in Worstead.[131] They were at work in nineteen of the twenty-four parishes. And they were doing rather well. While none of them was on a par with Henry Bonde of Salisbury or John Frythe of Canterbury, at least five in Worstead employed household servants. As a group, Tunstead Hundred weavers were, in 1381, more prosperous, or at least more heavily taxed, than those elsewhere (Figure 1.2). Over 40 per cent of them paid in excess of 12d. Until then, their principal overseas markets in northern Germany and the Baltic had not been blighted by the warfare

129 Sutton (1989), p. 201.
130 Miller and Hatcher (1995), p. 117. Tunstead Hundred enjoyed one of the highest rates of growth in assessed lay wealth between 1066 and 1334: Darby et al (1979), p. 251.
131 Fenwick (2001), 2, pp. 159–171.

that was endemic in much of Europe. Sadly it was all about to change.[132] Never again were they to have it so good. The evidence of the plea rolls indicates that cloth-making went into a sharp decline in the Hundred, from which there would be no significant recovery until the early-sixteenth century (Figure 1.4).

In 1334 the amount to be contributed by each town and village in the kingdom as tax (known as a tenth and fifteenth) was permanently fixed. In 1449, in recognition of the economic plight then suffered by much of the kingdom, Parliament allowed a reduction in the amount payable. Norfolk's overall allowance was 15.7 per cent, that of Tunstead Hundred was 20.2 per cent and that of North Walsham 26.7 per cent.[133] In his calculation of the 1524/25 tax, as a percentage of 1334 lay wealth, Darby put the district in the lowest quintile nationally. He wrote that ' the worsted area to the north-east of the city [ie Norwich] had dwindled from its peak in the fourteenth century'. Cornwall similarly referred to a fifteenth-century

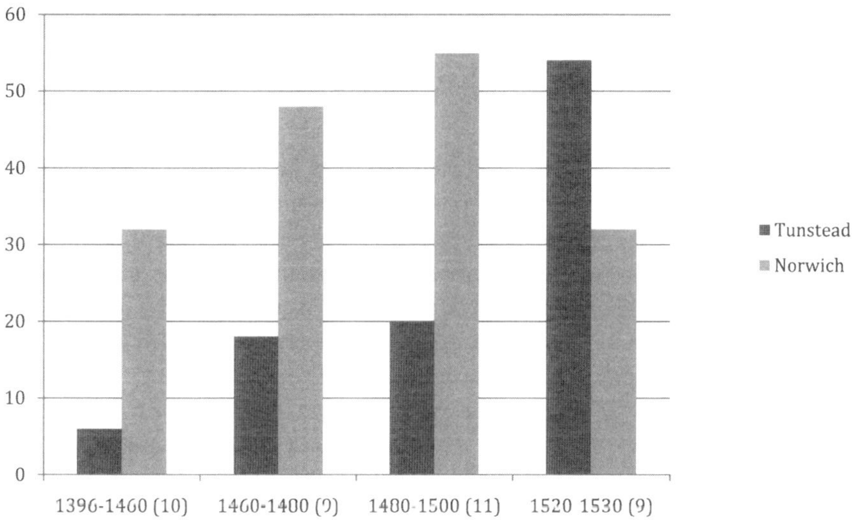

Figure 1.4: *Numbers of Weavers in Tunstead Hundred and Norwich*

Notes: This shows the number of weavers cited in the plea rolls during the periods stipulated in Tunstead Hundred and in Norwich. The figures for 1480–1500 include those weavers cited in the plea roll for the Hilary 1480 term (TNA CP 40/871) who were not cited in the plea rolls used to compile Appendix B. The index for the Hilary 1480 term, among others, went on-line very shortly before this volume went to press. Figures in parentheses are the number of different court terms.

Source: TNA CP 40

132 Munro (1999), pp. 139–40.

133 Hudson (1895), pp. 258, 285.

'recession' there.[134] Not until 1497 did any worsted weaver in Tunstead Hundred die sufficiently wealthy to have his will proven in the Prerogative Court of Canterbury. Even then, the deceased, John Deye of Worstead, had evidently prospered by cultivating close links with Norwich and London. Among bequests to Norwich, he gave 6s. 8d. to the cathedral, 13s. 4d. to the convent and 4s., for food, to the prisoners in both the castle and guildhall. His nephew and perhaps his agent, another John Deye of London, received 10 marks worth of worsted. His widow Katherine was the beneficiary of his looms.[135]

Why this long fall from grace? Tunstead Hundred may simply have suffered from the Great Slump that blighted much of England in the mid fifteenth-century. With the exception of Norwich's double worsteds, overseas demand for the product was muted. Moreover, as recounted above, Norwich came to dominate the textile industry in Norfolk to the detriment of other centres, but this only begs the question. Such a marked shift of cloth-production from countryside to town is unknown elsewhere in the country. Is there another, more complex explanation based on the involvement of weavers in the Peasants' Revolt? Hinton suspected their involvement, certainly in London and perhaps elsewhere, noting repeated attacks on Flemings.[136] Out of six local offenders killed in battle at North Walsham or summarily executed thereafter and specifically named in an inquisition at Worstead in 1382, only one can be confidently identified as a weaver, namely Robert Smyth of Ridlington. Very likely a thorough study of local documents would reveal others who were implicated. Their business success may have been on the wane, or alternatively given them the confidence to challenge an incompetent government. In any event they resented the repeated imposition of heavy taxation for reasons, mainly unsuccessful military campaigns, that they found increasingly unpersuasive. Did bishop Henry and his successors exact revenge on them for their temerity? Did he de-industrialise the locality or, at least, reduce the scale of cloth-making, so that the value of individual transactions fell below the 40s. threshold imposed by the common bench?[137] If he did, how did he do it? A possible explanation presents itself. According to Sutton, by the 1320s 'the Norfolk textile workers were sufficiently organized to be able to defeat the interests of London and Norwich dealers

134 Darby et al (1979), pp. 259–60; Cornwall (1988), p. 34. Oldland, whose views on medieval textile production are worthy of great respect, opines that 'there is no evidence for any decline in the country industry before the sixteenth century': Oldland (2011), p. 183. At least as regards Tunstead Hundred, the author disagrees.

135 TNA PROB 11/11/202.

136 Hinton (1973), pp. 195–8. The weavers had earlier, in 1329, successfully rebelled against the local alnager Robert Poley: Sutton (1989), p. 209.

137 If he did, he was not the first to try. In the 1330s the three leading Flemish towns of Bruges, Ypres and Ghent sought to suppress village and small town draperies in Flanders: Munro (1991), pp. 115–6.

[and] apparently continued to maintain their independence of outside control and even much internal control into the sixteenth century'. Yet Oldland observed that 'The Norfolk worsted industry was far more regulated and supervised, and at an early date, than the fifteenth-century woollens industry' and that Norwich exercised 'much greater control over regional production of worsteds than London ever enjoyed over broadcloths'.[138] Norwich's ascent as a centre of worsted production began at the time of the Peasant's Revolt, well before the advent of double worsted.[139] Was the control exercised by Norwich a mechanism not only by which the city promoted and protected its own industry, but suppressed that of Tunstead Hundred weavers? Perhaps a free market in the fourteenth century became, after the rebellion, a heavily regulated one in which the rules were fixed against them.

Both Britnell and Poos were convinced that the key figures in the Essex textile industry were the fullers. In Colchester they 'were more likely than any other group to be earning large trading profits'; and, in the countryside, 'virtually the only people [...] who might qualify as large scale investors and traders' with 'considerable wealth and wide ranging interests'.[140] So, let us test this by looking at the evidence. For Colchester a poll tax return survives for 1377, but it attributed occupations to very few taxpayers and there is no extant return for 1381.[141] During that town's golden period expiring in 1430, weavers marginally outnumbered fullers in the plea rolls and the value of their debt disputes was significantly higher. In time both groups were subjugated to the clothiers, of whom, in the last twenty years of the century, there were twenty-two – more than in Cranbrook, Lavenham or any other town in England – at least two of whom dealt direct with Hanseatic merchants even after the Germans had abandoned the town's outport at Hythe (Appendix B).[142]

The hundred of Hinckford is situated in the north-centre of the county, on the far side of the river Stour from Babergh (Map 1.4). It was within the cloth-making sector of rural Essex, north of a line running from Chelmsford to Maldon and east of a line running from Chelmsford to Saffron Walden. It provides the best extant occupational data from the 1381 poll tax in Essex. In the late fourteenth century the hundred included forty-six parishes, none of which, in 1381, had more than two hundred and fifty taxpayers and the largest of which included Braintree,

138 Sutton (1989), p. 214; Oldland (2011), p. 194.

139 The number of weavers admitted to the freedom of Norwich increased dramatically from twenty-four during the reign of Edward III (1327–77) to fifty-nine during that of Richard II (1377–99): Dunn (2004), p. 216.

140 Britnell (1986), p. 77; Poos (1991), pp. 68–9.

141 Fenwick (1998), 1, pp. 194–205.

142 Richard Aystone: CP 40/895, m. 16v (AALT IMG 574); and John Flyngaunt: CP 40/907, m. 366r (AALT IMG 692)

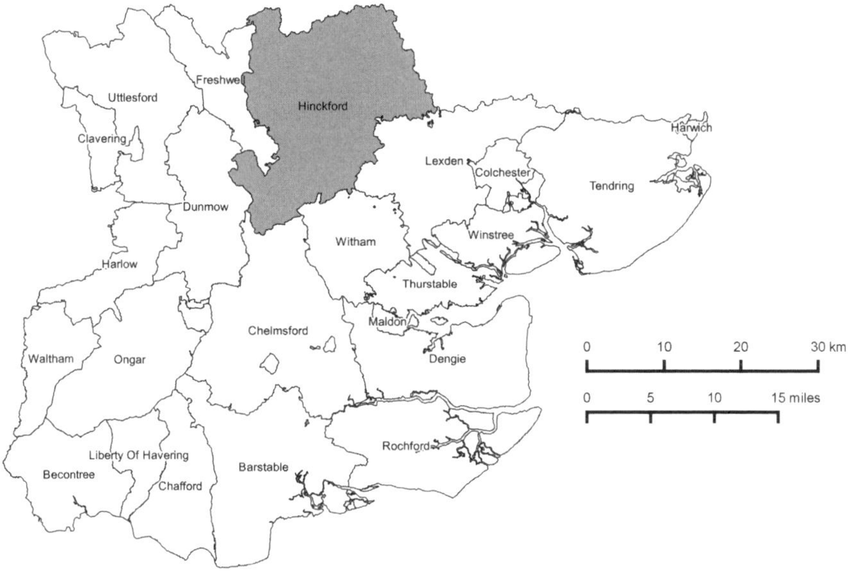

Map 1.4: *Essex Hundreds*
Map format attribution: David Addy

Bocking and Halstead. The total taxable population was just 3,976.[143] None of the settlements would grow to any great size over the course of the following hundred and fifty years. As Poos observed of the county's cloth-making centres, 'the great majority of these places were quite small rural communities'.[144] The 1381 poll tax returns enable us to observe several trends (Figures 1.2 and 1.5, and Table 1.3). A significantly greater proportion of Hinckford fullers than those in Suffolk paid more than the basic 12d. tax. Some may have owed their success to unscrupulous behaviour, since, shortly afterwards, two at Bocking and one at Halstead were indicted for overcharging.[145] They were marginally less heavily taxed than Hinckford weavers, but the two crafts were probably of similar economic standing. In 1381 they were outnumbered by weavers and were found in fewer centres, but later the opposite was true. Seven of the fifteen fullers in 1381 came from Bocking, on the river Blackwater, but only one is found in later records. Instead, the plea rolls suggest that Halstead came to the fore, as a main market town some dozen miles up the river Colne from Colchester. Of four fullers recorded there between 1396 and 1470, John Gauge was most prominent. He faced six claims in the common bench

143 Fenwick (1998), 1, pp. 180–1.
144 Poos (1991), p. 62.
145 Poos (1991), p. 69.

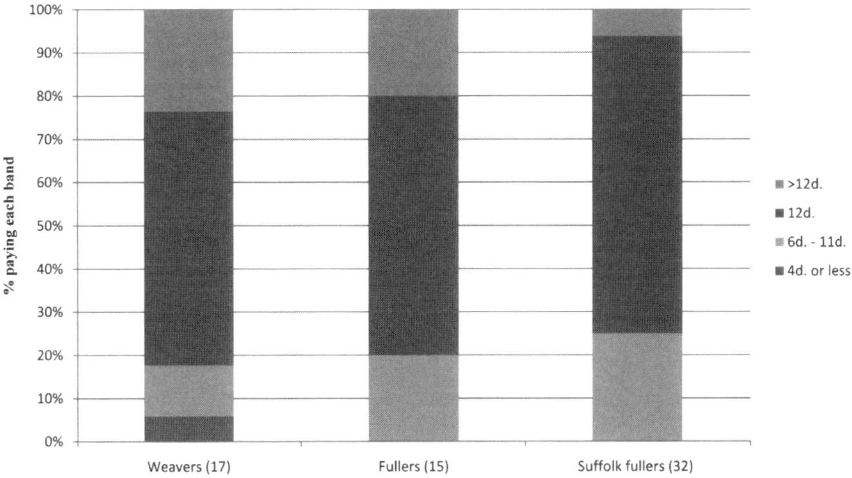

Figure 1.5: *Poll Tax Payable by Fullers and Weavers in 1381 in Hinckford Hundred, Essex; and Fullers in Suffolk*

Source: Fenwick (1998), 1, pp. 206–22; Fenwick (2001), 2, pp. 505–36.

TABLE 1.3: WEAVERS AND FULLERS IN HINCKFORD HUNDRED, ESSEX

	Fullers		*Weavers*	
Period	*No.*	*Centres*	*No.*	*Centres*
1381	15	8	17	11
1396–1470 (10 terms)	11	7	7	4
1480–1500 (10 terms)	4	3	3	2
1520–30 (9 terms)	2	2	3	3

Source: Fenwick (2001), 1, pp. 206–22. TNA CP 40.

between 1418 and 1430 with a substantial mean value of over £5, including one debt incurred in London and due to the City grocer John Dartre.[146] Another Halstead fuller and Gauge's contemporary, John Pykstone, dealt with the Sudbury draper Richard Othelesho. Slightly later, a Thomas Fuller put into trust and beyond the grasp of his creditors a 'teyntour gardyn' in the town in which he would have stretched and dried his fulled cloth.[147] Notwithstanding the apparent success of

146 TNA CP 40/629, mm. 9r, 27r, 211r (AALT IMG 20, 56, 425); CP 40/647, mm. 8v, 107v (AALT IMG 448, 642); TNA CP 40/677, m. 162r (AALT IMG 321).

147 TNA CP 40/656, m. 370v (AALT IMG 1707); TNA CP 40/776, m. 478v (AALT IMG).

John Gauge, this evidence does not support the view that fullers generally enjoyed in Essex the pre-eminence once bestowed upon them.

For a while, textile manufacture in Hinckford prospered. Among the eight cloth-makers in Braintree who produced 2400 dozens to the alnager in 1394/95, two produced 600 and a third 480.[148] Thereafter, decline is reflected both in the number of weavers and fullers recorded in the plea rolls (Table 1.3) and in the alnage accounts. In the 1390s gross textile production in Essex and Suffolk was almost identical, just over 2,700 cloths per annum each.[149] The first half of the fifteenth-century may, however, have been a period of stagnation in Essex cloth-making during which it fell well behind its northern neighbour. In 1468/69 only 2,627 broadcloths or equivalent were sealed, slightly less than seventy years before and half the output in Suffolk. Braintree's production had declined to 210 cloths.[150]

As the Hanseatic merchants moved their trade from Colchester and Ipswich to London in the 1460s, and as the clothiers began their rise to prominence as a link with the Capital, small town craftsmen struggled to maintain their independence. References to fullers and weavers in the plea rolls were replaced by those to clothiers, of whom twelve were recorded in Bocking and Braintree (Appendix B). This dozen did much of their business in London less than forty miles away. Out of fifteen disputes nine arose in the Capital, of which four were with mercers.[151] From this one may surmise that much of their output was unfinished broadcloth. The plea rolls suggest that Essex cloth-making enjoyed something of a renaissance in the latter years of the fifteenth century, and that the share of London trade enjoyed by Essex clothiers was second only to those of Suffolk (Table 1.3). Even if Essex's cloth-making areas did not keep pace with those of Suffolk, they still enjoyed outstanding levels of growth in lay wealth in the late Middle Ages.[152]

Evidence of cloth-making in Suffolk at Bury St Edmunds and Sudbury can be traced back to the thirteenth century, albeit on a more modest scale than the great cloth towns of the lowland eastern plains. Both paid fines for exemption from the assize of cloth in 1202.[153] Jocelyn de Brakelond, cellarer of the abbey in the time of Abbot Samson, referred to Bury's spinners and fullers.[154] Henry III bought cloth from the town.[155] At a time when personal names were still a reasonably reliable

---

148 Salzmann (1913), p. 157.
149 Gray (1924), p. 34.
150 Heaton (1920), p. 85; Britnell (1986), p. 191. For output trends in Colchester, see Britnell (1986), pp. 181–3.
151 TNA CP 40/883, m. 32r (AALT IMG 63); TNA CP 40/943, m. 68r (AALT IMG 137).
152 Darby made specific reference to the wealth of Braintree: Darby et al (1979), p. 258.
153 Miller and Hatcher (1995), p. 101
154 Unwin (1907), pp. 254–5.
155 Miller (1965), p. 70.

guide to occupation, John the Shearman rented from the sacrist William of Hoo a shop in the tailors' street.[156] A town rental mentions three men with the name *Fuller*, dwelling near the South Gate, and the same number with the name *Listere* (meaning dyer).[157] Later, the 1327 subsidy return included one taxpayer by the name of *le Fuller* and three by the name of *le Lestere*. A contemporaneous list of those who formed the Bury mob that destroyed the old Abbey Gate included a William de Burgh dyer, four named *le Lestere* and also William le Webstere (meaning weaver) who was probably not sufficiently wealthy to pay tax.[158]

Less is known about Sudbury in the thirteenth century, although its cloths were then recorded as an export item in the customs of the quay of Ipswich.[159] In 1327 four taxpayers in Sudbury bore the name *Textor* (meaning weaver) with two of them contributing the substantial sum of 5s. each.[160] In the decade before the Black Death at least a score of broad weavers and a dozen narrow weavers were at work in the town.[161]

Ipswich probably had no 'large-scale industry', although Masschaele's analysis of the taxation on moveable goods of 1283 revealed that cloth was, in aggregate, the fourth most valuable item in town. It was owned by only seventeen people, a small proportion (8 per cent) of taxpayers, marking it out as the product of 'specialised manufacture and heavy initial investments'.[162] Whether these seventeen had manufactured the cloth themselves, or held it purely for trade, we do not know. The 1327 Ipswich subsidy return certainly revealed little sign of textile production, including among 210 taxpayers only Hugh le Webbere and Gilbert Carde, possibly a weaver and card-maker respectively.[163] At that time very little cloth was recorded as being exported from the port, suggesting perhaps that Suffolk's early-fourteenth-century industry was targeted at domestic consumption.[164] By the beginning of the fifteenth century there was some evidence of textile production in the town, but nevertheless only ten cloth-workers were identified in the petty court rolls for 1400/01.[165]

Cloth-making in Hadleigh can be traced back into the thirteenth century, with numerous dyers and fullers appearing, from as early as 1276, before the court of Hadleigh Hall held by the priory of Christchurch, Canterbury. As well as a fulling

156 Gransden (2015), p. 106.

157 Redgrave (1908), pp. 198, 212–5, 217.

158 Hervey (1906), pp. xxx–xxxii, 215–7. Those with goods worth less than 5s. were exempted from paying the subsidy: Hervey (1906), p. ix.

159 Twiss (1873), p. 187

160 Hervey (1906), pp. 136–9.

161 Redgrave (1910), p. 102.

162 Miller and Hatcher (1995), p. 102; Masschaele (1997), pp. 21–3.

163 Hervey (1906), pp. 217–20.

164 Childs (1996), p. 130.

165 SROI C/2/1/3/34–37.

mill, in 1305 two dye-houses were to be found on the same manor, one of them let to Vincent the Fuller for 6d. a year, and the other to William Geffrey for 10d. At least five fullers rented property in the manor.[166] The 1327 subsidy return included the names of two men known as *le Webbe*, one as *le Walker* who was probably a fuller, and four as *Tinctor* who were probably dyers. The presence of their names in the return in itself indicates that they were not destitute, but since only two of them paid as much as 2s. tax, they were not particularly affluent either.[167]

The 1327 lay subsidy contains surname evidence associated with cloth-working in forty small towns and villages in Suffolk, including Clare, Kersey, Lavenham and Long Melford. These smaller centres were already concentrating on the production of relatively cheap and light fabrics known as kerseys, and after the Black Death were to challenge the supremacy of the older ones, with Lavenham in particular eventually outpacing them all.[168] Nevertheless, in contrast to so many other eastern lowland towns, cloth-making enjoyed a marked renaissance in Bury St Edmunds, Hadleigh and Sudbury, a revival that continued until the end of the Middle Ages despite competition from these upstarts. In the 1350s Suffolk cloth-makers were responsible for producing 3.7 per cent of English sealed cloth; in the 1390s 5.7 per cent; and in the late 1460s 13.2 per cent, by which time over five hundred of them were at work in the county.[169] Between 1480 and 1500 significantly more than twice the number of cloth-workers and clothiers were recorded in the plea rolls in Suffolk than in any other county in England (Appendix B), and were more widely distributed than in any save Norfolk (Figure 1.1). One in every four recorded clothiers resided in the county. Furthermore, Suffolk clothiers enjoyed the largest share of London trade (Figure 1.3). Chapters 3 to 7 consider in more detail how this came about.

Town v Country

Complaints by English cloth-makers in towns against competitors in the countryside date back at least to the thirteenth-century.[170] Those of the York weavers were particularly 'frequent and serious'.[171] Most textile historians have addressed the tortuous translocation from urban to rural textile production. Braudel

166 Hervey (1902), pp. 152, 154, 156, 161, 162, 166, 169–71.

167 Hervey (1906), pp. 154–5. He recorded the names, perhaps incorrectly, as *le Waller* and *Tinccor*.

168 Miller (1965), p. 78.

169 Gray (1924), p. 34; Heaton (1920), p. 85; Amor (2004), p. 417.

170 Miller (1965), p. 71.

171 Heaton (1920), p. 47.

expressed it succinctly 'the countryside was once more invaded by handicrafts'.[172] Various reasons have been advanced for this process. Push factors from the towns included industrial conflict, restrictive practices, over regulation, labour shortages, higher costs and more onerous taxes; while among pull factors to the countryside were more and less expensive industrial space, rural markets, outside capital, a cheaper workforce, the putting out system and, most famously, the fulling mill. These factors were not, however, always as persuasive or cogent as some historians would have us believe. Carus-Wilson regarded fulling mills as decisive, whereas Bridbury and Miller regarded them as incidental to the success and location of English production. But even this distinction is too crudely drawn because, for example, urban fullers could control rural fulling mills. William Cobald, a fuller of Bury St Edmunds, leased one at rural Lackford, while William Alysaundre, another fuller of the same town, may well have done so in Monks Eleigh.[173]

Much of the blame for the decline of urban cloth-making has been attributed to guilds. Miller believed that 'the very existence of craft guilds or endeavours to establish them might encourage merchants to transfer their entrepreneurial activities to the countryside'.[174] In this respect, they appear to have played two laudable, yet ultimately damaging roles: an earlier one as trade unions seeking to protect their members in dispute with urban elites; and a later one as employer associations in cahoots with the same elites to protect their industry from cheaper and substandard imports from the surrounding countryside. The late-thirteenth and early-fourteenth centuries witnessed violent conflict between merchant-employers and manufacturing craftsmen in, for instance, London and Leicester, that had a long-term deleterious impact on cloth production there. Swanson described the York weavers' guild 'from its very inception as a defensive organization against the power of the entrepreneurs'.[175] During the mid-fifteenth century Great Slump urban magistrates were keen to protect industries so as to minimize unemployment and social unrest. In Bury St Edmunds the abbot and the town's weavers' guild agreed new ordinances that were aimed at protecting them from competition by external clothiers.[176] As we shall see, such restrictive practices may have served their purpose in Bury, but more often they did not. Whatever social policy goals they achieved within towns, they invariably stifled economic activity too. Hadleigh may have eventually lost ground to its neighbours partly because its clothiers' guild concentrated cloth-making in few hands. By the late 1460s the town had been overtaken

172 Braudel (1982), p. 310.
173 SROB 262 Osbern; CCA-DCc-MA/33, f. 47r.
174 Miller (1965), p. 73.
175 Swanson (1989), p. 38.
176 SROB B9/1/2 published in Historic Manuscripts Commission Report xiv, App. pt. viii, pp. 133–8.

by Lavenham as a centre of textile production.[177] At about the same time in larger cities such as Lincoln, Salisbury and York, where guilds were long established, the industry was at a low ebb. In the mid-fifteenth century the master ('graceman') of the weavers' guild in Lincoln sought to exercise jurisdiction over, and collect membership dues from three weavers from Nottingham villages. In the early-sixteenth century the master of the York weavers' guild tried to do likewise from ten weavers from north Yorkshire villages.[178] We do not know whether either of them was ultimately successful, but in both cases they were swimming against the tide.

In exceptional cases urban and rural cloth production continued to thrive together. Suffolk experienced 'little of the marked antagonism between the older towns and the newer centres of industry.'[179] Elsewhere too the translocation could, at least for a while, be harmonious. In Somerset and Wiltshire, until the Great Slump, urban and rural cloth production developed in tandem.[180] Some historians have even questioned whether it ever happened at all. Bridbury wrote that 'the countryside never gained on the towns because every industrial advance meant that the villages making such advances took on first the colouring and then the forms and features of towns, so that, in the end, they became towns in all but name'. Hadleigh and Lavenham in Suffolk are good examples. Munro lent powerful support to this argument and concluded that the distinction between town and country was 'blurred'.[181]

Looking at the plea roll evidence, just over five-hundred of the cloth-workers and clothiers recorded between 1480–1500, representing 21.8 per cent of the total, lived in the twenty-four greater medieval towns of England (Appendix B).[182] Of these only twelve had a textile community of at least ten individuals and just three – London, Norwich and Bury St Edmunds – one of more than fifty (Figure 1.6). These bare statistics, however, hide some important detail. Just over a third of cloth-finishers – ie dyers and shearmen – lived in the greater towns; and just over a fifth of cloth-producers – ie spinsters, card-makers, weavers and fullers; but only one in twelve clothiers did so. Indeed, just two great towns, Bury St Edmunds and Colchester, were home to more than ten clothiers. This suggests that the clothier, with his 'more efficient business organization', was the central figure in any

177 Weller papers; Amor (2004), pp. 419–20.

178 TNA CP 40/788, m. 89v (AALT IMG 1222); CP 40/1013, m. 34r (AALT IMG 68).

179 Thornton (1928), p. 155.

180 Hare (1999), p. 5; Hare (2011), 182.

181 Bridbury (1982), p. 65; Munro (2003), p. 227.

182 As well as London, these were Beverley, Boston, Bristol, Bury St Edmunds, Cambridge, Canterbury, Chester, Colchester, Coventry, Exeter, Gloucester, Hereford, Leicester, Lincoln, Lynn, Newcastle, Norwich, Oxford, Salisbury, Shrewsbury, Winchester, Yarmouth and York: Kermode (2000), pp. 442–3.

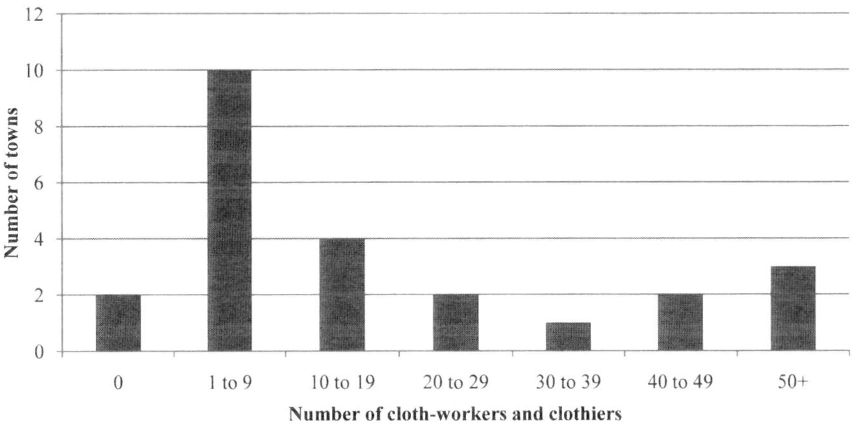

Figure 1.6: *Cloth-making in the Great Towns 1480–1500*

Source: TNA CP 40B (Appendix B)

migration of cloth-making from town to country.[183] Carus-Wilson summed up the advantages he enjoyed in living outside any municipality, he 'had not to suffer a constant drain on his purse for local taxation, pageants, liveries, and all the burdens of public office. Nor had he to submit to the industrial controls that bound the city manufacturer. He was free to employ what workers he pleased [and] to offer whatever wages would secure them'.[184]

Growth or Decline?

A reader coming to the end of this chapter might well be left with the impression of a thriving cloth industry in late medieval England. Not so according to the authors of a recent authoritative volume on *British Economic Growth 1270–1870*. Their surprising contention was that woollen textile production actually fell between 1270 and 1500. In other words, any growth in production in newer areas after the Black Death was never enough to compensate for the decline of production in the towns of the lowland eastern plain in the thirteenth century. In the absence of significant quantitative evidence about the output of those towns, definitive confirmation or refutation of this observation is impossible. Even more surprisingly, they argue that production peaked in the 1350s and then fell continuously thereafter, such that any growth in one part of the country must have been counterbalanced by decline elsewhere. The authors reached this conclusion by the simple method of deducting the volume of wool exports from aggregate wool

183 Oldland (2007), p. 110.

184 Carus-Wilson (1987), p. 684.

output at any given time. They calculated wool output primarily by reference to Campbell's Medieval Accounts Database and contended that it fell sharply after the Black Death. Even though the volume of wool exports also fell sharply, not enough was left to support growth in textile manufacture.[185]

These arguments sit uncomfortably with the local evidence surveyed in this chapter. There is little sign in the great cloth-making towns of eastern England in the thirteenth century for large numbers of cloth-workers or for widespread division of labour. Yet there is ample evidence of both in fifteenth-century England, which implies a thriving and larger industry than the spatially confined one of the thirteenth century. So do the poor sources of the thirteenth century and the better ones of the fifteenth century create a mirage?

Moreover, how do we square this alleged decline with the evidence of the alnage accounts that show a three-fold increase between the 1350s and the 1390s in the number of cloths sealed each year;[186] or with Phythian-Adams and Britnell's view that the period 1380 to 1430 was 'a golden age of English urban civilization';[187] or with the work of Schofield and Darby that strongly linked growth in lay wealth to the cloth industry?[188] The only way in which the apparent increase in cloth production between the 1350s and the 1390s could, in fact, conceal a fall in output is if over two thirds of production in the earlier period was narrow cloths and exempt from alnage. Such a prospect is possible, but intrinsically unlikely. Why bother to regulate and tax less than one third of an industry? A downward trend after 1400 is incontestable. The alnage accounts document a fall in aggregate production between the 1390s and the late 1460s/early 1470s from 49,308 to 39,345 sealed cloths. A Parliamentary report from the officers of the exchequer suggests that in 1433 output was below even the lower figure.[189]

A crude method of assessing the credibility of the assertions in *British Economic Growth 1270–1870* is to create a balance sheet of places where cloth production was rising and falling after 1300. Among large textile towns and cities, several

185 Broadberry et al (2015), pp. 81–83, 138, 146–7. Campbell's Medieval Accounts Database is, by his own admission, not conclusive, but he has a powerful ally in Powell who also believed that wool output fell sharply: Power (1965), pp. 36–7. They have recently been challenged by Oldland who concluded that 'wool production remained remarkably strong': Oldland (2014), pp. 43–4.

186 Gray (1924), p. 34, but see the comment in note 58 above.

187 Britnell (1986), p. 159.

188 Schofield (1965), p. 507; Darby et al (1979), pp. 247, 256, 258–61.

189 Gray (1924), pp. 13, 34; Heaton (1920), p. 85. In 1432–3 the revenue raised 'from the farms of subsidies and alnage' was £720 10s. 1d., which equates to alnage and subsidy on 38,427 broadcloths: *Rot. Parl.*, iv, p. 433. However, since the collection of these moneys was farmed out, the total represented the amount paid by the farmers to the Crown, not that paid by cloth-makers to the farmers.

appear to have enjoyed some recovery by the 1350s from their industrial crisis, notably Bristol, Canterbury, Lincoln, Winchester and York. But only in Bristol and York would cloth-makers continue to prosper, at least until the mid-fifteenth century. In 1400 Salisbury was probably England's leading textile city, but by 1450 it had lost many of its weavers. Colchester in Essex also enjoyed early-fifteenth-century industrial and commercial growth. Nearly all textile towns struggled during the Great Slump, but perhaps Bury St Edmunds, Newbury, Northampton and above all Norwich fared best. By 1500 Bury St Edmunds, Norwich, and possibly Coventry, were the main success stories, with the first two dependent to different degrees on worsted production.

Among smaller towns and villages, in the second half of the fourteenth century those in Tunstead Hundred in Norfolk stood out. To a slightly lesser extent, the cloth-makers of Hinckford Hundred in Essex, and of Hadleigh in Suffolk also did well, although only those in Hadleigh went on to enjoy largely uninterrupted success. In the early 1400s Castle Combe in Wiltshire was a small boom town, but that boom lasted little more than fifty years, and the settlement was later eclipsed by Devizes. Cloth-makers of Babergh and Cosford in Suffolk and, further west, of Frome/Mells in Somerset, south Devon and Stroud in Gloucestershire prospered through most of the fifteenth-century. Cranbrook and its satellites in the Kentish Weald came to the fore shortly before 1500. Weighing this all up, the author would tentatively suggest that the English textile industry grew between 1350 and 1400, declined between 1400 and 1470 and then, with the end of the Great Slump, enjoyed some export-led growth from 1470 until 1500 without ever again reaching its 1400 peak. Suffolk was probably the only county in which the growth of cloth production was more or less continuous.

In the next chapter we shall consider the factors that influenced the fortunes of these textile centres and the reasons why their fortunes waxed and waned.

– 2 –

Influential Factors

The success of cloth-making in one place but not another, at one time but not earlier or later, cries out for an explanation. There was no *prima causa* for the development of textile manufacture anywhere in England. Many factors were in play. Very often it is impossible to assess which were more influential than others, or how they interacted or intertwined. In this chapter we explore them under the general headings of physical geography, economy and technology, and society. The influence of some of these factors is considered in greater depth throughout chapters 3 to 6.

With her seminal paper on *Industries in the Countryside,* Joan Thirsk became the midwife of an academic debate over proto-industrialisation, or 'industrialisation before industrialisation'.[1] The protagonists were in search of a universal explanation for the transition from feudalism to capitalism and seized on the putting out system of textile production as a key causal stage in that transition. Their theory embraced some of the factors addressed below, but suffered a major weakness. None of the rural textile centres considered in the previous chapter ever experienced significant factory-based industrialisation; on the contrary, all suffered de-industrialisation in the early modern period.[2] So, in considering the influence of any of the factors explaining where and when textile production was established, one must remember Thirsk's wise words that 'there is no certainty or finality in any explanation for the growth of a rural industry in one district rather than another'.[3]

---

1 This debate is admirably summarized by Clarkson in *Proto-Industrialisation: The First Phase of Industrialisation?*

2 In this chapter, these are referred to as the featured textile centres.

3 Thirsk (1984), p. 217.

Physical Geography

Climate and Weather

In the late spring or summer of 1257 Samalas, in far-off Indonesia, erupted, emitting at least 40 cubic kilometres of material.[4] It was the largest volcanic eruption in the Common Era and had a profound impact on world climate. Together with later eruptions elsewhere and a reduction in solar radiance, it may well have contributed to changes in sea level temperature and atmospheric circulation that moved northwestern Europe from the period of the Medieval Climate Anomaly to the Little Ice Age. The first half of the fourteenth century was 'the most prolonged episode of unstable climatic conditions during the entire second millennium'; while the first half of the fifteenth century was 'the negative tipping point' that experienced 'the single greatest change in atmospheric circulation of the last 10,000 years'.[5] In the mid-fourteenth century William Merle, the rector of Driby in Lincolnshire, kept a diary in which he recorded heavy snow fall and hard frosts. Such cold weather was blamed by those paying the lay subsidy of 1341 (the *Nonae*), in various parts of central England, for crop destruction and outbreaks of murrain among sheep, as well as poor quality or low numbers of lambs.[6] As we shall see in chapter 3, a longer-term drop in winter temperatures and more frequent severe winters helped reduce the mean length of the wool staple and the mean weight of fleeces, so limiting the volume of wool available to cloth-makers both in England and continental Europe. Furthermore, the bad weather increased the demand for warm, heavy broadcloths over the lighter fabrics that had been the mainstay of the thirteenth-century textile industry. Darker, heat-absorbing colours came into fashion, increasing demand for woad with which they were dyed.[7]

Undulating Landscape

Some academics would have us believe that proto-industry was concentrated in 'barren mountain regions' and 'harsh mountainous areas'.[8] Yet, however picturesque

---

4 http://www.bbc.co.uk/news/science-environment-24332239

5 Campbell has written and lectured extensively on environmental change based on the evidence of dendrochronology and ice core analysis: Campbell (2000), p. 369; Campbell (2010), pp. 305–6; Campbell (2012), pp. 122–3; http://www.econsoc.hist.cam.ac.uk/podcast-campbell.html

6 Baker (1966), p. 523; Livingstone (2015), pp. 332, 335–6.

7 The rationale behind the fashion for dark, heat-absorbing cloth was kindly explained to the author by Bruce Campbell in conversation on 3 April 2016. The will of the prominent Bury St Edmunds draper John Odeham was replete with references to black cloth: SROB 129 Hawlee.

8 Coleman quoting, sceptically, from the work of Kriedte, Medick and Schlumbohm: Coleman (1983), pp. 438, 441.

they might be, mountains were too remote from markets and transport networks. In fact, for good reasons, rural cloth-making evolved in the undulating landscapes of south Suffolk and north Essex, the Cotswolds, and the Mendips. Suffolk is generally thought of as a flat county. Anyone who climbs 136 metres to the top of Great Wood Hill on the Newmarket Ridge stands at its highest point. Yet the county's medieval cloth industry was concentrated in one of its most undulating districts where the relative relief often exceeds 36 metres per kilometre.[9] A bird's eye view of Lavenham illustrates this well (Plate 2.1). The centre of this small town stands at an elevation of about 65 metres, above all but nine of the featured textile centres (Figure 2.1). It is higher than Salisbury in Wiltshire and Stroud in Gloucestershire, and less than 7 metres lower than Frome in Somerset. In an undulating landscape the land drains easily and, in doing so, creates rivers and river valleys. Nearly, all the English textile centres that developed after 1300 were situated in a river valley, be it that of the Brett, By-Brook, Colne, Exe, Frome, Kennet or Stour. In Stroud almost every section of the stream was leased out, as was the case in Long Melford and also possibly Monks Eleigh.[10] Thirsk considered that the location of Suffolk's textile

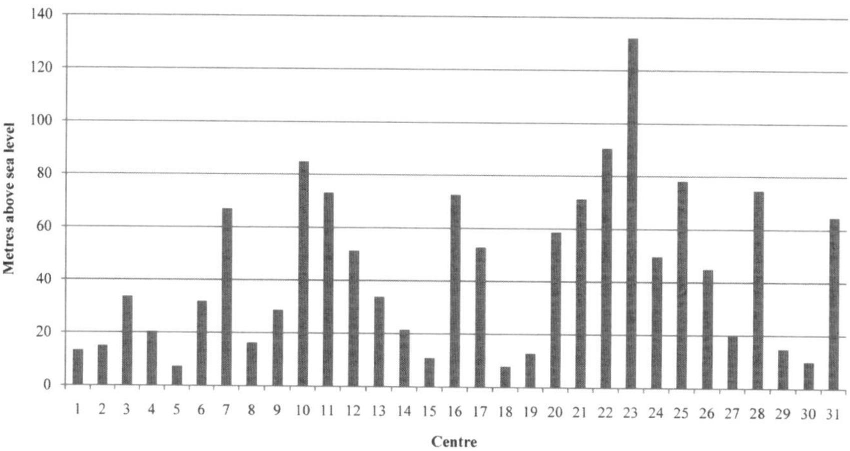

Figure 2.1: *Elevation of Textile Centres*

Notes: 1. York 2. Beverley 3. Louth 4. Lincoln 5. Boston 6. Stamford 7. Leicester 8. Norwich 9. North Walsham 10. Coventry 11. Northampton 12. Bury St Edmunds 13. Sudbury 14. Hadleigh 15. Ipswich 16. Braintree & Bocking 17. Halstead 18. Colchester 19. Bristol 20. Stroud 21. Frome 22. Castle Combe 23. Devizes 24. Salisbury 25. Newbury 26. Reading 27. London 28. Cranbrook 29. Canterbury 30. Exeter 31. Lavenham.

Source: http://www.mapdevelopers.com/elevation_calculator.php

9 Personal communication from Edward Martin, with a copy of Rosemary Hoppitt's map of relative relief, 13 December 2015.

10 Darby et al (1979), p. 259.

manufactures was 'due entirely to the network of rivers in the southern half of the county emptying fast-flowing waters into the estuaries of the Stour, the Orwell and the Deben'.[11]

River valley locations were critical to the success of the textile industry because the presence of flowing water was almost always a prerequisite of cloth-making.[12] Fulling and dyeing depended on it. The safe disposal of waste trade material was difficult if not impossible without a river to wash it away. In thirteenth-century France it took approximately two cubic metres of water to produce a standard measure of coloured cloth.[13] In England it was no different. Most London dye-houses were established close to the Thames.[14] Norwich dyers were to be found working on both banks of the Wensum.[15] In Exeter they carried on their business on Exe Island, an area of marshland between the city walls and the river Exe, where they had ready access to plentiful supplies of water and were far enough away from residential sectors to avoid creating a nuisance.[16] Of the thirty-one featured centres, only Devizes stands more than 100 metres above sea level (Figure 2.1). Since rivers grow more mature as they travel downstream, the town suffered from a paucity of flowing water. The price was an absence of fullers and dyers, the crafts most dependent on it (Appendix B). Only the tiniest quantities of woad and madder were ever carried to Devizes from Southampton.[17]

Arable and Pastoral Husbandry

The management of the English landscape changed, as a consequence of the demographic collapse caused by the Black Death and subsequent epidemics, in a way that ultimately encouraged the development of rural cloth-making. The demand for land fell, pushing down rents and, in time, encouraging engrossment of holdings and field enclosure. The demand for labour rose, pushing up wages, raising standards of living and creating new markets for agricultural and manufacturing produce among people who previously could not have afforded it. Aggregate demand for grain fell, while per capita demand for meat and dairy products rose. All these economic forces combined to turn many large open fields, once dedicated to labour intensive arable cultivation, into smaller pastures that favoured the rearing of livestock. The impact of these general economic trends is observable in

11 Thirsk (1984), p. 222.
12 Zell (1991), p. 236.
13 Rawcliffe (2013), p. 203.
14 Crowfoot (2001), p. 20.
15 Rawcliffe (2013), p. 209.
16 TNA CP 40/717, mm. 82r, 425r (AALT IMG 168, 858).
17 Distribution maps in Hicks (2015), pp. 204–8.

most of the major cloth-making counties. On the heavy clay soils of Suffolk and north Essex 'decline of cereals production was severe' and gave way to stock fattening and dairying.[18] In the Kent Weald 'the area under the plough sharply contracted', while more substantial landowners concentrated on the raising, fattening and sale of cattle.[19] In the sheep-corn country of Somerset and Wiltshire agriculturalists used 'land in ways better suited to its inherent qualities, using it with great flexibility, and sometimes by increasing the numbers of livestock, and especially sheep'.[20] As the area of land under pasture steadily rose, so did the proportion of pasture-fed sheep as against those folded on arable, so making long staple wool suitable for worsteds more readily available.

These changes should, in theory, have increased the number of smallholders available to work in non-agricultural employment; thus creating a rural workforce ready to learn new skills in spinning, weaving and fulling, and providing the labour necessary for the putting out system of production.[21] However, in their respective studies of the textile industry in Essex and Norfolk, Poos and Whittle both challenged this hypothesis. Poos argued that in Essex the 'marked shift towards pastoral production' in the 1400s occurred after the county's industry 'had already achieved its full medieval development', although he later conceded that 'remarkably little is known about the structure, organization and participants in the cloth industry in rural and small-town Essex in the later middle ages'.[22] It will be argued in this volume that the Suffolk cloth industry did not reach full maturity until the final third of the fifteenth century, once the Great Slump had run its course and Hanseatic merchants had deserted the ports of Ipswich and Colchester. Only then did the clothier emerge as a major figure and the putting out system become of critical importance. It seems very unlikely that Essex was markedly different. The earlier analysis, in chapter 1, of cloth-making in Hinckford Hundred would suggest that this district at least followed the Suffolk pattern and, indeed, no clothier was recorded there before 1480 in the plea rolls that have been studied for this volume.

Whittle pointed out that the agrarian regime of north-east Norfolk could not be described as pastoral, but 'was a mixed farming region with a highly productive arable sector'.[23] However, Norfolk was a county of worsted weavers, not clothiers (Appendix B). Furthermore, the earlier analysis of cloth-making in Tunstead Hundred strongly suggests that it was depressed for much of the fifteenth century.

---

18 Britnell (1991), p. 57; Bailey (2007), pp. 225, 297.

19 Mate (2006), pp. 173, 176.

20 Miller (1991), p. 150. In Wiltshire Hare identified 'signs of the emerging pastoral economy': Hare (2011), p. 59.

21 Bolton (1980), pp. 269–70; Mate (2006), p. 176.

22 Poos (1991), pp. 63–4.

23 Whittle (2000), p. 248.

It would not have been prime territory for the evolution of a putting out system even if husbandmen had had the time to learn weaving. If the challenge made by Poos and Whittle can be thus rebuffed, then the link between land management, surplus labour and cloth-making becomes very difficult to refute. The evidence suggests that textile production was more likely to grow in those regions that experienced a significant switch from arable to pastoral husbandry.

Engrossment and Enclosure

This link between land management, surplus labour and cloth-making may perhaps be re-enforced by studying the process of engrossment and enclosure of holdings. Engrossment was the accumulation in the hands of a few tenants of land that had once been divided among many. Enclosure was the creation of a hedge or fence around land in single ownership. In the early sixteenth century agricultural commentators recognized that enclosure was beneficial to animal husbandry, but the process had already been underway for many years before then. Fitzherbert opined that 'though a man be but a farmer, and shall have his farm XX yeres, it is lesse coste to hym and more profyte to quyckeset, dyche and hedge, than to have his cattell goo before the herdeman'.[24] Engrossment and enclosure were more likely to happen in villages where fields were already fragmented, as was the case in much of East Anglia. The enclosure rate between 1455 and 1489 was at its highest, so that by 1500 as much as 45 per cent of the English countryside was enclosed.[25] Engrossment increased the number of residents who were landless or held only small parcels of land. Enclosure often denied them the opportunity to pasture their own beasts and so curtailed a source of valuable income. For them, work in the textile industry was particularly welcome.

Evidence of engrossment and enclosure in textile counties is widespread. An analysis by Martin of the complex land holdings in Worstead, Norfolk showed that as early as the thirteenth century the engrossment of holdings was well underway. In 1270 about 60 per cent of arable land belonged to nineteen holdings measuring between 20 and 90 acres, while nearly two-thirds of holdings were less than 5 acres in size.[26] In the Kennet valley of Berkshire tenants were 'rationalizing their holdings, enclosing, engrossing and creating larger farms'.[27] In Wiltshire the clay vales and the Cotswolds in the west, where cloth-making flourished, experienced more enclosure than the chalkland in the east.[28]

---

24 Fitzherbert (1534), pp. 76–7.
25 Wordie (1983), pp. 492, 494.
26 Martin and Satchell (2008), p. 80.
27 Yates (2007), p. 155.
28 Hare (2011), pp. 16, 27.

The same process can be seen on some Suffolk manors by studying rentals naming tenants at one date, in which the entry for each parcel of land was interlineated with the name of the tenant at a later date. In Monks Eleigh such a rental was prepared in 1379/80 and interlineated at some time around 1430. Between these two dates there was a sharp fall in the recorded number of tenants from ninety-two to forty-two.[29] Furthermore, there were early signs of acquisitive individuals consolidating in their own hands adjacent parcels that had once belonged to several tenants. For instance, John Reve held nearly all of the lands known as *Goodynggesquarter and Gaywodequarter* that included eight crofts and had previously been in the names of three different tenants. Austin Dunton, probably the Hadleigh draper of that name, held a dozen different parcels of land comprising about 76 acres, including *Salmanneslond Le Ryche, Gambyslond, Manefeldlond* and *Boytones* which appear one after the other in the rental and had previously been in four different hands.[30] Although Monks Eleigh had, by 1430, evolved into a moderately successful textile centre, only two cloth-workers can be identified in the interlineations, namely the fullers Roger Reyver and Robert Talbot. Neither was among the leading tenants. Reyver rented two cottages and another nine acres of land, and Talbot just over 15 acres. For them textile work provided invaluable income.

Engrossment and enclosure occurred on the manors of Thorney just outside Stowmarket, a town where cloth-making boomed in the second half of the fifteenth century. Three rentals were prepared, probably in the first quarter of the fifteenth century, and interlineated probably in the third.[31] The evidence of these rentals, when combined with that of contemporaneous court rolls, indicates a decline in the number of tenants on the manors and a process of engrossment of holdings. This was then followed by enclosure. The Kebbyl family took a lease of the demesne land and, in the face of hostility from some of their neighbours, quickly enclosed it. Thereafter, the tenanted land too was gradually enclosed.[32] Here also, those rendered landless or deprived of grazing welcomed the opportunity of work in textiles.

These examples strongly suggest that cloth-making flourished in many of those regions where land holdings were engrossed and enclosed.

---

29 CCA-DCc-Register/B: Monks Eleigh Rental 1379/80, ff. 149v–157r. The c. 1430 date was not written on the rental, but ascertained by reference to the dates on which some of the tenants were cited in the plea rolls. Five of those named in the original rental were cottagers, for whom there were no interlineations. For a few other parcels there was no interlineation. At each date some of the tenants may have sub-let their land, but there is no evidence that sub-letting was more prevalent at one date than another. See Yates (2007), pp. 159–61.

30 In 1474 William Dunton, who had been executor of Austin Dunton and so may have been his son, held the manor of Boytones in Monks Eleigh: *CCR 1468–1476*, p. 405.

31 SROI HA/CC2/1/3, 4; HA/CC2/2/1.

32 Amor (2006), pp. 177–183, 185, 190.

Economy and Technology

Standards of Living

After the Black Death peasants 'could buy the more costly clothes. They could indulge in the greater comfort of the linings and dress more colourfully, with blues and greens to some extent replacing the old whites and russets.'[33] As a Parliamentary petition declared in 1363 'the commons of this your said realm, both men and women, have worn and daily wear extravagant and inappropriate clothing.'[34] Thereafter, the government introduced a string of sumptuary laws, aimed at imposing, on each of the various classes, a dress code appropriate to their social standing, but this legislation was not successful. The fashionable doublet and hose replaced the plain, shapeless tunic and this form of attire was so esteemed that it was often mentioned as a valued item in wills. For example, Roger Ropkyn of Thrandeston bequeathed a red doublet, and John Parmenter of Great Finborough one of green worsted with a hood.[35] In Ipswich items of clothing were mentioned in wills more frequently than any other chattel.[36]

The per capita demand for textiles rose, as did the need for skilled craftsmen to turn them into the latest fashions, providing the industry with a host of new domestic customers. Historians have speculated that about ⅛ of consumer spending was on clothing.[37] Dyer reckoned that 'in 1500 a million-and-a-quarter adults were buying annually an average of 3 yards of cloth each', generating aggregate sales of about 160,000 broadcloths.[38] They bought from men such as the Hadleigh clothier William Forthe who, as we shall see in chapter 6, created a sales network in the small towns and villages of East Anglia. They did not aspire to the finest broadcloth that would have cost them in excess of 9s. a yard, equivalent to at least eighteen days' wages, but set their sights on low-to-middling-quality fabrics.[39] Robert Joure, a husbandman of Great Finborough, bought from Edward Wymark, a draper of Bury St Edmunds, 6¼ yards of red cloth, 6 yards of '*blak lynyng*', an ell (45 inches) of worsted and an ell of kersey called '*hosecloth*' all for 35s. 1d., or 2s. 4d. a yard. Wymark's other customers included John Whepsted, a labourer of Brettenham; John Grene, a husbandman of Ixworth Thorpe; and Robert Smyth, a butcher of Rickinghall Inferior.[40]

---

33 C. Dyer (1989), p. 177.
34 *Rot. Par.*, v, p. 504.
35 SROB 108v, 308 Baldwyne.
36 Amor (2011), p. 34.
37 Munro (2009), p. 52.
38 C. Dyer (2005), p. 150.
39 Munro (2003), p. 319.
40 TNA CP 40/951, mm. 5v, 392v (AALT IMG 10 dorse, 748 dorse).

This evidence of affluence is supported by real wage indices, collated by various economic historians, indicating that during the fifteenth century the lower orders enjoyed a 'golden age in the history of standards of living' that would remain unmatched for the next four hundred years.[41] Farm labourers could provide for an entire family on less than two hundred days' work a year.[42] This gave time for leisure and play that had once been denied them by the rigour of subsistence living. According to Bailey, 'a leisure culture emerged among the lower orders of rural society [and] games and pastimes became increasingly popular'.[43] Despite the passage of legislation against football and tennis, camping closes sprang up in many parts of East Anglia on which could be played a game that combined football and handball.[44] In 1414 two hundred tennis balls were unladed from the ship *Maryknyght* in Ipswich.[45]

How does one square this merry world of football and tennis with one in which men and women were willing to subject themselves, as outworkers, to a local clothier in the hard labour of distaff, loom and fulling trough? In a revisionist paper, *Unreal Wages*, Hatcher challenged the real wage indices on which this rosy view of peasant life is based. In his view, they exaggerated the amount that an agricultural labourer might hope to earn outside busy harvest time, and wrongly assumed that he could find work if and when he wanted to, when, in fact, employment was 'intermittent and piecemeal'.[46] During the Great Slump of the mid-fifteenth century, when agriculturalists were adversely affected by poor harvests and loss of livestock to disease, unemployment and low pay were particularly serious problems for their seasonal workers.[47] It was at this very time that the clothier emerged in Suffolk (Figure 6.1) and, slightly later, in Essex and Kent. He may well have seized the moment to recruit smallholders and landless peasants who were hungry and desperate for work. Once they had become dependent on him for much of their income, it would have been a brave decision to quit work in the textile industry,

---

41 Britnell (1986), p. 1. Real wage indices are collated by comparing wage rates against prices of an appropriate basket of consumables at any given time: Hatcher (1977), p. 71; Farmer (1991), pp. 520–4; G. Clark (2007), pp. 99–100, 109.

42 Allen and Weisdorf (2011), pp. 720–1.

43 Bailey (1994), pp. 163–4.

44 *Statutes*, 2, pp. 57, 163; Dymond (1999), pp. 154–5.

45 TNA E 122/51/39.

46 A recent estimate of the average number of days worked per person in 1433 arrived at a figure of 165: Broadberry et al (2015), p. 264.

47 Hatcher (2011), pp. 8–12. Hatcher has won support from Broadberry et al (2015), pp. 449–50, 263–4; and from Humphreys and Weisdorf in a paper presented to the Economic History Society conference in 2016: https://jacobweisdorf.files.wordpress.com/2012/04/unreal-wages_humphries-and-weisdorf_jan-2016.pdf

even when animal husbandry enjoyed some recovery in the final third of the fifteenth century.[48]

A growing body of evidence now suggests that ordinary people in post-Black Death England did not consistently enjoy full employment or as high a standard of living as historians once thought. Certainly, relatively few of them possessed semi-luxury or luxury goods such as pewter or silver.[49] However, clothing represented a higher need than such items and, notwithstanding the decline in national population, almost certainly aggregate domestic demand for commercially produced textiles was higher after 1349 than it had been before.

Export Trade

The best estimates of the population of late-medieval England are between 2.2 and 3 million in 1377, and between 2.2 and 2.3 million in the first half of the 1520s. In the 1450s numbers almost certainly fell below 2 million.[50] So, however well peasants were doing and however much cloth they were buying, there was still a chronic lack of aggregate demand in the domestic economy. Much of continental Europe had been as badly affected as England by plague. Nevertheless, the surviving population and GDP of those lands that now comprise Belgium, France, Germany, Italy and the Netherlands, all key overseas markets for English cloth-makers, were ten times larger than England's.[51] Although very few English cloth-makers engaged directly in exports, the success of commercial textile manufacture depended ultimately on prising from traditional competitors, principally the Low Countries drapers, an ever increasing share of international sales.[52] The growth in the export trade after the Black Death was not uninterrupted, but growth there undoubtedly was, from almost nothing in most years during the first half of the fourteenth-century to nearly eighty thousand broadcloths in 1499/1500.[53] Furthermore, as discussed elsewhere in this volume, over that period the quality of the exported cloth steadily improved.

---

48 Britnell (1997), pp. 238–9.

49 Seer (2013), pp. 133–7. See also Table 4.1.

50 Hinde (2003), pp. 64, 68–73; Broadberry et al (2015), p. 205. Clark gave significantly higher population figures, so, as yet, there is no final consensus: Clark (2010), p. 120.

51 Maddison (2003), pp. 34–5, 46–7; Broadberry et al (2015), p. 201. Maddison provided population and GDP estimates for West European countries as at 1500, but his unrealistically high estimate of population and low estimate of GDP for the UK leaves one with more confidence in his relative figures than his absolute ones.

52 The names of a few cloth-makers, such as William Debenham of Ipswich in the 1390s and Roger Crytott of Lavenham in the 1460s, appeared in both the alnage and the customs accounts, but they were exceptional: Amor (2011), p. 132.

53 Carus-Wilson and Colman (1963), pp. 138–9.

The reasons for this growth have been oft rehearsed. The fiscal and bullion policies of Edward III and his successors made wool extremely expensive, and at times unavailable in the Low Countries and so gave English manufacturers a competitive edge over their old rivals. Munro considered that 'English cloth enjoyed a significant advantage in being produced from the same fine wools tax-free, while bearing only a minimal 2 per cent export duty'.[54] Bridbury contested the importance of fiscal policy on the grounds that the decisive downturn in wool exports occurred many years after the hike in export duty.[55] Furthermore, if Broadberry et al are correct, then the English industry was at its medieval peak in the late 1200s when the export duty on wool was much lighter.[56]

Although most countries in western Europe experienced some political and social upheaval, Flanders suffered worst. There 'internal and external troubles, acting and reacting upon each other shook the industry to its foundations'.[57] During the Anglo-Burgundian War of the early 1430s, for instance, 'the Flemish drapery towns suffered mass unemployment and famine'.[58] So, whoever is correct about the importance of fiscal policy, the Flemings as main rivals to English cloth-makers were enfeebled. The 'final victory of the English cloth trade', that witnessed a spurt of exports in the closing years of the fifteenth century, was the result of a revival in the overland trade route from northern Europe to Italy, the opening of new silver and copper mines in south Germany and the debasement of the English currency. These events combined to suck into Antwerp, from opposite directions, English cloth and the bullion necessary to buy it.[59] The London–Antwerp axis became paramount in the textile trade.

Of course, it was not quite so straightforward. Various other factors at different times, in different parts of the country influenced the volume of cloth exports. During the period of their English occupation, Normandy and Gascony were prime markets for the textiles of southern and eastern counties and the loss of these provinces was a severe blow. Berkshire cloth-makers prospered while Italian galleys continued to depart regularly from Southampton, but struggled when the revival of the overland route marked the decline of this Mediterranean trade. An *entente commerciale* with Spain benefited the cloth-makers of south Devon who specialized in medium-priced, finished fabrics that were much in demand in Iberia. The same products sold well in the Baltic, attracting Hanseatic merchants to the ports of Colchester and Ipswich from whence they could ship them in bulk. The

54 Munro (1970), p. 226.
55 Bridbury (1982), pp. 89–93.
56 Carus-Wilson and Coleman (1963), pp. 194–6.
57 Carus-Wilson (1987), p. 657.
58 Munro (1970), p. 240.
59 Munro (2003), pp. 292–4.

peak in Hanseatic exports from East Anglia coincided with the maximum number of recorded Suffolk weavers and fullers, and the subsequent decline paralleled the rise of the county's clothiers (Figures 2.2, 4.2 and 6.1). For a critical period during the mid 1400s this German trade cushioned the Suffolk textile industry against the worst effects of the Great Slump and gave it the momentum necessary to outpace all other counties by 1500.

In the late Middle Ages the background noise of war was unremitting. Conflict was disruptive and economically debilitating with a profound affect upon cloth exports. Although a contemporary poet might advise the king that "the trewe process of Englysh polycye" should be to "cheryshe marchandyse" above foreign conquest, such advice often went unheeded.[60] In the early-fourteenth century the great international fairs of Champagne declined and overland routes across the Alps were cut off. Bigger ships with larger crews were necessary to provide protection against foreign navies and privateers. Transport and transaction costs increased and, according to Munro, priced out of international trade the cheaper fabrics that had once been the mainstay of English towns.[61] International trade became dominated by better quality broadcloths that could bear these higher costs. Overseas markets repeatedly opened and closed as the nation's alliances with its neighbours shifted and its military fortunes waxed and waned. Particular low points were the

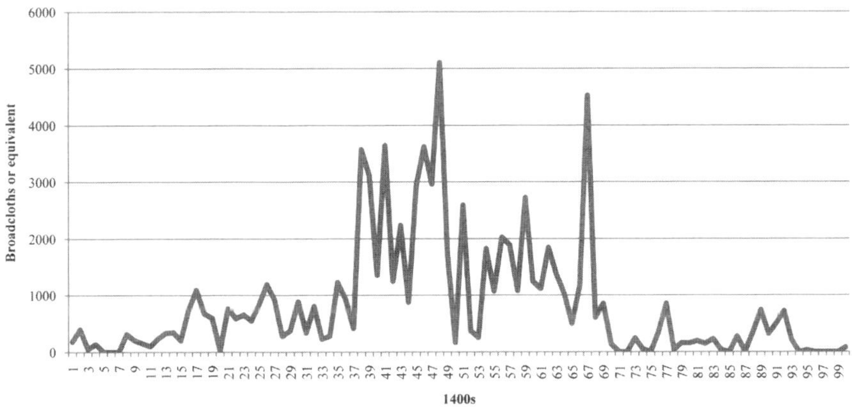

Figure 2.2: *Hanseatic Exports of Cloth from the Headport of Ipswich (includes Colchester)*
Source: Carus-Wilson and Coleman (1963), pp. 87–112.

60 Warner (1926), lines 1, 6 (p. 1).

61 Munro (2003), pp. 234, 237–9. The export of cheaper cloth did not cease altogether as the case of John Sutton, referred to in chapter 4, exemplified. Indeed, Chorley contended that by the mid-fourteenth century 'English cloth competed by its cheapness at the lower end of the international market': Chorley (1988), p. 10.

loss of nearly all territories in France at the end of the Hundred Years' War, and quarrels with the Hanseatic League that ultimately shut English merchants out of the Baltic.[62] During the fifteenth century an ear for news abroad and an 'inventiveness at finding and exploiting new markets' were critical to success.[63]

Credit

Credit is the lubricating oil of the business world. Overdrafts facilitate day-to-day trading while term loans enable investment for growth. In the late Middle Ages its importance was enhanced by a bullion crisis. The total value of silver coin in the country fell from between £1½ and 2 million in 1331, to one-tenth of that amount by 1422. It may well have fallen further still before beginning to recover by 1470.[64] Anyone without access to credit at that time would have found the business of cloth-making very difficult indeed.

Much credit was made available by suppliers accepting, voluntarily or otherwise, delayed payment for their goods and services. As we shall see, the wills of some Suffolk clothiers expressly referred to the book debts due to them. When he died in 1499, John Reynham of Nayland's personal property (excluding land) was valued at £377 11s. 9½d., of which his stock comprised £70 and his book debts £260. So, his book debts were worth nearly four times as much as his stock and more than two-thirds of the total probate value (Appendix D).[65]

Another popular, but more formal form of credit was the *scriptum obligatorum*, now known as a bond or more colloquially as an IOU. It usually contained the full record of a transaction, including a summary of any conditions and stipulations, the names of parties, the date of sealing, the amount of debt and the date or dates of re-payment. Sometimes it also stipulated the place and actual mode of re-payment.[66] The great advantage that a bond enjoyed over a simple record of debt was that it left little scope for argument. The contents could not easily be challenged in court and the only defence was to produce a release ('acquittance'), or to show that the bond had been entered into under duress or was a forgery.[67] A high proportion of all disputes before the common bench were over bonds. Indeed, the ''Londoners and the law' project team found that they comprised 44 per cent of

62 The author explored these subjects in much greater detail in his volume *Late Medieval Ipswich: Trade and Industry.*

63 Bailey (2007), p. 299. See also Coleman who opined that 'adaption was the key to the survival' of textile producing regions: Coleman (1983), p. 445.

64 Allen (2001), p. 607; Britnell (2004 (A)), p. 25.

65 TNA PROB 2/87.

66 Postan (1973 (B)), p. 30.

67 Postan (1973 (B)), p. 33.

Londoners' cases that reached the pleadings stage.[68] City merchants were willing to loan money or supply goods on credit, but insisted on a bond to secure the debt.

Two groups, in particular, still had sufficient bullion to play the role of banker. One was a corps of wealthy goldsmiths, many from overseas. Thus in 1477 the Italian Damiano de Podio lent to the Ipswich merchant Richard Ryle the sum of £18 12s. 4d. and, in return, took as security a bond entered into in the ward of Lombard Street in London.[69] A few years later, John of Rotterdam, who was based in Ipswich, was probably providing a similar service to other merchants of the town.[70] A second group was the leading London grocers, men such as Richard Lee whom we shall meet again in chapter 5. Their involvement in the international wool trade gave them the financial resources necessary to invest in the cloth industry.[71] These men were not, however, inclined to make risky and speculative loans to people they did not know. As Sutton wrote, 'ultimately credit relied on personal knowledge and trust, and worked best when limited to a small number of persons who knew each other'.[72]

Suffolk clothiers and dyers could persuade lenders to open their purses for two reasons. Firstly, many of them were themselves well known in the City by reason of family connection or membership of a merchant company. For instance, John Kyng was born and bred in Hadleigh, based his cloth-making in Shelley, but lived in London and belonged to the grocers' company. Robert Fen of Stoke-by-Nayland was buried in the church of St Mary Woolnoth in Lombard Street and numbered among his beneficiaries the grocer John Hunt.[73] Secondly, the county's textile industry was growing faster than anywhere else. Although this growth itself created solvency risks, it meant that a clothier or dyer was likely to find buyers for his wares and so be able to repay a loan. In this sense, success bred success and enabled Suffolk cloth-making to continue on its upward trajectory. The close connection with London merchants is reflected in Figure 1.3 and discussed in more detail in chapters 5 and 6. Elsewhere in the country some established clothiers and dyers were also well known in the Capital and could, no doubt, still secure credit. However, as we have seen in chapter 1, during the Great Slump of the mid-fifteenth-century, their fortunes and future were far less secure. In those areas where the textile industry was much younger, such as the Kentish Weald, the clothiers might

68 Stevens (2012), p. 240.

69 TNA CP 40/871, m. 312v (AALT IMG 1521). Podio's occupational status was not stated in the pleadings, but his Italian nationality and his presence in Lombard Street suggest that he was probably a goldsmith.

70 SROI C/2/10/3/1, ff. 281, 282, 287, 288; C/2/10/3/2, f. 33.

71 Nightingale (1995), pp. 485–6; Lee (2015), pp. 334–5.

72 Sutton (2005), p. 310.

73 TNA PROB 11/5/126, 11/6/204.

not yet have had an opportunity to establish a good credit rating. This might explain why so few of them did business with Londoners and why the likes of Stephen Karkeregge had more modest operations than their later sixteenth-century counterparts.

Markets and Fairs

Both English peasants and overseas merchants bought much of their cloth in formal markets. Carus-Wilson argued that the existence of a tight framework of urban and rural markets in medieval England was essential to the development of a rural textile industry, because it drew producers 'into the orbit of the *grande industrie* [it] was the proliferation of markets that was so striking a feature of the English economy in the late twelfth and thirteenth centuries. This brought every countryman within easy reach of a place where he could dispose of surplus produce to dealers in touch with the cities or with merchants at the ports'.[74] Her view has not gone unchallenged. Thirsk and Zell stressed the importance of 'social structure' and downplayed that of markets in encouraging the development of cloth-making.[75] Indeed, the number of formal weekly markets declined during the fifteenth century. Nevertheless, their contribution to the success of any trade or industry is undeniable. Davis summarized their benefits admirably: 'Regular market places and times, enforcement of contracts and debts, quality controls, supervision of prices, and protection against stolen goods [all] helped lower transaction costs and expedite trade'.[76] Fairs were held less frequently, generally once a year, but would expect to attract buyers and sellers from further afield.

Textile manufacture succeeded in those counties that enjoyed the densest network of local markets – Essex, Gloucestershire, Kent, Norfolk, Somerset, and Suffolk. These fora had invariably been founded before commercial cloth-making took off, so were a cause rather than a consequence of industrial development. Of the thirty featured textile centres, only Stroud lacked a market, but, even here, one was to be found at Minchinhampton three miles away. Only Cranbrook, North Walsham, Stroud and Sudbury lacked their own annual fair.[77] More than half of William Forthe's many claims were against people who resided in market settlements. Some, such as Colchester and Ipswich, were large towns; others, such as Pirleston, Tunstead and Wicklewood in Norfolk, were much smaller; while Blythburgh, Dunwich, Kessingland and Southwold were strung like beads along the Suffolk

74 Carus-Wilson (1987), p. 669.
75 Thirsk (1984), pp. 217–8; Zell (1991), p. 235.
76 Davis (2012), p. 3.
77 http://www.history.ac.uk/cmh/gaz/gazweb2.html

coast.[78] Indeed, in Norfolk and, above all, in Suffolk, markets were spread more thickly than in any other counties. Suffolk had a hundred and three, with a density exceeding ten per hundred square miles.[79] Most of the county's textile centres had their own. The cloth market and cloth hall in Kersey were probably the scene of a fracas in the town between John le Cook the Younger of Long Melford and his alleged assailants Richard Brasyere, Hugh de Rushbroke, John Masoun and James Brice among others. Brasyere, Rushbroke, and Masoun were all mercers and Brice was a draper. Brasyere came from Long Melford and Masoun from Sudbury.[80] These examples show a correlation between, on the one hand, success in textile production and, on the other, the density of local markets and fairs, which makes it difficult to deny that the first was often dependent on the second.

Local markets and fairs stood on the lowest rung of a mercantile ladder that included, on the step above, those of large towns and major ports and, at the top, those of London. Only Newbury, Reading and Cranbrook had neither a very large town nor a headport within a distance of thirty miles (Figure 2.3). Furthermore, although Coventry was more than a hundred miles from Southampton, according to Hare it was still 'a key player' in that port's trade.[81] In the fifteenth century the gradual silting up of the east coast ports of Boston and King's Lynn drove East Midlands wool merchants towards Ipswich, bringing with them raw material for Suffolk's textile industry.[82]

The county's clothiers also benefited from the proximity of the fair at Stourbridge outside Cambridge, held in late August and September each year after harvest. Merchants were attracted from throughout England and continental Europe. Stourbridge Fair experienced fluctuating fortunes during the late Middle Ages, prey to the vicissitudes of international relations and also to the anti-fair policy adopted by London's merchant companies, such as the grocers and mercers, who intermittently prohibited their members from attending.[83] Nevertheless, according to Lee 'the fair drew upon two particularly dynamic areas of the later medieval English economy – the textile trade and London. Cloth dealers from manufacturing and marketing centres like Coventry and Suffolk provided Stourbridge Fair with a wide range of textiles to be sold retail and wholesale'. Clothiers who rented

78 TNA CP 40/834, m. 214r, 226v (AALT IMG 381, 1286); TNA CP 40/895, m. 167r (AALT IMG 305); CP 40/911, m. 4v (AALT IMG 10 dorse); CP 40/931, mm. 5r, 172r (AALT IMG 11, 358).

79 Farmer (1991), p. 331. The Gazetteer of Markets and Fairs includes sixteen markets in Suffolk that were not included in Farmer's calculations. See also Scarfe (1999), pp. 76–7; Dymond (1993), pp. 76–7.

80 TNA SC 6/1001/3; CP 40/360, m. 63v (AALT IMG 9394).

81 Hare (2015), pp. 108–9, 202–5.

82 Amor (2011), pp. 56, 74–5.

83 Nightingale (1995), pp. 368–9, 396–7, 448, 568; Sutton (2005), pp. 214–7.

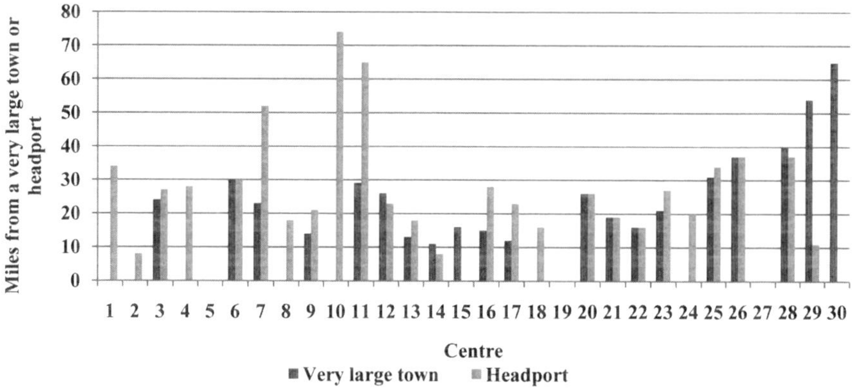

Figure 2.3: *Minimum Distance from Textile Centres*

Notes: The numbering of the centres is the same as in Figure 2.1. Very large towns (by medieval English standards) means those with populations of over 5,000 in 1377 (applying a multiplier of 1.9 to the number of taxpayers[84]) ie London, York, Bristol, Coventry, Norwich, Lincoln, Salisbury, Lynn, Colchester, Boston, Beverley and Newcastle-upon-Tyne. Headports were Newcastle-upon-Tyne, Kingston-upon-Hull, Boston, Lynn, Yarmouth, Ipswich, London, Sandwich, Chichester, Southampton, Melcombe/Weymouth, Exeter and Dartmouth, Plymouth and Fowey, Bridgwater and Bristol. Some textile centres were themselves very large towns and/or headports in which case no column is shown. Distances, in this figure and in and later figures and tables, are measured as the crow flies using the website http://www.freemaptools.com/how-far-is-it-between.htm

Sources: Kowaleski (2000), p. 473; A. Dyer (2000), p. 758.

stalls there came from Bury St Edmunds, Hadleigh, Monks Eleigh, Nayland, Sudbury and the Waldingfields.[85]

The relative importance to the textile trade of private sales, smaller rural markets, larger urban ones and fairs is difficult to assess. The fastest growth in Suffolk cloth production occurred during the fifteenth century when smaller rural markets and fairs suffered intermittent downturns in business.[86] This might suggest that it was the larger markets that were crucial. The increasing concentration in London of the overseas cloth trade made the City a magnet for clothiers from all the eastern and southern textile counties (Figure 1.3). Indeed, their 'crucial responsibility lay in the delivery of the cloth to the London market and finding a purchaser there'.[87] The pivotal role of London merchants has already been explored in chapter 1 and will be addressed in more detail in chapters 5 and 6. In the fifteenth century, the drapers' Blackwell Hall in London, which was to become the single most important forum

84 Choosing the correct multiplier is far from straightforward: Rigby (2010), pp. 401–4.

85 Lee (2005), pp. 124–5, 140.

86 Bailey (2007), p. 265.

87 Ramsey (1982), p. 23.

for the sale of provincial cloth, was already attracting much of the cloth trade. Yet more was conducted informally in the Capital with grocers and, above all, mercers.

Transport

The evolution of a hierarchy of markets and fairs in England was made possible by a domestic transport network linking communities by boat along rivers and coasts, and by horse and cart along roads. Without such a network the output of any textile manufacturer would have been limited to his immediate locality. Carriage by water was generally preferred on the grounds of cost, while carriers were less exposed to the risk of ambush and robbery.[88] According to Langdon, 'the importance of inland water transport in helping to explain the development of various economic patterns in medieval England has been broadly accepted' and 'the eastern part of the country was particularly favoured with effective navigable rivers'. By careful combination of primary and secondary navigation with trans-shipment points, where goods could be transferred from smaller to larger boats, substantial loads were carried downstream from towns and villages far from the sea.[89] In Yorkshire the Ouse was described in 1394 as 'a highway and the greatest of all the King's rivers, and is for the use of merchants in ships with divers merchandise from the high sea to the city of York'.[90] In East Anglia William Forthe of Hadleigh conducted longer distance trade with many people who lived in riverside settlements. Two hailed from Wisbech in Cambridgeshire on the river Nene; another from Greetwell in Lincolnshire on the river Witham; and three more from Eaton near Norwich at the confluence of the rivers Yare and Wensum.[91]

Edwards and Hindle published a map on which they marked the relatively small area of English land that, in the Middle Ages, was more than fifteen miles from navigable water.[92] However small that area might have been, within it lay some of the most important fifteenth-century textile centres in south Suffolk and north Essex that were closely tied into the London market (Figure 1.3). So, either rivers were unnecessary for the transportation of a substantial proportion of the nation's cloth, or more rivers were navigable, and higher upstream than once believed. Unger

88 Rackham (1986), p. 264; Edwards and Hindle (1991), p. 129.

89 Langdon (2007), pp. 111, 120–1, 126, 129.

90 Edwards and Hindle (1991), p. 126.

91 TNA CP 40/834, m. 214r (AALT IMG 381); CP 40/883, m. 173v (AALT IMG 311 dorse); CP 40/911, m. 227v (AALT IMG 449 dorse).

92 Edwards and Hindle (1991), p.130. The Cambridge Group for the History of Population and Social Structure has published a similar map as at 1600–60, but marking more extensive areas as being outside the 15-mile zone. It seems unlikely that the total length of navigable river in England was greater in the Middle Ages than it was in 1600: http://www.campop.geog.cam.ac.uk/research/projects/occupations/transport/waterways/waterways1600

suggested that late medieval design improvements in vessels used in coastal and riverine transport enabled them to access creeks and rivulets.[93] The use of the Suffolk rivers for transportation of cloth is considered in more detail in chapter 4 and suggests that he was right.

Far more ships hugged the coast than ever went to open sea. This was as true in Devon and Hampshire as it was in East Anglia.[94] Once Hanseatic merchants and ships abandoned Colchester and Ipswich for the port of London, large volumes of cloth had to be carried down the east coast and up the Thames from north Essex and south Suffolk; while dyes and dyestuffs had to be brought back in the opposite direction. A sharp increase between 1460 and 1470 in the number of recorded mariners and shipmen resident in the east coast towns and villages of Suffolk cannot be directly linked to the carriage of cloth, dyes and dyestuffs, but it occurred in the very decade of the Hanseatic exodus (Figures 2.2 and 2.4). Moreover, in 1500 at least three mariners were in dispute with London grocers.[95] Although there was some subsequent fluctuation in numbers they remained consistently above their pre-1470

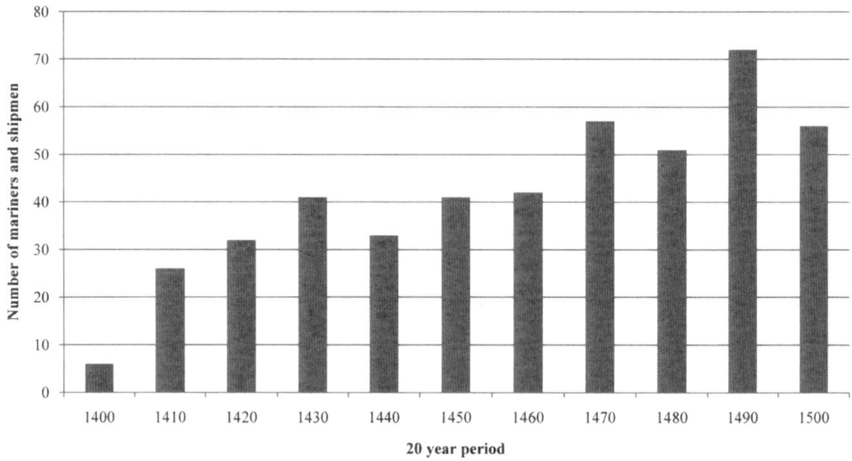

Figure 2.4: *Number of Mariners and Shipmen from East Coast of Suffolk*

Notes: Each column records the number of Suffolk mariners and shipmen named in the five Hilary term TNA CP 40 plea rolls for the date given, five and ten years before the date, and five and ten years after the date. The same methodology is used in Figures 4.2, 5.1 and 6.1. The term 'seaman' generally meant shipmaster, and it is submitted that the term 'mariner', that appears to have superseded 'shipman' in the final third of the fifteenth century, meant the same.[96]

Source: TNA CP 40A.

93 Unger (2007), pp. 373–4.

94 Kowaleski (1995), p. 4; Kowaleski (2011), p. 168; Amor (2011), pp. 87–8.

95 TNA CP 40/951, mm. 224r, 37v (AALT IMG 456, 74 dorse).

96 Kowaleski (2011), p. 180.

level. The seamen of Covehithe, Pakefield, Southwold and Walberswick figured most prominently in the plea rolls. To those living in such relatively small and isolated settlements seafaring was one of a limited number of career options. By the late-fifteenth century safer and more settled seas meant that ships and crews no longer needed to be big enough to defend themselves from armed attack. A 20-tonne coastal vessel might cost £40 and fall within the price range of enterprising seamen.[97]

The extensive use of navigable water in the Middle Ages did not make roads redundant. For instance, Newbury clothiers took their products south along the old carting route to Southampton and later east along the highway to London.[98] Coventry was another major textile centre that 'relied on its position on the road network'.[99] According to Farmer, 'by the mid-fifteenth-century, professional hauliers were travelling routinely between the midlands, London, Oxford, Bristol, Salisbury and Southampton'.[100] Wool from Lincolnshire may have arrived in Suffolk by boat, but from elsewhere in the East Midlands it almost certainly came by cart. It was common to combine the two methods of transport in one trip. Both roads and rivers are depicted with emphasis on the medieval Gough map.[101] According to the antiquarian who first studied it, and after whom it is named, 'the greatest merit of this map is, that it may justly boast itself the first among us wherein the roads and distances are laid down'.[102] Stenton was in no doubt that it had been intended to be 'of practical value to travellers', and that the road system drawn by Ogilby in 1675 was 'already in being in the later Middle Ages'.[103] Of the thirty textile centres, seventeen are linked by the red lines on the Gough map that depict these roads.[104]

Later historians agree that the main highways were metalled and far from the deplorable state into which they would fall, as a result of much greater use, in the early modern period. Furthermore, in the countryside the network of roads and lanes was even denser than it is now.[105] The upkeep of roads and bridges was a charity

97 Kowaleski (2011), p. 167.

98 Platt (1973), pp. 159, 161; Millea (2007), pp. 72, 85.

99 Edwards and Hindle (1991), p. 129.

100 Farmer (1991), p. 352.

101 For another interpretation of the medieval river and road network, see Edwards and Hindle (1991), p. 132.

102 Millea (2007), p. 27.

103 Stenton (1936), pp. 7, 15. The Cambridge Group for the History of Population and Social Structure have digitized Ogilby's road maps of 1675: http://www.campop.geog.cam.ac.uk/research/projects/occupations/transport1379 1729/

104 Millea (2007), pp. 85–6. York, Beverley, Lincoln, Boston, Stamford, Leicester, Norwich, Coventry, Northampton, Bury St Edmunds, Bristol, Salisbury, Newbury, Reading, London, Canterbury and Exeter.

105 Bolton (1980), pp. 151–2; Rackham (1986), pp. 263–4; Edwards and Hindle (1991), p. 124; C. Dyer (2005), p. 23.

recognized by the Catholic church to which clothiers and other merchants gave generously both during their life times and in their wills.[106] Hauliers traversed these roads throughout the year. Indeed, the Southampton brokerage books indicate that they were busiest during the winter months.[107] Two wheeled horse-drawn carts were the vehicles most commonly seen en route. They were generally of a standard size and capable of carrying nearly a ton weight. Ox-drawn wains could carry loads twice as heavy, but at half the speed. Pack horses would struggle with a tenth of the load of a cart which made them more expensive and rarely used outside mountainous areas.[108]

Raw Materials

Local sources of raw materials, particularly coal and iron, were key to the location of industry during the Industrial Revolution, was this also the case in the Middle Ages? Some medieval cloth-makers enjoyed ready access to raw materials. Those in the Cotswold towns of Stroud or Castle Combe could rely on good quality local wool and beds of fuller's earth – a clay material that can decolorize liquids without chemical treatment. Carus-Wilson referred to fuller's earth diggings in Minchinhampton, and Zell to those at Boxley, fourteen miles north of Cranbrook, that served the textile industry of the Kentish Weald.[109] Timber provided fuel for heating vats of dye and was also the raw material from which potash could be made for use in dyeing. In 1086 the Weald was the most densely wooded region of England and, although it was subsequently cut back at a ferocious rate, enough remained to service late medieval dyers. In south Suffolk Rackham found as many as thirty ancient woods.[110] Ash, elm and oak were recorded in Higham and Hitcham, and oak in Little Waldingfield.[111]

Nevertheless, Thirsk correctly challenged the idea that the local 'availability of certain natural resources' was necessary to the success of cloth-making.[112] In fact the opposite was often the case, with these materials having to be carried long distances. Suffolk was the prime example. Local wool was poor and, as we shall see in chapter 3, better-quality fleeces were sucked in from throughout England. Fuller's earth was sourced from Kent, while dyes and dyestuffs were only cultivated in the county in tiny quantities and were generally imported from abroad. It is a tribute to

106 Spufforth (2002), p. 176. For the long list of bequests for local roads made by Lavenham testators see Betterton and Dymond (1989), pp. 113–5.

107 Hicks (2015), pp. 46, 49.

108 Langdon (1984), pp. 49, 58–63, 65; Hicks (2015), pp. 43–4, 49. Ox hauling may have remained more important in the south-west.

109 Carus-Wilson (1959), p. 193; Zell (1991), p. 180. See also Hunt (1855), p. 148.

110 Rackham (1986), pp. 78, 88, 106.

111 TNA CP 40/853, mm. 53r, 53v (AALT IMG 116, 1028); TNA CP 40/911, m. 312r (AALT IMG 613).

112 Thirsk (1984), p. 218.

the networks of medieval markets, fairs, rivers, coastal routes and roads discussed above that these handicaps did not prevent Suffolk's cloth-makers from becoming the most successful in the land.

Technology

Postan once characterized technical development as being 'remarkably static for the whole of the Middle Ages'.[113] Whether or not this was true of agriculture, it was not the case in textiles. Several technological advances were made in cloth-making during the Middle Ages that are described briefly below and more fully in chapter 4. The distaff and spindle were gradually superseded by the spinning wheel, although this transition had made only limited progress by 1500. The upright loom was replaced, almost universally, by the horizontal narrow loom. This in turn gave way in many larger workshops to the broadloom. It may have been easier to house broadlooms in rural areas where properties were not so cramped. Munro described the invention of the horizontal loom as 'the single most important innovation in the development of the true woollen and the medieval woollen industry itself'.[114]

Nevertheless, the innovation that has attracted the most interest from historians was the fulling mill. In the dark days of 1941 Carus-Wilson published a seminal paper on fulling mills that, she argued, had given rise to 'An Industrial Revolution of the Thirteenth Century'. They had re-orientated cloth-making 'to concentrate on the swift, clear streams of the north and west, in remote valleys far beyond the bounds of the ancient chartered cities of the plains'.[115] These mills used hydraulic energy for the first time in English industry. The faster the flow of water, the greater the energy output. Cams and trip-hammers converted the rotary power of the water-driven wheel into reciprocal power, pounding cloth, not with feet or clubs, but with a pair of heavy blocks of oak (about 24 kg in weight), known as stocks, up to forty times a minute. With just one attendant, a fulling mill could scour, felt and full a standard-sized good-quality woollen in about 20 hours. The result was a 70 per cent saving in costs and 3.3 fold productivity gain.[116] First recorded in England in the late-twelfth century, fulling mills were soon widespread. Carus-Wilson's study of cloth-making in Stroud and Castle Combe re-enforced her view. In those particular centres fulling mills were pivotal to the rapid growth of textile manufacture.[117]

---

113 Postan (1973 (A)), p. 17.
114 Munro (2003), p. 197.
115 Carus-Wilson (1967), p. 198.
116 Munro (2003), pp. 205–7; Munro (2009), p. 6.
117 Carus-Wilson (1959), pp. 192–202.

Nevertheless, they may have been exceptions. Miller was among those who gently deflated the full force of her argument. In his paper of 1965, *The Fortunes of the English Textile Industry during the Thirteenth Century*, he argued that rural cloth-making pre-dated the diffusion of the fulling mill and continued in regions, such as Suffolk, that lacked swift, clear streams to drive them. The cost advantage of mill fulling was less important in the thirteenth than in the fourteenth century. Only then did production of heavily fulled broadcloth begin to overtake that of lighter cloths that required little or no fulling. Furthermore, labour became more scarce and wages higher.[118] Even Miller may have overstated the true impact of the fulling mill in the Middle Ages. Langdon discerned a sharp increase in their numbers only after 1480.[119] As we shall see in chapter 4, this was probably the case in Suffolk. By that time the number of master fullers was in decline leaving the field clear for clothiers to introduce this labour saving, but job destroying plant and machinery.

The invention of the broadloom triggered a gradual transition from narrow to broadcloth production and may have contributed to the translocation of the industry from town to countryside. At the very end of the fifteenth-century, greater use of the spinning wheel and mill-fulling enabled cloth-makers to increase productivity and turn-out heavier and better-quality cloth. Nevertheless, technology probably had little influence on the regional location of textile manufacture.

Society

Population Pressure

A heavily populated region might be well suited for industrial development because, as discussed above, unemployed or under-employed peasants were available as a pool of cheap labour. For the proto-industry school these peasants were the forerunners of the industrial proletariat. Population growth could occur through immigration, or by reason of inheritance customs. Where partible inheritance was the rule, each son received a portion of his father's estate. Rather than being forced away to find land or work elsewhere, each son would stay and survive, albeit on an ever-diminishing plot. Any additional industrial income would be a godsend. The link between partible inheritance and rural industry gave rise to the phrase 'scratch a weaver and find a parcener'.[120] This neat analysis is, however, rather muddied by the early existence of an active market in both free and villein land.[121]

---

118 Miller (1965), pp. 72–4, 78, 81.
119 Langdon (1984), p. 42.
120 Thirsk (1984), p. 231.
121 Clarkson (1985), p. 43.

A peasant, whose plot might be just about sufficient to support him and his family in normal times, could be forced to sell in the event of harvest failure in order to buy food. He would find himself in the same position as a younger son without an inheritance. Furthermore, it was after the Black Death, once population pressure had deflated, that rural cloth-making made its greatest strides.

In 1290 Norfolk was the most populous county in England.[122] The village of Worstead, in the north-east of that county, may be the best exemplar of industrial development fuelled by high population. Partible inheritance was probably the rule there. As explained above, by the late-thirteenth century a large number of its tenants struggled to make a living from small plots of land. Although arable husbandry was particularly intensive, an additional non-agricultural income might well have been necessary for survival.[123] Cloth-making became well established in Worstead and neighbouring villages before the Black Death, at a time when population pressure was still intense. So why did the industry not also evolve on the far side of the Wash in the south Lincolnshire Fens which experienced similar over-population and intense subdivision of holdings? Maybe the local organisation of wool growing and the prodigious outward flow of wool exports through Boston deprived prospective cloth-makers of yarn.

Thirsk speculated that, owing to partible inheritance and immigration, the clothing district of Suffolk might have grown more populous between the eleventh and fourteenth centuries and posed similar 'problems of unemployment and over-population which industries alone could solve'.[124] In absolute terms, population undoubtedly did increase, as it did so almost everywhere.[125] However, the district was not more heavily populated in 1327 than many other parts of the county. Indeed, over those three hundred years there had been either no change or even a modest fall in its population relative to neighbouring districts. Some cloth-making settlements certainly enjoyed population growth between 1327 and 1524, but such growth was not uniform across the whole of south-central Suffolk.[126] This all suggests that rather than population growth being the cause of cloth-making in the district, the reverse was true.

What was true in Suffolk, was equally the case in the nation's other rural textile centres. Cloth-making in Stroud was certainly not the result of population pressure. In 1381 the town itself had not even existed as a nucleated settlement. There

122 Broadberry et al (2015), p. 25.

123 Sutton (1989), p. 205; Campbell (1993), pp. 48–9; Martin and Satchell (2008), p. 80.

124 Thirsk (1984), p. 222.

125 The population of England grew from 1.4–1.9 million in 1086 to 4–5.6 million in 1348: Hinde (2003), pp. 25–6. During the same period that of Suffolk grew from about 100,000 to about 225,000: Bailey (2007), p. 67.

126 Todd and Dymond (1999), pp. 80–83; Bailey (2007), pp. 69–70, 284.

had been only 64 poll taxpayers in Bisley, 111 in Minchinhampton and 62 in Rodborough.[127] In 1377 just across the border in Wiltshire, Castle Combe had had only 138 poll taxpayers.[128] Cause and effect should not be confused.

Lordship

Thirsk considered that rural industries were more likely to evolve in communities, such as those in central Suffolk, with a weak manorial framework. Such weakness meant little control over inward immigration or the cultivation of the village fields.[129] Her view might be taken to imply that strong lordship was inimical to cloth-making, but was this the case? After all, powerful lords possessed the resources and ability to create towns and markets, and to invest in fulling mills. In the late-thirteenth and early-fourteenth centuries lords did erect fulling mills in the hope of encouraging local industry and generating an income stream from fulling cloth. As we shall see in chapter 4, such initiatives were often unsuccessful, but sometimes perhaps they worked. The priory of Christchurch, Canterbury erected a fulling mill at Hadleigh and also, possibly, at Monks Eleigh and in each of these communities textile manufacture flourished.

After the Black Death the various textile centres experienced contrasting modes of lordship. Castle Combe had, in Sir John Fastolf, a benign but absentee lord who bought cloth from the town and did his best to protect its industry. When Walter Hungerford challenged the right of Castle Combe to hold its own market, and accused four local clothiers, two weavers and a chapman of impeding the collection of tolls at his own Saturday market in Chippenham, Sir John stepped in to meet most of the cost of obtaining from the Crown a new charter for a market and fair.[130] Across the county border in Gloucestershire, the manorial structure of Stroud also favoured industrial growth. It stood on the dividing line between two manors, Bisley and Minchinhampton, both of which had absentee lords with no axe to grind with local cloth-makers.[131] In the east, the cloth-making villages of Tunstead Hundred in Norfolk may not have enjoyed such disinterested lordship. The manorial structure there was very complex and fragmented, but no-one could doubt the power of the bishop of Norwich.[132] After the Peasants Revolt there is reason to believe that he did not look kindly on the local weavers, once so prosperous, but soon to be impoverished.

127 Fenwick (1998), 1, pp. 254, 256.
128 Fenwick (2005), 3, p. 12.
129 Thirsk (1984), pp. 223, 231; Zell (1991), p. 3.
130 TNA CP 40/717, m. 384v (AALT IMG 1761); Carus-Wilson (1959), p. 204.
131 Carus-Wilson (1959), p. 192.
132 Campbell (1993), pp. 52–3

Powerful lords were not always enemies of cloth-makers. During much of the fifteenth century Walter Hungerford and his descendants were among the leading Wiltshire families.[133] They had extensive property interests in the west of the county in and around Heytesbury in the upper Wylye valley where 'manorial structures remained intact'. Under their stewardship the region's textile manufacture enjoyed considerable growth and in 1469 Heytesbury Hundred was responsible for 19 per cent of the county's production.[134] In Suffolk Bailey wrote that most textile towns 'were settled on the large manors of major landlords [...] who may have encouraged their tenants by buying their wares, but who certainly offered patronage and security in uncertain times [as well as] more effective manorial courts and a greater degree of stability than could lesser lords'.[135] Betterton and Dymond considered that 'the more liberal attitude towards new industrial developments' of the earl of Oxford may well have encouraged cloth-making in Lavenham in its formative years.[136] In Long Melford local fullers were frequently locked in litigation with local gentry, namely the Cloptons and Martyns, suggesting that such families took an interest in the industry. In nearby Stoke-by-Nayland, John, the first Howard duke of Norfolk, had an excellent commercial brain with wide business interests. He enjoyed a close relationship with the local clothier Robert Sergeaunt, selling him wool and woad and fulling Sergeaunt's cloth in his mill.[137] Howard even dabbled in cloth-making himself. In 1466 a man bearing his name presented twenty broadcloths to the alnager in Ipswich.[138] By then he owned his own fulling mill valued at 66s. 8d. per annum.[139] In November 1482 he paid the fuller John Taylour 3s. 2d. to finish seven broadcloths and, over the following few weeks, made him several more payments for dressing his cloth.[140]

Other in those exceptional cases where a local lord was positively hostile to cloth-makers, the contrasting examples given above do not allow us to determine with any certainty whether strong or weak lordship was better for textile manufacture. Further study of local sources may shed more light.

Skills and Knowledge

In his search for explanations for the success of cloth-making in south Suffolk, Bailey noted that 'a pool of expertise already existed there, mainly within the three

133 http://www.oxforddnb.com/view/article/14181?docPos=14
134 Hare (1999), p. 15; Hare (2011), pp. 35, 194.
135 Bailey (2007), p. 300.
136 Betterton and Dymond (1989), p. 3.
137 Crawford (1992), p. II, pp. 302, 320.
138 TNA E 101/342/25.
139 Crawford (1992), p. xlv.
140 Crawford (1992), II, pp. 320, 322, 325–6.

urban centres of Bury St Edmunds, Clare and Sudbury'.[141] How skillful were cloth-workers and what expertise did they need?

The time, attention and sheer effort required of medieval spinsters, weavers and fullers was considerable, but their technical skill was generally limited. It was gruelling, but not intricate work. Most woollens had a simple plain weave, known as tabby, that was forgiving of errors. Cloth finishing was a different matter. Dyeing demanded a good working knowledge of organic chemistry and shearing a good eye and steady hand. Competent cloth finishers could hide mistakes that had occurred earlier in production, while incompetent ones could quickly ruin a cloth. This may explain why the greater towns, with their stricter regime of regulation and inspection, retained such a significant proportion of finishing work.

Adam Smith was certainly dismissive of the need for prolonged training in cloth-making. He poured scorn on the institution of apprenticeship and had no confidence in its ability to ensure quality standards in textile manufacture. Writing in *The Wealth of Nations,* he stated that 'long apprenticeships are altogether unnecessary', being simply a restrictive practice that maintained high wages and prices. The skills of most trades, he argued, could be learned in weeks, if not days.[142] Recently, Professors Ogilvie and Epstein have reignited the debate, with Ogilvie sharing Smith's view and Epstein stoutly defending craft guilds as 'sustained systems for the transmission of skills'.[143] Whatever the merits of apprenticeship, it was a widespread and enduring institution in medieval cloth-making. In 1381 in Canterbury four weavers had apprentices, including John Frythe who also had six servants.[144] Such was the demand for labour in Castle Combe that, to the consternation of John Gilys and other local farmers high up in Over Combe, and in breach of a 1406 statute, cloth-makers recruited as apprentices the children of agricultural labourers.[145] In Bury St Edmunds the card-maker John Candeby and the weaver John Couper both remembered former apprentices in their wills.[146] Nevertheless, not all training was through the formal apprenticeship system. In 1478 John Sele of Ipswich engaged David Brewer to teach him the art of dyeing for an agreed tuition

141 Bailey (2007), p. 298.

142 Adam Smith (2000), pp. 91–2.

143 Ogilvie (2004), pp. 302–304; Epstein (2008), pp. 155, 160–2. Their debate focused on the early-modern, rather than medieval period.

144 Fenwick (1998), 1, p. 420.

145 TNA CP 40/647, m. 161v (AALT IMG 751). The statute provided that 'he or she that useth to labour at the plough or cart, or other labour or service of husbandry, til he be of the age of twelve years' was debarred from being an apprentice, and also anyone whose parents did not have land or rents worth at least 20s. a year: *Statutes,* 2, pp. 157–8. Carus-Wilson referred to other cases of enticement of apprentices heard in the manor court: Carus-Wilson (1959), p. 203.

146 SROB 187 Osbern, 217 Hawlee.

fee of 40s. Brewer alleged that he had 'well and faithfully taught, serviced and informed Sele so that he learned all things relating to the art', but his pupil nevertheless failed to pay up.[147] Furthermore, formal apprenticeship training, that was far more common in towns than in the country, did not protect urban craftsmen from the competition of rural weavers and fullers.

Another type of skill altogether was one that enabled cloth-makers to maintain a stake in the industry and turn out products that were saleable to sophisticated consumers in unpredictable markets here and abroad. A clothier with extensive commercial and intelligence networks was well equipped to do this, particularly if he had regular contact with Hanseatic traders or members of London's great merchant companies. An alternative for those of lesser means, was to befriend a cloth-worker who had fled the continent and brought with them a knowledge of European fashions. Here may lie the true importance to the domestic textile industry of the Flemish weavers who established themselves in England in the mid- to late-fourteenth century.

Flemish Weavers

In 1336/37, at the bequest of Edward III, Parliament enacted a statute that, inter alia, promoted the immigration of foreign cloth artisans into England 'under the king's protection and safe conduct to dwell in the same lands choosing where they will'.[148] This was not the first measure aimed at attracting skilled textile workers from overseas, but it is the best known.[149] Flemish weavers, in particular, were encouraged, and sometimes forced to escape the turmoil of their homeland and bring with them to England much-needed skills.[150] Earlier historians interpreted this exodus as giving a fresh kick-start to an English textile industry that had become moribund by the early-fourteenth century.[151] It was an interpretation that went out of favour with the work of Heaton and, in Suffolk, McClenaghan. Heaton famously quipped that, in Yorkshire, 'there were more cloth workers from Lincoln than from the whole of the Low Countries'.[152] McClenaghan searched the later-fifteenth-century alien subsidy returns in vain for Flemish weavers. More recently Betterton and Dymond found 'no evidence' of their having any connection with fourteenth-century Lavenham, while Poos considered them 'difficult to detect in

147 SROI C/2/10/1/7, m. 11r.
148 Statutes, 1, pp. 280–1.
149 Munro (1999), pp. 122–3.
150 Van Bavel (2010), pp. 271–3.
151 Lipson (1921), pp. 12–4.
152 Heaton (1920), p. 16.

the smaller production centres' of Essex and ascribed little importance 'to their presence in the region's industry'.[153]

And yet the Flemish weaver doggedly refuses to disappear from the medieval history of the English cloth industry. Carus-Wilson found isolated individuals in the West Country and the West Riding, but concluded that 'most of them seem to have settled in eastern and southern England'.[154] Is it then more than coincidence that those textile regions became the most successful? In the poll tax returns, an Andrew Fleming was a weaver in Bocking, two John Flemings worked their looms in Braintree and Stebbing respectively, as did a Giles Fleming in Worstead. Godfrey Braban and John Brabon of York, and Giles Brabon of an unidentified parish in Suffolk all bore a name suggestive of a Brabantine origin.[155] During the Peasants Revolt Flemings fell victim to several fatal attacks in London, Norfolk and Essex and some of them were almost certainly weavers.[156] In 1400/01 Adam Duchman was one of two weavers named in the Ipswich petty court rolls.[157] In 1436 aliens resident in England were obliged to swear an oath of allegiance to the English Crown. The record of such oath-takers, who were predominantly from the Low Countries, included weavers living in Cambridgeshire, Devon, Essex, Kent, London, Norfolk, Oxfordshire and Suffolk.[158] The two from Suffolk were Peter Porter in Bury St Edmunds and John More in Ipswich. In 1483 a Flemish weaver named Powes was a householder in Framsden, while the clothier John Stanesby employed eleven alien cloth-makers, who were reputedly Italian nationals, in his workhouse in Bildeston.[159] Elsewhere in the county, Isebraven Ducheman worked as a weaver in Beccles, as did John Ducheman in Raydon and Roger Eppe, whose name suggests a connection with Ypres, in Weybread.[160]

Lambert and Pajic's recent study of Colchester between 1351 and 1367 identified 126 immigrants from the Low Countries settling in the town, of whom at least twenty-seven were Flemish, and very probably engaged in textile production and trade. Other influxes of similar numbers, during a similar period, were recorded in London, York and Winchester. They concluded that 'by providing manpower, expertise and financial resources, the Flemings made a vital contribution to the development of an industry that made Colchester's future look distinctively

153 Betterton and Dymond (1989), p. 2; Poos (1991), p. 70.

154 Carus-Wilson (1987), p. 677.

155 Fenwick (1998), 1, pp. 206–7, 208; Fenwick (2001), pp. 166, 535; Fenwick (2005), 3, pp. 142, 154.

156 Hinton (1973), pp. 195–8.

157 SROI C/2/3/1/34 (7 Oct 1400).

158 *CPR 1429–36*, pp. 541–88.

159 TNA E 179/180/111.

160 TNA CP 40/629, m. 406r (AALT IMG 822); CP 40/656, m. 341r (AALT IMG 689); CP 40/736, m. 300r (AALT IMG 587).

different from that of many other English towns [and that] government policy to attract foreign workers in order to stimulate key sectors of the economy in fourteenth-century England did make a difference'.[161]

So where does the truth lie? It may depend on where and when you look. In the fifteenth century the majority of Suffolk aliens gravitated towards Ipswich and other east coast settlements where there was only limited cloth-making.[162] This helps to explain the paucity of Flemish weavers recorded in the alien subsidy returns. Arrivals en masse were probably unusual. Nevertheless, small communities and even individuals, with skills and knowledge brought from the continent, could exert influence far beyond their numbers. After all this was an industry that was targeted at the export market and at ultimate customers with whom Flemish weavers might still have contact and with whose preferences and tastes they would be familiar.

Why Did the Suffolk Clothier Triumph?

In his best-selling volume of world history *Sapiens*, Harari wrote that 'one of the distinguishing marks of history as an academic discipline [was that] the better you know a particular historical period, the *harder* it becomes to explain why things happened one way and not another'.[163] What is true of world history, is also true of textile history.

Some of the factors discussed above influenced the development of cloth-making in most of the featured centres, while others were more important in one place than another, or at one time than another. All the centres experienced a similar climate, much colder than it had been before 1300. A sufficient number of rural ones were situated in river valleys to suggest that an undulating landscape favoured cloth-making. In those regions where pastoral husbandry and field engrossment and enclosure were prevalent, surplus labour was more likely to be available to man the putting-out system of production. Higher standards of living after the Black Death probably increased aggregate domestic demand for commercially produced textiles, although this did not necessarily favour one region over another. Export demand gave further impetus to English cloth-making. At times during the fifteenth-century the ports of Colchester, Exeter, Ipswich and Southampton were important outlets, but by 1500, London had secured the bulk of overseas trade. Those counties whose clothiers enjoyed the closest relations with, and could secure the most credit from, London merchants did best. The correlation between counties that produced cloth commercially and those with a high density of markets and fairs was sufficiently close

---

161 Lambert and Pajic (2014) pp. 739, 743–4, 752–3.

162 TNA E 179/180/92, E 179/180/111; Amor (2011), pp. 32–3; Kowaleski (2012), p. 107; http//www.englandsimmigrants.com/page/sources/alien-subsidies/east-anglia/suffolk

163 Harari (2014), p. 266.

to support Carus-Wilson's view that one benefited from the other. A network of rivers, coastal routes and roads made possible the long-distance carriage of cloth and the raw materials necessary to produce it. This meant that even when the export trade became focused on London, textile manufacture continued unabated in regions far from the City. The presence or otherwise of local sources of raw materials was relatively unimportant. Technology and, in particular, the fulling mill was once regarded by historians as the main determinant of the location of cloth-making in England. More recent research has shown this was not so. By 1500, however, technology was beginning to increase the productivity of cloth-making and the quality of the end product. Population pressure rarely provided the initial spark for the textile industry, but once manufacture was underway it tended to draw in more people in search of work. The nature and strength of lordship appear to have been neutral factors. Skills and knowledge were essential to successful cloth-making, but formalized apprenticeship training did not protect urban craftsmen from rural competition. The Flemish weaver was an enduring figure in English cloth-making perhaps because he was familiar with continental tastes and could pass on fashion tips to his colleagues.

With an eye to the title of this volume, one should try to explain what gave Suffolk's clothiers the edge over their contemporaries elsewhere and made the county the leading producer of cloth by 1500. Certainly it was not local supplies of raw materials, nor any technological advantage. Suffolk had, in parts, an undulating landscape with several river valleys. After the Black Death, peasants increasingly preferred pastoral over arable husbandry, and the more successful engrossed and enclosed land. Timber was readily available. Most people enjoyed a higher standard of living than their forebears and could afford to buy their cloth rather than depend on homespun. A substantial number were keen to supplement their agriculture earnings with income from cloth-making. Their lords were generally supportive, and rarely hostile to the development of textile industries on their manors. In the towns, the residents of older established textile centres had learned skills that they could pass on to their country cousins. However, while all these factors favoured cloth-making, as we have seen, they were not unique to Suffolk.

What gave the county's cloth-makers the real edge were the following – burgeoning overseas demand for their products at a time when most of the country was mired in the Great Slump; an unusually dense network of markets in many towns and villages that facilitated their meeting and developing relationships with suppliers and buyers; relatively fast transport links by sea, river and land to London and the Low Countries; good credit ratings with the City's merchants; and the local presence of just enough Flemish weavers and other overseas cloth-workers to help them keep tabs on the latest continental fashions.

In the next four chapters we shall examine the way in which they turned wool into cloth – analyzing the Suffolk wool trade; cloth production and finishing in the county; and the role of the clothier.

– Part Two –

Suffolk

– 3 –

Wool and the Wool Trade

One day in 1444 a group of friends set out to do what country boys have done since time immemorial. They went poaching in Eriswell for 'hares, rabbits, pheasants and partridges' and so infringed Sir Thomas Tudenham's right of free warren. Sadly, for them, they were apprehended and hauled before the court, which is why we know about their adventure. Among them were six shepherds, all working in the Suffolk Breckland – John Hey, John Mayhewe, and John Deye from Icklingham; and John Eryswell, John Grove and Thomas Symond from Eriswell.[1] These men lived with their beasts and knew them best. They watched over them in their folds to protect them from rustlers and dogs. They tarred or greased them to ward off disease.[2] In early summer, they toiled by running streams to wash, clip and shear their fleeces. If each of them was looking after four hundred sheep, then they were collectively responsible for flocks of 2,400 animals. If each of them washed, clipped and sheared ten sheep per day, then together they devoted more than 240 man-days to the task.[3] In doing so, they added to the stock of English wool that clothed much of Europe.

Sheep and their Fleeces

Owing to advances in animal husbandry during the eighteenth century, we cannot be sure what medieval sheep looked like. They would probably have resembled more closely those small beasts found today on Orkney and Shetland than those that now graze the Breck. They may have had black faces and short tails and the ewes probably had no horns.[4] Certainly, they grew a light, coarse ('slight') fleece

1 TNA CP 40/736, m. 167r (AALT IMG 326).

2 C. Dyer (1991), p. 235. In 1487–8 the Prior of Ely spent 5s. on a barrel of tar for his flock at Lakenheath: CUL EDC 7/16B.

3 These two assumptions are based on the work of C. Dyer (2005), p. 108; and Iredale (1972), p. 67 respectively.

4 Ryder (1972), p. 62; Ryder (1984), pp. 14–5, 28; Baines (1979), p. 30.

and, in textile production, that was of great significance. The quality of wool was determined by various factors. In any single fleece it varied considerably. Since sheep habitually turn their backs on bad weather, wool from an animal's shoulders was top grade, while that from its britches and belly was generally unusable.[5] Regional variations in quality were caused by local climate, soil type, pasture and husbandry. Sheep reared on windswept hillsides in the Welsh Marches grew fleeces that were fine, close and warm. Like human hill walkers they insulated themselves with thin but multiple layers, so enduring cold winters and surviving on a meagre diet. Those that basked in the sunshine of lowland England grew coarser coats, but, if folded on poor soils, still only produced short staple wool. The growing number fattened in rich pastures produced longer wool fleeces. Careful estate management, flock control and selective breeding also mattered. Supplementary feed, winter shelter in purpose-built cotes and close supervision enhanced the wellbeing of any flock, and minimised the risk of liverfluke, scab or other disease that could have a devastating effect on the quality of an animal's fleece.[6]

The weight of a fleece also varied with sex and age. Castrated male sheep, known as wethers, were favoured for wool production because they invariably grew the heaviest coats, twice those of hoggs (two-year olds) and yearlings and a third heavier than ewes. The weight of any given wether's fleece depended on similar factors to those that determined its quality.[7] In the later Middle Ages mean fleece weights fell sharply, generating much debate over the reasons why. Stephenson argued that the climate was to blame. In the late-fourteenth and early-fifteenth centuries, as the northern hemisphere moved into the Little Ice Age, average temperatures dropped and very severe winters became more frequent. The late 1430s, the years of the so-called 'Sporer Minimum' when solar radiance was particularly weak, proved to be perhaps the hardest for England in the second millennium.[8] These harsh winters inflated mortality and reduced the health of flocks, exposing them to a greater risk of disease. Stone, while acknowledging the adverse climate, gave greater weight to human decision making. In his words 'demesne agriculture was in fact not in perpetual thrall to the environment'.[9] The Black Death and subsequent epidemics of plague sharply reduced England's population in the later Middle Ages. So, the demand for agricultural produce fell and the available labour pool shrank, deflating prices, inflating wages and slashing seigneurial profits. Neither lay nor ecclesiastical lords could any longer find or afford the staff to look after their sheep so well. They deliberately sacrificed the quality of their wool in the

5 Lipson (1921), p. 129; Miller (1991), p. 399; Ryder (1995), pp. 13–6; Crowfoot (2004), p. 15.
6 Bowden (1962), pp. 25–6; Munro (1978), p. 118; Crowfoot (2004), p. 15.
7 Stephenson (1988), p. 733; Bailey (1989), p. 126; Campbell (2000), p. 155.
8 Stephenson (1988), pp. 382–5; Campbell (2012), p. 122.
9 Stone (2005), p. 19.

search for economies. Stone offered the example of the nourishing of lambs at Hinderclay. During the fourteenth century they were rarely, if ever, weaned on ewes' milk, which was reserved for human consumption. Before the Black Death, however, they were given cows' milk instead. Afterwards, owing to its cost, this supply of cows' milk ceased with an inevitable adverse affect on the lambs' health and productivity in later life.[10] So, while the deterioration in the climate lowered the bar above which the health of medieval sheep could not rise, graziers took decisions about the management of flocks that kept them well below that bar. Owing to a combination of environmental, human and genetic factors, the weight of fleeces fell from an average of about 1.8 lbs. in the 1350s to 1.24 lbs. a hundred years later, and in Suffolk often even lower.[11]

English sheep farmers were famed throughout Europe for the fine quality of their short, thin wool. The fleeces of sheep reared on the hillsides of Herefordshire and the Cotswolds enjoyed the highest repute and were in great demand from the cloth-makers of the Low Countries. As the general quality of English wool fell, these premium brands became ever more prized.[12] By contrast, coarse Suffolk wool was poorly regarded and lowly valued. In a list drawn up for Parliament in 1454, as an unsuccessful attempt to fix minimum prices, it was near the bottom at 52s. per sack (Plate 3.1). In the Staple lists of 1475 and 1499 it was not included at all.[13] Neither English nor Continental cloth-makers could use Suffolk wool in the production of high quality woollens. It was, however, fit for use in less expensive cloth, and often had the merit, per unit area, of returning a higher yield at less cost than finer wool.[14] It was, in the words used by John Deye the Elder of Long Melford in his will, 'cogware wool'.[15]

The challenges faced by Suffolk wool growers were exacerbated by the unpredictable nature of the wool market. Munro tracked the relative prices of wool, grain, meat and dairy for the period 1331 to 1450 and found a fluctuating pattern that would have made very difficult the advance planning necessary to maximise agricultural returns.[16] According to Campbell, sheep were 'one of the most complex components of medieval agriculture', while Lloyd opined that 'English wool prices were probably susceptible to the influence of more factors than the prices of any other commodity in the middle ages'.[17] Most of these factors affected demand,

10 Stone (2003), p. 9.
11 Broadberry et al (2015), p. 109; Bailey (1989), p. 127.
12 Munro (1979), p. 218.
13 *Rot. Parl.*, v, p. 275; Munro (1978), p. 148.
14 Campbell (2000), p. 161.
15 SROB 111 Baldwyne. Cogware was a coarse and inexpensive type of cloth.
16 Munro (2012), p. 307.
17 Campbell (2000), p. 157; Lloyd (1973), p. 13.

rather than supply, and were the consequence of domestic fiscal policy and international relations quite outside the control of the grower. The middle decades of the fifteenth century, the period of the Great Slump, were particularly difficult for pastoralists in Suffolk as elsewhere.

Despite all these uncertainties, and notwithstanding an overall decline in the total number of animals and the prices of their wool, sheep farming continued and indeed enjoyed a 'distinct revival' from the 1460s.[18] In regions of low population, cheap land and poor soil, where intensive cereal husbandry was impossible, it was often the best, if not the only option. By keeping labour costs low, reducing administrative burdens to a minimum, and achieving economies of scale, graziers could extract some profit from their flocks. They preferred a modest return from wool to none at all. So began, in many parts of England, a process of engrossment and enclosure that created the great estates of the early modern period and the fortunes of families such as the Townshends of Norfolk.

Breckland

The Suffolk Breckland lies in the north-west of the county. Its parishes, some of which are among the largest in Suffolk, extend in aggregate to just over 150 square miles and account for about one tenth of the total landmass of Suffolk. Several small towns grew up in medieval Breckland, of which Mildenhall became the largest, but outside these towns, the countryside was sparsely populated.[19] Much of the land was gorse-covered heath, beneath which lay a light, dry, sandy soil of limited use for arable, and better suited for pastoral farming. This low-lying land merged with the wetlands of the Fens, and part of the boundary between them is marked by the remains of hythes, which were small landing places for boats, to the north-west of the main road running through Lakenheath. In an age before large-scale drainage and land reclamation, these wetlands extended to the Isle of Ely and beyond as far as the North Sea.[20] They provided a link with King's Lynn in Norfolk.[21] Direct trade between that great port and Breckland towns and villages, including Ampton, Brandon, Icklingham, Lakenheath, and Wangford contributed significantly to the region's economy. The occupational status of litigants in court of common pleas cases, and the relatively high value of their disputes provide some clues to the nature of this trade. John Wareyn a merchant of King's Lynn sued William Edward a wool chapman of Brandon for ten marks (£6 13s. 4d.); Henry

18 Bailey (1989), p. 293.

19 Harvey (1906), pp. xi, 196–203; Harvey (1910), pp. 218–30; Middleton-Stewart (2011), pp. xix–xxi; Todd and Dymond (1999) pp. 82–3.

20 Bailey (1989) pp. 3, 42–3; Martin (1999), pp. 20–1.

21 Langdon (2007), pp. 113, 130.

Baxter of King's Lynn sued the Brandon boatman, Edward Bonyfaunt, for £10; and John Coket, merchant and wool grower of Ampton, sued Thomas Carter of King's Lynn for £21.[22]

Today, Breckland is an ornithological wonderland, famous for some of Britain's most elusive resident birds. In the Middle Ages it was better known for other reasons. Bailey referred to 'the remarkably diverse resource base of the breck-fen edge'. In his seminal work *A Marginal Economy? East Anglian Breckland in the Later Middle Ages* he described an economy based on barley, fishing, rabbits, and sheep, as well as the cutting of reeds, sedge and turf.[23] Rabbits arrived with the Normans and thrived in purpose-built warrens. William Galyon, Esquire who, as we shall see, grazed two hundred sheep at Culford, also kept two hundred rabbits at Barnham and another three hundred at Mildenhall; while the bishop of Ely had six hundred at Brandon, protected from a flint warrener's lodge.[24] Richard Smere of Brandon bought 2,500 rabbits from Edmund Lychfield, prior of Flitcham, for £20, at just under 2d. per head.[25]

Good little earners as they often were, rabbits were never as important as sheep to the Breckland economy. As early as the Domesday Book, the hundred of Lackford had the highest concentration of sheep in East Anglia.[26] In those common bench cases which recorded rustling and dog worrying of sheep, the largest flocks were always found in Breckland – three hundred belonging to John Bernard Esquire at Mildenhall, two hundred to William Galyon in Culford, and six hundred to Sir Edmund Bedyngfield in Eriswell. In 1420 Galyon pursued the woolman Thomas Bredman for payment of £16, a sum that would have bought more than eleven hundred fleeces.[27] Other sources recorded that Henry Harlegray of Mildenhall bequeathed seven hundred sheep to his two sons; the abbot of Bury St Edmunds kept two thousand sheep in Culford; and the prior of Ely sold five hundred fleeces from his Lakenheath manor.[28] Later in the century, John Coket of

22 TNA CP 40/736, m. 354r. (AALT IMG 695); CP 40/834, m. 48r (AALT IMG 96); CP 40/871, m. 224r (AALT IMG 414); PROB 11/7/118.

23 Personal communication, summer 1992. In 1360 the Prior of Ely claimed compensation for the theft from his Lakenheath marsh of rushes worth a hundred marks (£66 13s. 4d.): TNA CP 40/402, m. 111r (AALT IMG 1315).

24 TNA CP 40/616, m. 401r (AALT IMG 800); CP 40/656, m. 49r (AALT IMG 101); Bailey (1989), pp. 254–5.

25 TNA CP 40/951, m. 206r (AALT IMG 421). The price of rabbit remained fairly steady during the fifteenth century at just under two-pence per head: Bailey (1989), pp. 296–9

26 Bailey (1989), p. 117.

27 TNA CP 40/656, m. 49r (AALT IMG 102); CP 40/616, m. 401r (AALT IMG 800); CP 40/891, m. 184v (AALT IMG 1049); CP 40/636, m. 450v (AALT IMG 1844). This assumes that a fleece weighed 1.18 lbs.: Bailey (1989), p. 127; and that a stone of wool sold for 3s. 4d.

28 SROB 243 Baldwyne; Bailey (1990 (A)) p. 46; CUL EDC 7/16B.

Ampton bequeathed four hundred wethers out of his flocks at Honington, two hundred more from his flocks at Icklingham and all his sheep and lambs at Little Livermere.[29] Their flocks needed full-time, skilled care and more than half the shepherds recorded in the pleas rolls came from those parishes that Bailey described as central or peripheral to Breckland (Map 3.1).[30]

Before the Black Death peasants had been active in sheep farming. For example, in 1283 in Barnham, ownership of 2,525 sheep had been divided between forty-three taxpayers; in Euston 1,103 between twenty-six; and in West Stow 1,097

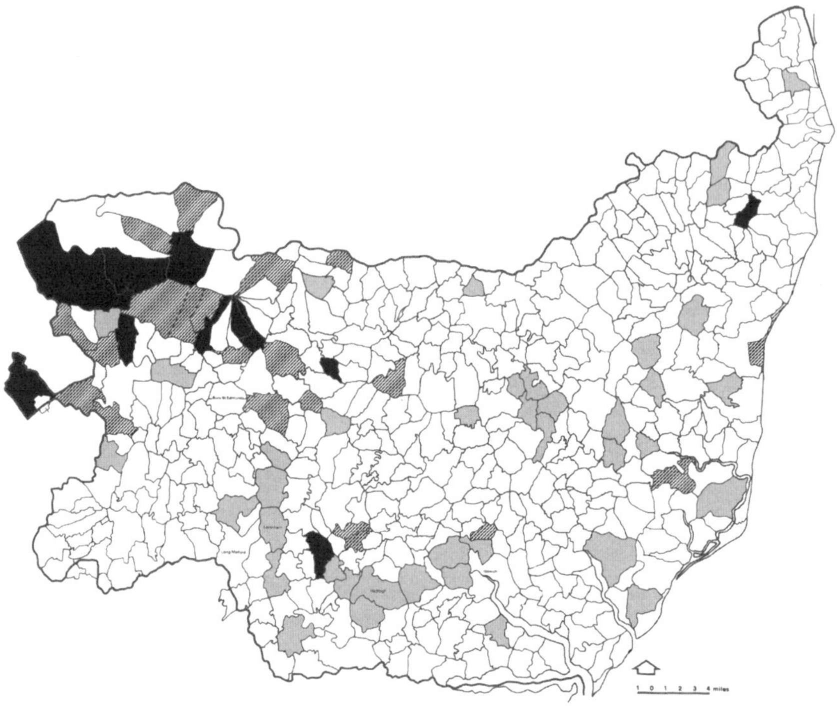

Map 3.1: *Suffolk Sheep and Shepherds 1350–1500*

Notes: Grey marks a parish in which a court case refers to sheep. Hatched dark grey marks a parish in which a court cases refers to shepherds. Black marks a parish in which a court case refers to both sheep and shepherds.

Source: TNA CP 40.

Map format attribution: David Dymond and Edward Martin.

29 TNA PROB 11/7/118.

30 Bailey (1989), pp. 35–6.

between twenty.[31] Thereafter, although Breckland did not experience enclosure in medieval times, the engrossment of holdings and the growth of seigneurial power squeezed out smaller farmers and denied them access to sheep folds. Even if peasants were willing to hazard the uncertainties of wool growing, in Breckland they were rarely given the chance, and, when they were, they were not allowed to fold many beasts.[32] The Register Baldwyne includes all 2,324 wills proven in the court of the archdeacon of Sudbury between 1439 and 1474.[33] Just one of them referred to a sheep-fold. Thomas Hyve of Lakenheath bequeathed 'to Cecily my wife 20 sheep, as they run out of my sheep-fold; to Agnes Brownwyn 12 sheep; to John Lacy, my shepherd, 12 sheep'.[34] In 1497 the substantial estate of Thomas Man of Lakenheath included a 'foldright called Bowes'.[35] The fortunes of his family were on a rise that would lead to their taking a lease of the Lakenheath manor within twenty years.[36]

Central Suffolk

Other flocks were scattered across the county in more than fifty towns and villages (Map 3.1). Within a group of parishes around Debenham, on the eastern edge of the wood pasture lands of High Suffolk, Henry Quyncy had a hundred sheep at Ashfield, while Robert Wingfield Esquire lost thirty-three at Cretingham.[37] In a survey of 144 wills made between 1374 and 1530 by testators from the same area, Peter Northeast found bequests of a relatively modest aggregate number of 181 sheep.[38] Further south, in the Suffolk industrial belt of Babergh and Cosford, clothiers themselves were known to keep animals, as no doubt did those of their neighbours who practised husbandry.[39] The woolman John Wyndecole leased a seven-acre croft in Monks Eleigh in which he may well have grazed sheep.[40] Nevertheless, the size of flocks was also relatively small (Table 3.1). Some of these sheep were reared locally, but others may have been brought from distant parts to service the cloth industry. John Porter claimed that his sheep were worth 2s. 6d. each, 25 per cent more than the then going

31 Powell (1910), pp. xxx, xxxi and tables.
32 Bailey (1490), pp. 45–9.
33 Northeast (2001) and (2010), passim.
34 SROB 263 Baldwyne.
35 SROB 83 Boner.
36 CUL EDC 7/15/11/48; Bailey (2007), p. 217
37 TNA CP 40/437, m. 378v (AALT IMG 3776); CP 40/796, m. 61r (AALT IMG 123).
38 Dymond (2012), p. 445.
39 C. Dyer (2002), p. 326. Small flocks of sheep were similarly kept in the cloth-making region of the Kentish Weald: Zell (1994), p. 202.
40 CCA-DCc-Register B: Monks Eleigh Rental 1379/80, f. 156v.

TABLE 3.1: REFERENCES TO FLOCKS OF SHEEP IN SUFFOLK INDUSTRIAL BELT 1350–1500

Town/Village	*Number of sheep mentioned*	*Value per sheep*
Assington	20	2s.
Great Waldingfield	40	2s. 4d.
Hadleigh	20	2s. 6d.
Kersey	20	2s.
Lavenham	60	2s.
Layham	80	Value unascertainable
Lindsey	56	No value given
Monks Eleigh	6	No value given
Shimpling	No number given	No value given
Stratford St Mary	20	2s.

rate for Breckland beasts.[41] John Cooke was involved in a fight in Lavenham with six husbandmen from Herefordshire. Unless they had driven sheep across country, it is difficult to understand why they would have been in town.[42] A possible drove way has been identified at the east end of Water Street by which 'herds of sheep were being brought in to Lavenham (rather than just their fleeces)'.[43] John Howard engaged in the wool trade, as in many other commercial activities. He sold to William Howet and the local clothier Robert Sergeaunt a substantial quantity of wool for £21 7s. 11d.[44]

The Sandlings

On the far side of the county, along the east coast, a region known as the Sandlings shared many of Breckland's ecological and economic features. Heathland, marsh, turbary, alder, sedge and rushes were important resources to many manors. Ghostly graffiti, with lace-like images of ships, can be found on the columns of local churches, such as that at Parham. They are a reminder of the importance of fishing

41 TNA CP 40/636, m. 343v (AALT IMG 1720). We know that Breckland beasts were worth about 2s. per head because James Wymbyssh of Kentford was alleged to have sold one hundred sheep for £10: *CPR (1461–67)*, p. 253.

42 Long distance droving of animals was not uncommon in the Middle Ages: Stenton (1936), pp. 18–9.

43 TNA CP 40/636, m. 343v (AALT IMG 1720); CP 40/756, m. 4v (AALT IMG 919); Fradley (2007), p. 8.

44 Crawford (1992), II, p. 302.

and coastal trade.[45] As we saw in chapter 2, mariners and shipmen emerged from virtually every sea-side parish, commonly sailing to London and sometimes as far as Southampton.[46] Some combined voyaging with pastoral farming. William Herman of Covehithe unlawfully rescued his beasts from the pound where they had been imparked by the servant of the lords of the manor, while Henry Andrew and Henry Cotman of Pakefield allowed theirs to trespass on a neighbour's land.[47] The sea was not an unmitigated blessing. It was also a threat. From the thirteenth century, coastal erosion began to eat away at the land.[48] The story of the slow disappearance of the town of Dunwich is well known, but it was a process that continued along much of the shoreline. John Cokerell complained that his neighbour Richard Spisour had failed to maintain his own sea defences, and so allowed the flooding of 20 acres of his marsh at Orford.[49]

What were surprisingly less evident in the late medieval Sandlings were sheep. In the late 1370s the Prior of Holy Trinity Ipswich kept at least two hundred animals at Sotterley, while in 1477/78 the prior of Blythburgh paid 6s. 8d. to pasture a small flock in the great close called *le Haughe* at Blythburgh.[50] Simon atte Fen in Lound, John Giles in Sudbourne and Thomas Warde in Alderton all sought compensation for the loss of flocks, but only Warde's numbered as many as a hundred beasts.[51] The bailiffs of Dunwich charged a penny per sheep per annum for pasturing on the town's marsh. An account of 1428/29 provides the names of twenty graziers who, between them, owned 214 sheep and seventy-three lambs. The biggest flock, of just forty-seven sheep and two-dozen lambs, belonged to Robert Genewe.[52]

---

45 http://www.medieval-graffiti-suffolk.co.uk See also Bailey (1990 (B)) pp. 102–14; Kowaleski (2012), pp. 87–118.

46 In 1480 the Southampton merchant William Gunter sued three mariners from Blythburgh each for payment of a debt of £10 that had been incurred in his city: TNA CP 40/871, m. 282v (AALT IMG 1464). Mariners and shipmen were recorded from Gorleston in the north, through Lowestoft, Kirkley, Pakefield, Kessingland, Benacre, Covehithe, Easton Bavents, Southwold, Walberswick, Dunwich, Westleton, Sizewell, Aldeburgh, Orford, Bawdsey to Felixstowe in the south of the county. For a discussion of Norfolk's coastal trade, see Rutledge (1993), pp. 78–9.

47 TNA CP 40/636, m. 460v (AALT IMG 1958); CP 40/736, m. 295v (AALT IMG 1621).

48 Bailey (1991) pp. 195–8.

49 TNA CP 40/576, m. 111r (AALT IMG 226–7).

50 TNA CP 40/477, m. 410v (AALT IMG 839); Richmond (1981), p. 36.

51 TNA CP 40/516, m. 394v (AALT IMG 957); TNA CP 40/636, m. 219v (AALT IMG 1473); TNA CP 40/834, m. 423v (AALT IMG 1653).

52 Bailey (1992), pp. 132, 137–8

In the middle decades of the fifteenth century John Hopton had an enduring interest in sheep farming in Blythburgh and Easton Bavents near Southwold, but he is never known to have owned more than seven hundred and fifty animals. In Richmond's words 'he was not a large-scale sheep farmer'.[53] Having leased his demesne at Easton Bavents to his shepherd John Deynes, he paid him for pasturing three hundred sheep. He also bought a small flock from the executors of Nicholas Nel. The challenges he faced in making wool pay are well illustrated by his experience on acquiring, in 1435, the manor of Easton Bavents from the widow Ela Shardlow. With her intimate knowledge of the land and close relationship with her staff, she had been able to generate a reasonable profit from farming some two hundred sheep. By contrast, Hopton, then a newly arrived gentleman from Yorkshire, barely broke even in the first seven years and blamed his shepherd Nicholas for losing an eighth of his flock through negligence.[54] As Hopton had acquired the manor on the eve of the Sporer Minimum, his losses may well have been as much the fault of the climate as of the hapless shepherd. If Hopton struggled to make sheep farming pay in the fifteenth century, there is little sign of other lay or ecclesiastical lords even trying.[55] It had not always been so.

Sibton Abbey

Before the Black Death the Cistercian abbey at Sibton had been a big player in sheep farming in the Sandlings and south Norfolk. The Institutes of its order's General Chapter provided that:

> Monks of our order should derive their means of subsistence from the work of their hands, from farming, and from animal husbandry. Accordingly, we too are permitted to own for our own use [...] animals other than those which more ordinarily provoke curiosity and show vanity rather than serve any useful purpose.[56]

The Cistercians pursued this objective by acquiring, clearing and enclosing vast acres of intractable countryside to graze sheep, and by consolidating their lands into granges administered from agricultural centres that were staffed by lay broth-

---

53 Richmond (1981), p. 98.

54 Richmond (1981), pp. 79–82.

55 There was no shortage of small religious houses along the east coast: Northeast (1999), pp. 70–1. Butley priory owned over 1,000 acres of sheepwalks on the heathland, but there is little extant evidence of wool growing there in the late Middle Ages: Fenwick and Harrup (2009), pp. 8, 23; personal communication with Valerie Fenwick 29 November to 2 December 2015.

56 Burton and Kerr (2011), p. 149.

ers and later hired labour. They became pre-eminent not only in growing the wool, but in dressing, washing, stretching, drying, sorting, packing and marketing it for direct sale to merchants. Henry I began the issue of exemptions from tolls throughout England that gave them a further competitive edge in transporting wool to ports for export. Although Dyer has suggested that their granges were not technically more sophisticated than other demesnes, and that their progressive farming is a 'modern myth', for a while they were undoubted masters of the wool trade.[57]

Founded in 1150, Sibton Abbey was Suffolk's only Cistercian house.[58] The size of its flocks can be tracked through the centuries. Abbot Samson of Bury St Edmunds (1135–1211) granted the abbey the right of passage for 1,500 sheep along the Tuddenham to Kennett road in Breckland.[59] A hundred years later, at the end of the thirteenth century, three price lists of wool were prepared, of which the best known and most comprehensive was that of the merchant Francesco Pegolotti. His was the only one to include any Suffolk wool and it was the wool of Sibton Abbey that he mentioned. He recorded that the abbey exported ten sacks per annum, at a price of £9 6s. 8d. each, which suggests a flock of about two thousand sheep. Move forward again to the 1360s and the abbey's own accounts indicate a similar number of animals, albeit that the value of their wool had fallen by more than half to 3s. 4d. per stone, equating to £4 6s. 8d. per sack.[60] Sibton's wool growing business, like that of its sister houses, around the country, was fatally undermined by various adverse factors. Starting in the mid-thirteenth century, sheep were struck down by a series of pandemics, perhaps liver fluke, but more generally described as murrain.[61] The abbey's flocks suffered particularly badly in 1363, 1365 and again in 1372, losing about a third of total numbers each time. By the early sixteenth-century the monks had abandoned all sheep farming. Neither the abbey's accounts for 1508/9, nor its valuation of 1535, disclose any income from the sale of wool.[62] Today, all that remains above ground to remind us of past glories are the ruins of the refectory, probably the oldest in the world, and the stunning arcade transplanted to the parish church (Plates 3.2 and 3.3).[63]

---

57 Farmer (1991), p. 395; C. Dyer (2002), pp. 134–7; Burton and Kerr (2011), pp. 167–75, 184–6.

58 Knowles and Neville Hadcock (1953), p. 115.

59 Davis (1954), p. 134.

60 Denney (1960), p. 25; Munro (1978), p. 161. In the 1420s wool was sold elsewhere in Suffolk for 3s. 4d. per stone: TNA CP 40/656, m. 229v (AALT IMG 1423); CP 40/676, m. 241v (AALT IMG 1449).

61 Philip Slavin 'The beginning of the end: sheep panzootics and the fortunes of the wool industry in England 1250–1330', presentation to Economic History Conference, 30 March 2014.

62 Denney (1960), pp. 25, 131–48; *Valor Ecclesiasticus* (1817), pp. 434–6.

63 Personal conversation with Dr Glyn Copack, 5 September 2014.

In addition to disease, at least two other problems confronted the abbey in its efforts to continue sheep farming. The health of any flock could be maintained, and even improved by importing fresh breeding stock from elsewhere, so it was important for growers to be able to move their sheep about.[64] In 1415 the abbot was defending an alleged right, by virtue of long-term use, to drive beasts between his manors at Rendham and Mendham over land belonging to John Yong – we do not know the outcome of the contest. Before the Black Death, labour had been plentiful and cheap, afterwards it was neither – good people became hard to find and even harder to keep.[65] The abbot engaged John Cote and Ralph de Danbere, and, later, Adam Hert to work for him, but presented with a better offer, each of them absconded without completing their term of service.[66] Although the Statute of Labourers was intended to prevent such misconduct, in practice it rarely did.

Woolmen

Woolmen (otherwise known as woolmongers, wool broggers or wool chapmen) were key figures in the late medieval wool trade. Dyer has painted a vivid portrait of one of their number, John Heritage of Moreton-in-Marsh, Gloucestershire. They were middlemen balanced 'precariously between the provincial growers and the metropolitan tycoons'. They spent long hours on the road, visiting markets and fairs, building and maintaining the trust of the graziers who were their suppliers and the cloth-makers who were their customers. They worked within a 'tangled web of credit', making and receiving payments by instalment.[67] In juggling cash-flow, the trick was to ensure that the former did not outrun the latter. They sorted, weighed, packed and stored their produce in woolhouses, and arranged its onward carriage. They needed an expert eye for the quality of, and defects in the fleece, and, if they were to turn a profit, an up-to-date knowledge of the market and prevailing prices.[68] If they could not find a ready buyer then the wool would soon dry out and, within a year, lose nearly a third of its weight.[69]

In the days when religious houses, such as Sibton, dominated the trade, the monks sold direct to exporters not only their own wool, but the *collecta* that they had gathered from the small growers around them. Such monopolistic practices

64 As between Hundon and Clare: Britnell (1991), p. 208. See also Stone (2005), p. 153.

65 Burton and Kerr (2011), pp. 159–60.

66 TNA CP 40/402, m. 181v (AALT IMG 1954); CP 40/736, m. 5v (AALT IMG 1043).

67 C. Dyer (2012 (A)), p. 120.

68 They used a metal wool probe to test the quality of the fleece, an example of which is carved on the tomb of Robert de Selby of Kingston-upon-Hull: Badham (2015), pp. 146–7.

69 Stone, (2005), p. 117.

cut woolmen out of the market. Their protests to Parliament in 1262, and again in 1302, were heard sympathetically, but were largely ineffective.[70] So, they operated on only a parochial scale, buying small quantities of wool and selling door-to-door to spinners or at local markets.[71] Power referred to a Suffolk villager in the reign of Henry II, described as a 'wolmonger', 'who reaped on his lord's land four times a week in harvest time'.[72] At the end of the thirteenth century Robert the Woolmonger rented a small messuage in Redgrave.[73] Thirty years late Nicholas le Wollemongere of Sudbury paid 7d. in subsidy, a modest sum which no doubt reflected the size of his business. Two of his contemporaries, Robert and Thomas le Wollemongere, were implicated in the Bury riot that destroyed the old Abbey Gate and caused so much other damage in 1327.[74]

After the Black Death, as the great religious houses gradually withdrew from sheep farming and the direct link between grower and exporter was broken, woolmen assumed a new importance and their fortunes began to rise.[75] According to Power 'in the fifteenth century they formed one of the richest and most important trading groups in the country'.[76] As early as 1381 a woolmonger paid the highest recorded poll tax in Salisbury of 13s. 4d.[77] In 1395 Richard II pardoned sixty-six of them for 'all contempts, trespasses, frauds, deceits and unjust and excessive weighings and purchases of wools contrary to statute'.[78] Most of these came from Lincolnshire whose wools they were selling to Italian merchants. When, very shortly afterwards, the Italians switched their focus of attention from the east to the south coast, in response to their new granted right to ship direct to the Mediterranean, woolmen were happy to supply them with Cotswolds wool through the two great annual fairs at Burford and Northleach.[79] Memorial brasses, such as that of William Grevil in the church of Chipping Camden, Gloucestershire, bear witness today to the success of their enterprise.[80] In the North woolmen were also beginning to prosper, appearing in York's Register of Freemen from 1413.[81]

70 Munro (1978), p. 127.

71 Sutton (1989), p. 214; Swanson (1989), p. 31; Davis (2012), p. 7.

72 Power (1941), p. 46.

73 Gransden (2015), p. 193.

74 Hervey (1906), pp. xxxii, 137.

75 Bowden (1962), pp. 77–8. Dyer challenged any impression of their general affluence, saying of John Heritage that 'in his twenty years of trading he seemed to have lived a hand-to-mouth existence', and that his profit margin was no more than 9.3%: C. Dyer (2012 (A)), pp. 121, 128.

76 Power (1933), p. 52.

77 Fenwick (2005), p. 124.

78 *CPR 1391–96*, pp. 626–31.

79 Fryde (1999), pp. 89, 90, 96–8; Lee (2012), p. 441.

80 Power (1941), p. 50.

81 Ponting (1971), p. 26; Kermonde (1999), p. 202; Swanson, p. 140.

Middlemen are never popular, particularly when they are doing well. Woolmen began to attract the enmity of growers, who thought they were paying too little, and of cloth-makers who accused them of charging too much. They were the 'caterpillar of the Commonwealth', making their 'insidious profit between bark and tree'.[82] It was high time for the state to rein them in. The first legislation targeted at woolmen was passed in 1465, and re-enacted in 1489 and 1531, forbidding them from purchasing or contracting to purchase unshorn wool between 18 March and 24 August in each year.[83] Nevertheless, since they remained an essential link in the chain of supply, woolmen became enduring figures within the industry. What of their fortunes in Suffolk?

Suffolk Woolmen

We know the names of eighty-nine woolmen distributed widely across the county (Map 3.2). The vast majority lived in the west, twenty-one in Bury St Edmunds alone, and many others in towns and villages that straddled major roads radiating out from Bury to Thetford, Ipswich, Sudbury, Haverhill, Newmarket and Mildenhall, which could serve as convenient collecting points.

Icklingham became their most important centre in Breckland. William Gardener pursued a long and successful career starting no later than 1470, when he was locked in litigation with the Hadleigh merchant Robert Forthe, and concluding no earlier than 1505, making the rounds of a territory that included both Suffolk and Norfolk.[84] Another eight woolmen are recorded in villages north of Bury St Edmunds, namely the Fornhams, where the abbot of Bury St Edmunds had large flocks, Ampton, Great Livermere and Timworth.[85] Thomas Medewe of Great Livermere did business with Abbot William Cadenham (1497–1513).[86] Bury St Edmunds woolmen also had family ties and owned land in the Fornhams, Timworth and Ingham.[87] A case in point was the Coket family, at least three of whom were woolmen in the mid-fifteenth century. Walter of Ingham appointed John of Bury as his executor in 1462, who, in turn, appointed his brother John of Timworth in the following year and bequeathed land at Fornham to his widow

82 Power (1941), p. 48.

83 *Statutes*, pp. 410, 535–6. In the re-enactment the proscribed period was changed to run from 1 March until 15 August.

84 TNA CP 40/837, m. 173r (AALT IMG 276); CP 40/971, m. 180r (AALT IMG 371).

85 Bailey (2007), p. 218.

86 TNA CP 40/943, m. 276v (AALT IMG 1359).

87 SROB 77 Hawlee, 185 Pye.

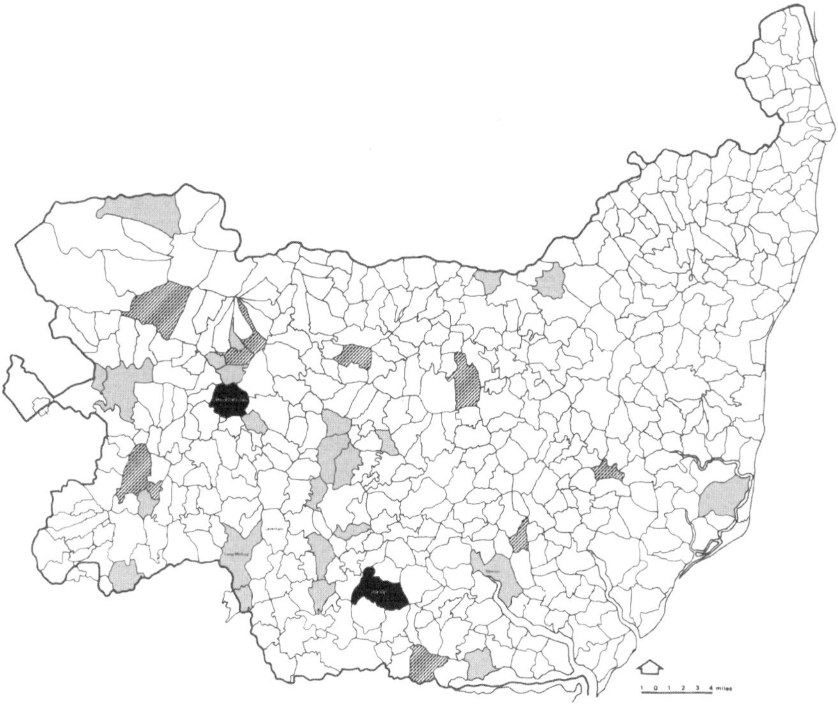

Map 3.2: *Suffolk Woolmen 1350–1500*

Notes: Grey marks a parish in which one woolman was recorded. Hatched dark grey marks a parish in which between two and four woolmen were recorded. Black marks a parish in which not less than five woolmen were recorded.

Source: TNA CP 40.

Map format attribution: David Dymond and Edward Martin.

Joan.[88] John Coket of Ampton, mentioned above, probably carried on the family business in the next generation.[89]

In the east Richard Dysse was the most notable of Ipswich woolmen. He lived with his wife Margaret and two children in the parish of St Mary Tower, and died and was buried in the churchyard there in 1438. He numbered, among his neighbours, some of the town's leading burgesses. His business interests were extensive. He had suppliers or customers in Boxford, Bury St Edmunds, Combs, London,

88 SROB 294 Baldwyne, 77 Hawlee.

89 TNA PROB 11/7/118. Coket made bequests for the repair of the highways from Bury St Edmunds to Lavenham and Long Melford which suggests that he may have supplied wool to the cloth-makers of those towns.

Nayland, Stratford St Mary and Watton in Norfolk. One was a spinster, another a tailor and a third a fellow woolman. Some, such as Sir Robert Ponynges, were probably graziers. Dysse owned a shop and an inn where he transacted his business. He achieved civic recognition as a tithingman. In his will he left silver plate to his local church as well as gifts to the poor in greatest need, and made a bequest of £10 for an honest chaplain to go on pilgrimage to Rome.[90] Thomas Champenays was slightly older and had a lower profile. He was described as a woolman when he appeared before the Ipswich petty court in 1400, and it may have been the same man who re-appeared in Wickham Market fifteen years later.[91] After Dysse and Champenays there is no later record of an Ipswich woolman. By the time Dysse died the overseas wool trade of Ipswich was in decline and the town had only a peripheral role in domestic textile production.[92]

Most of John Heritage's dealings were within eight miles of his home at Moreton-in-Marsh, Gloucestershire where he bought wool to sell on his annual trip to London.[93] Some Suffolk woolmen were similarly able to work close to base. John Fytches the Younger of Sudbury sought payment in the common bench from five men all of whom lived within six miles of his own town – one in Belchamp Otten, Essex, and two in each of Glemsford and Long Melford.[94] Of the twelve disputes recorded in the plea rolls involving Hadleigh woolmen, at least five were with fellow townsmen. These included two claims against Richard Hacche, and one against Ralph Good.[95] As a group the woolmen of Bury St Edmunds conducted less long-haul business than those resident outside the town, perhaps because they had plenty of local custom (Table 3.2).

Although only one Suffolk woolman was recorded doing business in London, many of them had to spend long hours on the road, with two thirds of the suppliers and customers with whom they litigated being more than ten miles distant.[96] Walter Hardhed of Rattlesden, who was active in the middle decades of the fifteenth century, is the best recorded of all Suffolk woolmen with an extensive operation. Living in a village with little textile production of its own, he had few customers on his doorstep. Although some of his suppliers were closer to home in Bildeston, Buxhall and Elmsett, 70 per cent of his business was at a distance of more than ten

90 TNA CP 40/616, m. 284r (AALT IMG 563); TNA CP 40/656, m. Atts. r (AALT IMG 963); SROI IC/AA1/20; SROI C/2/8/1/2, 3, 5, 6, 7, 8, 9, 10.

91 SROI C/2/3/1/34 (21 December 1400); TNA CP 40/616, m. 94v (AALT IMG 1270).

92 Carus-Wilson and Coleman (1963), p. 135; Amor (2011), pp. 121–3, 163–71.

93 C. Dyer (2012 (A)), p. 100.

94 TNA CP 40/853, m. 286v (AALT IMG 1475).

95 TNA CP 40/616, m. 603v (AALT IMG 602); CP 40/616, m. 401v (AALT IMG 800); CP 40/837, m. 5v (AALT IMG 308).

96 Robert Bryght, a woolman of Bildeston, went to London in the early 1480s: TNA CP 40/891, m. 28r (AALT IMG 56).

TABLE 3.2: DISTANCE BETWEEN PARTIES TO DEBT LITIGATION WITH SUFFOLK WOOLMEN 1350–1500

	All woolmen		*Bury St Edmunds woolmen*		*Walter Hardhed of Rattlesden*	
Distance	*No. of claims*	*Percentage of claims*	*No. of claims*	*Percentage of claims*	*No. of claims*	*Percentage of claims*
Same town or village	10	11.5	6	20.7	0	0
Less than 6 miles	14	16.1	2	6.9	1	5
More than 6, but less than 10 miles	4	4.6	3	10.3	5	25
More than 10, but less than 20 miles	27	31.0	13	44.8	13	65
London	2	2.3	0	0	0	0
Norwich	6	6.9	0	0	0	0
Overseas	0	0	0	0	0	0
Other more than 20 miles	24	27.6	5	17.2	1	5
Total	87		29		20	

Notes: This Table (and also Tables 4.3, 5.2 and 6.2) includes those cases in which the place of residence of both the claimant and the defendant was stated and records the distance between those two residences.

Source: TNA CP 40.

miles (Table 3.2). He supplied cloth-makers in Hadleigh, Ipswich and Little Waldingfield, together with others working, nearly twenty miles away, along the Suffolk-Essex border. In Brantham, Dedham and East Bergholt he numbered among his customers three weavers and a fuller.[97] John Dunch was Hardhed's neighbour in nearby Stowmarket.[98] He travelled even further than Hardhed on business and included John Hopton in Easton Bavents among his suppliers.[99] As we shall see, some woolmen were ready to venture well beyond the county boundary in search of wool.

The scale of business undertaken by woolmen is reflected in the value of their disputes recorded in the plea rolls (Table 3.3). The mean value of longer distance transactions, not less than twenty miles, was more than 40 per cent higher

97 TNA CP 40/717, m. 271v (AALT IMG 1538); CP 40/756, m. 50v (AALT IMG 1011); CP 40/776, m. 198v (AALT IMG 392).

98 TNA CP 40/696, m. [torn]v (AALT IMG 870).

99 Richmond (1981), pp. 79–80.

TABLE 3.3: VALUE OF DEBT LITIGATION OF SUFFOLK AND BURY ST EDMUNDS WOOLMEN 1350–1500

	All Suffolk woolmen		*Bury St Edmunds woolmen*	
Value of claim	*No. of claims*	*Percentage of claims*	*No. of claims*	*Percentage of claims*
£20 or more	23	17.7	5	14.7
£10 or more but less than £20	24	18.5	6	17.6
£5 or more but less than £10	21	16.2	8	23.5
Less than £5 but more than 40s.	36	27.7	9	26.5
40s.	26	20.0	6	17.6
Total	130		34	

Notes: Many writs of debt were issued against more than one defendant and in most of these the enrolment alleged that each of them should pay the sum in dispute ('quod quilibet eorum reddat'). In such cases, for the purposes of this and later similar tables, each of the defendants is assumed to have been liable for the whole of the sum. In a few other cases the enrolment alleged that collectively the defendants should pay the sum in dispute. In these, each of the defendants is assumed to have been liable for an equal share of the full sum in dispute, and in some cases this brings the defendant's personal debt below the statutory minimum of 40s.

Source: TNA CP 40.

(£13 8s. 2d.) than that of transactions less than ten miles from home (£9 8s. 10d.). Generally, the cost and risk of long-haul transport, marketing and sale could only be justified by the prospect of a large return. The largest debt was the £97 owed, in 1418, by John Sterre, woolman of Stoke-by-Nayland, to John atte Hachche of Fobbing in Essex. Sterre clearly struggled to pay because he was pardoned from outlawry in 1420 for not appearing before the Justices of the Bench to answer the claim.[100] He may have overstretched himself during a period in the late 1410s, when wool prices had taken a serious downward turn.[101] Four of the five debts owed by Wickhambrook woolmen were for £10 or more, with Richard Broun (alias Worlych) and John Smyth each defending a claim for £40. These debts may have been incurred at one or other of several annual fairs that were held in south-west Suffolk, including one in each of the two adjacent parishes of Cowlinge and Swilland.[102] A third was held, under the auspices of the borough of Clare, at Wentford, less than

100 TNA CP 40/629, m. 225r (AALT IMG 453); *CPR 1416–2*, p. 287.

101 Lloyd (1973), p. 49; Farmer (1991), p. 514.

102 Scarfe (1999), pp. 78–9.

six miles from Wickhambrook, and is known to have been a forum for the sale of wool.[103] Woolmen saw these fairs as good opportunities to do business.[104] While the general value of disputes was relatively high (compare Table 3.3 with Table 4.2), nevertheless many were for more modest sums. The mean value of Walter Hardhed's twenty recorded disputes was £5 10s. 5d. More than 40 per cent were for even smaller amounts, less than £5, and 20 per cent for the minimum sum of 40s. Dyer was almost certainly correct in suggesting that, for small-scale operators like his John Heritage, the wool trade did not provide a ready path to riches.

Woolmen's common bench disputes can also tell us about the occupations of those with whom they dealt (Table 3.4). The disputes with the highest mean value were those with fellow woolmen. Two of the three cases involved members of the Stubbard family of Bury St Edmunds whose fortunes are considered below. Nearly half of disputes were with parties who were probably suppliers – clergy, lords and husbandmen/yeomen. Of the three, the mean value of disputes with the clergy was highest. Both debts due to the abbot of Bury St Edmunds, perhaps Suffolk's pre-eminent grazier, were for sums of £20 or more. The parsons of Eriswell, Framlingham and Lidgate all sued woolmen for significant amounts, perhaps for payment for the tithe wool they had sold. The prior of Thetford brought three claims in 1470, one against William Lovelond of Stoke for £20 and two others against each of John and Thomas Brounsmyth of Hadleigh for the more modest sum of £4.[105] Of the eight woolman disputes with lords, seven pre-dated 1450, while, of the eleven disputes with husbandmen/yeomen, ten were after that date. This surely marks the well-known retreat from seigniorial agriculture that saw Bailey's 'active, dirty boot farmers' taking leases from, and stepping into the cleaner shoes of the often absentee lords of the manor who had once governed Suffolk.[106] The mean value of the husbandmen/yeomen disputes was less than half that of the lords', reflecting the smaller scale of their pastoral operations – the sum of £4 5s. 4d. represented the price of perhaps four hundred fleeces.

Both the number and the mean value of the disputes with clothiers is surprisingly low. In all four cases, the claimant was William Copynger whose career is discussed below; none of the claims was earlier than 1485. It may be that he was supplying the smaller operators and that their wealthier counterparts were now sourcing wool direct from growers.[107] As mentioned above, in 1483 the clothier

---

103 Davis (2012), p. 287.

104 C. Dyer (2012 (A)), pp. 100–1, 103.

105 TNA CP 40/834, m. 191v (AALT IMG 1221).

106 Bailey (2007), p. 210. Farmer stated that the English sheep farmer was 'most likely by now to be a *firmarius* renting the demesnes and pastures the lords no longer bothered to work directly': Farmer (1991), p. 401.

107 In Yorkshire, too, wealthier clothiers sourced their wool direct: Heaton (1920), p. 118.

TABLE 3.4: OCCUPATIONS OF PARTIES IN DEBT LITIGATION WITH SUFFOLK WOOLMEN 1350–1500

Occupation	*No.*	*Percentage*	*Mean value of claims*
Clergy (inc. Clerk)	14	20.6	£18 1s. 5d.
Clothier	4	5.9	£6
Cloth-worker	8	11.8	£7 17s. 6d.
Husbandman/Yeoman	11	16.2	£4 5s. 4d.
Member of London company (Grocer)	2	2.9	£19
Lord/Gentry	7	10.3	£9 15s. 7d.
Merchant/Mercer	10	14.7	£9 4s. 6d.
Other	9	13.2	£8 1s. 5d.
Woolman	3	4.4	£26 11s. 1d.
Total	68		

Notes: This Table (and also Tables 4.5, 5.3 and 6.3) includes only those cases in which the occupation of both parties was expressly stated, not those in which such occupation is probably known, but was not stated as such. The cloth-workers were one dyer, three fullers, one fuller alias shearman, one shearman and two weavers.

Source: TNA CP 40.

Robert Sergeaunt of Stoke-by-Nayland bought John Howard's clip.[108] More than 40 per cent of the debt disputes of the great Hadleigh clothier, William Forthe, were with husbandmen and yeomen from many of whom he probably bought wool.

The ten years either side of 1420 were not vintage ones for wool growers. Furthermore, the Suffolk textile industry had not yet grown to full maturity. Nevertheless, during that period, thirty-two different Suffolk woolmen were recorded in seven plea rolls, compared with just seventeen woolmen in twelve plea rolls in the final twenty years of the century. While it is dangerous to draw firm conclusions from relatively small samples of data, by 1500 the golden age of the Suffolk woolman was probably drawing to a close.[109]

Bury Fair

Bailey correctly contended that Bury burgesses 'largely controlled and dominated' Breckland's wool trade.[110] Indeed, as should by now be apparent, Bury St Edmunds

108 Crawford (1992), II, p. 302.

109 Not one woolman was recorded in the Babergh Miltary Survey of 1522: Pound (1986), pp. 132–3.

110 Bailey (1989), p. 178.

was the centre of that trade in Suffolk. The town's winter fair held outside the east gate in November and December of each year had once been of international importance and, in the fifteenth century, still attracted business.[111] May we assume that Edmund Clerk, a Norfolk woolman, was attending the fair in the late 1430s when he found time to break into Robert Burgeys' house, steal his goods and abduct his servant Joan?[112] Several of the town's woolmen owned adjacent property that may have served as the centre of their operations and prominent local families made their fortunes there.[113]

The longest-lasting dynasty was that of the Stubbard (alias Grygge) family. Robert first appeared in the records in 1415 buying 303 fleeces from the abbot of Bury St Edmunds' flocks in Fornham.[114] Despite the tardiness with which he sometimes paid his debts, his relationship with the abbot appears to have survived for some years. In 1420 he faced a claim by the abbot for £24.[115] John Stubbard was a litigant in the common bench in a series of cases between 1430 and 1460.[116] In 1450 he claimed £4 each from John Herry, a weaver of Lavenham, and John Coo, a mercer of Boxford. He may have been the same man who died in 1473. If so, his business life was both a long and reasonably successful one. In his will he bequeathed silver plate, but more importantly, left his son William his best horse, saddle with bridle, pack cloth and 16 stone of wool so that he could continue his trade. He appointed his friend William Berewey, a fuller, to act as joint executor with the son, and together they set about collecting book debts not only from Richard Broun of Wickhambrook, but also from Gregory Belcham, a fuller of Hadleigh, and John Wryte a fuller and shearman of Lavenham. This proved no easy task since the debts remained outstanding ten years later.[117] The son had, in turn, passed away by 1490, and was succeeded by another, William, who too was proud to call himself 'woolman'.[118] As we shall see, family dynasties, such as the Stubbards, were not uncommon in the wool trade and textile industry. Children growing up in a wool or cloth household would not only imbibe the mysteries of the craft, but would also

111 Gransden (2015), pp. 179–82.

112 TNA CP 40/716, m. 364v (AALT IMG 732)

113 Adam Babyngton's dwellinghouse was in Eastgate Street: SROB 52 Hawlee. Together with his kinsman Walter, William Copynger owned a tenement on Angel Hill: Scott Thomas (1950), p. 156.

114 Bailey (1989), p. 290.

115 TNA CP 40/636, m. 388v (AALT IMG 1830).

116 TNA CP 40/676, m. 155v (AALT IMG 1278); CP 40/756, m. 209v (AALT IMG 1327); CP 40/796, m. 48v (AALT IMG 939).

117 SROB Hawlee 180; TNA CP 40/885A, m. 5r (AALT IMG 9).

118 Bailey (1989), 290; TNA CP 40/911, m. 280v (AALT IMG 554 dorse).

enjoy an early opportunity to build relationships outside their immediate family with those who might help them in their later careers.[119]

At about the same time as Robert Stubbard, Adam Babyngton also did business with the abbot in Fornham, and proved to be no quicker at paying his debts. In 1410 he bought four hundred of the abbot's fleeces and five years later had still not paid the full sum due.[120] In 1415 he himself pursued payment from two other woolmen in Bury St Edmunds and Rushbrooke respectively, as well as from Roger Shaldford, a Bury weaver. He may have lived until 1456 when a will in his name records an unusually large bequest of 26s. 8d. to the parish church of St James for tithes forgotten. He remembered his two daughters, Margaret and Anne, and appointed as executors, among others, his widow Rose and his son Hugh.[121] A Hugh Babyngton, son or perhaps grandson, was trading as a woolman as late as 1485.[122]

The most prominent of all Bury woolmen was William Copynger. He was appointed by John (alias Jankin) Smith as one of the original Guildhall Feoffees in 1470, and served as an alderman of the town. He enjoyed a close connection with the prominent Gardener family, serving as executor for Stephen Gardiner and, together with his kinsman Walter Copynger, a clothier, transferring to John Gardener a tenement called *The Bole* on Angel Hill. As mentioned earlier, William numbered among his customers four clothiers who hailed from Great Waldingfield, Hadleigh, Long Melford and Stratford St Mary, as well as two fullers of Long Melford. He was appointed as executor by his brother Walter in 1506, but was dead by 1514 when his own executors, Thomas and William Copynger, were seeking recovery of further debts.[123]

An Insatiable Demand

In the 1390s Suffolk cloth-makers produced the equivalent of about 2,500, and, by the late 1460s, 5,000 broadcloths each year. In addition, they also made worsted cloths that escaped the attention of the alnager. One sack of wool was needed for every four to four-and-a-half broadcloths and about 260 fleeces for every sack of

119 For contemporaneous examples of such cloth-making dynasties in the Kentish Weald, see Zell (1994), pp. 191–2.

120 He is described as 'Adam Woolman of Bury St Edmunds', but almost certainly this was Babyngton: Bailey (1989), p. 271n.

121 TNA CP 40/616, m. 87v (AALT IMG 1257); SROB 52 Hawlee.

122 TNA CP 40/891, m. 162v (AALT IMG 1013).

123 Statham, pp. xx, 283–4; Lobel (1934), p. 28; SROB 125 Hawlee; Scott Thomas, p. 156; SROB 182 Pye; TNA CP 40/1005, m. 71r (AALT IMG 143 front).

wool.[124] So, by 1470, the output of the woollen textile industry had grown to nearly 30,000 fleeces a year, creating a quite unprecedented demand. Where did all this wool come from?

Undoubtedly, much of it was grown locally. If each of the large Breckland parishes generated between five hundred and 1,000 fleeces a year, and the remainder of the county a few thousand, then the total might be a sizeable fraction of what was required. It would not have made export grade cloth, but it was good enough for domestic consumption. The remainder must have come from beyond the county boundary. Betterton and Dymond suggested that Suffolk cloth-makers drew on the wool of 'Lincolnshire, Northamptonshire, Buckinghamshire and Leicestershire'; and, more recently, Dymond re-affirmed that Suffolk broadcloth was 'made of short-fibred wool imported from limestone areas in, for example, Northamptonshire and Lincolnshire'.[125] A close study of those who brought wool to Suffolk, both great merchants and more humble woolmen, confirms that much did come from these counties, but that there were other important sources.

Merchants of the Staple made good use of the port of Ipswich in the fifteenth century for the export of wool. Many of them came from East Midlands towns, from Coventry in the west, through Leicester, Northampton, to Boston in the east.[126] Very likely, if they could sell some of their clip at the right price in England, then they would take the opportunity, rather than risk it to the high seas. Indeed, as the overseas wool trade declined during the course of the fifteenth century, the opportunity to off load in Suffolk must have become ever-more welcome. William Browne of Stamford was among them. He had ready access to high quality wool of Lindsey, lesser quality Kesteven wool, and poor quality Fenland wool.[127] In the late 1460s he traded with three merchants of Ipswich and a fourth of nearby Higham.[128] John Coote, a draper of Stoke-by-Clare, counted on Henry Noke, a yeoman of Boston, to act as his surety, suggesting a close relationship.[129] Meanwhile, East Midlands woolmen traded with Suffolk cloth-makers. We hear of John Pery of Biggleswade in Bedfordshire, and of John Stoghton of St Neots in Cambridgeshire who sued John Talbon the Younger, a weaver of Long Melford. Walter Hardhed, the

124 Carus-Wilson and Colman (1963) pp. 13, 15–6. Oldland challenged the presumption that, in the Middle Ages, a sack of wool produced only 4.33 broadcloths. Although this figure was used in a 1548 petition to Parliament, he argued that 'the amount of wool required for broadcloth, or its equivalents, increased significantly over time': Oldland (2014), 29–30. John Munro informed the author that he supported this view: personal conversation 5 July 2013.

125 Betterton and Dymond (1989), p. 26; *Suffolk Threads*.

126 Amor (2011), pp. 180–1, 270–1.

127 Rogers (2012), p. 135.

128 TNA CP 40/834, m. 69r (AALT IMG 134).

129 TNA CP 40/776, m. 538v (AALT IMG 2176–7).

well-travelled woolman from Rattlesden, dealt with John Glovere a yeoman of Olney in Buckinghamshire.[130]

The wool of the East Midlands was not enough to keep all the looms running in Suffolk. The county's cloth-makers were happy to work with Essex wool. The marshes in the south of that county, on the north bank of the Thames estuary, provided rich grazing for sheep.[131] John Cole, a mercer, and Richard Debenham and Richard Halk, two weavers, all of Stoke-by-Nayland, relied on Henry Rede, a woolman of Prittlewell.[132] The Stoke-by-Nayland clothier Nicholas Hatche did business with Roger Hatche, perhaps a kinsman, who was a woolman in Billericay.[133] The Suffolk woolman, John Sterre dealt with John atte Hachche of Fobbing.[134] Prittlewell was situated in the south Essex hundred of Rochford, and Billericay and Fobbing in that of Barstable (Map 1.4). On the far side of the Thames estuary, the north Kent marshes provided an equally good habitat for sheep whose fleeces were also sold to Suffolk cloth-makers. Two weavers and two fullers from the Shotley Peninsula transacted business with the woolman Thomas Carter of Maidstone.[135] By boat down the Stour, along the Essex coast, across the estuary and up the Medway the travel distance was little more than sixty miles.

Looking further afield, those Suffolk clothiers focusing on the quality end of the market may have procured premium grade wool from the Cotswolds and the Welsh Marches. We know that in 1509 Thomas Spryng III of Lavenham bought wool in London from John Heritage.[136] By the early years of Elizabeth I, cloth-makers of Coggeshall and Dedham procured, either direct or through London dealers, the wools of Shropshire and Staffordshire to produce their fine white broadcloths.[137] Intriguing references suggest that, more than a century earlier, Suffolk men did the same. The wealthy shearman Richard Davy of Nayland bequeathed a cross of silver and gilt worth £24 to the church of Ellesmere in Shropshire, betraying a close connection with that town.[138] We have already met the Herefordshire husbandmen who were caught up in a brawl in Lavenham. It was not only hot blood that brought the two ends of England together, but also trade. The Hadleigh woolman, Richard Baxtere sought payment from John Dane of

130 TNA CP 40/576, m. 228r (AALT IMG 474); CP 40/883, m. 250r (AALT IMG 448); CP 40/776, m. 198v (AALT IMG 392).

131 Gascoigne and Medlycot (2014), pp. 12–15 <http://content.historicengland.org.uk/images-books /publications/essex-historic-grazing-marshes-project/Final_Grazing_Marshes.pdf/>

132 TNA CP 40/776, m. 140r (AALT IMG 276).

133 TNA CP 40/907, m. 58v (AALT IMG 945).

134 TNA CP 40/776, m. 140v (AALT IMG 1391).

135 TNA CP 40/788, m. 241v (AALT IMG 1521).

136 C. Dyer (2012 (A)), p. 117.

137 Bowden (1962), pp. 64, 92.

138 TNA PROB 11/18/50.

Herefordshire, while the clothier John Wodhous of Stratford St Mary faced a claim for £20 by Sir Gilbert Talbot of Shropshire.[139]

Of all the counties that yielded up their wool to Suffolk, by the late-fifteenth century Norfolk was probably the most important. By then, the Townshends had begun to build their flocks on the lighter soils of north-west Norfolk, around Raynham. They carefully managed their sheep, relied on their own breeding and won control of local fold-courses. They created a compact group of estates in a manner reminiscent of the Cistercians. Despite early setbacks, caused by murrain and severe winters, they went on to enjoy the same success. By 1507 they kept nine flocks, comprising 8,418 sheep. They supplied relatively low-quality short-staple wool for use in the domestic textile industry and numbered among their customers cloth-makers in Hadleigh and Lavenham.[140] Suffolk bed weavers relied on longer staple wool of pasture fed sheep for which eastern Norfolk was well known and looked to Norwich to meet their requirements. In the 1420s the executors of the city woolman Robert Southfield pursued payment from five Suffolk men, including the hapless John Sterre. Later, two prominent merchants, Richard Brasyer, alderman and twice mayor, and Richard Broun, alderman, engaged in the same trade.

Some deals were done over the farm gate, some at one of the markets or fairs that dotted the county. Some wool arrived in Suffolk overland, the rest came by river and sea. The cost of transporting wool by cart was not prohibitive, although it was inflated if tolls had to be paid en route or on arrival. Bolton states that in the mid-fourteenth century, wool could be carried over a distance of fifty miles at about 1.5 per cent of the value of the cargo, but more when labour costs rose and wool values fell.[141] Dyer calculated that John Heritage paid 3s. 6d. per sack for four to five days' transport from Moreton to London, or a little over 2 per cent of the price he achieved on arrival.[142] Carriage by boat, along either the coast or inland waterway, was significantly cheaper, even if the route was much longer, which explains why the boatmen of Breckland and the mariners and shipmen of the Sandlings were so busy.[143]

Even before the spinster picked up her spindle, or the dyer filled his vat with woad, a great deal of hard work had been necessary to grow, collect and make the

139 TNA CP 40/596, m. 224r (AALT IMG 457); TNA CP 40/951, m. 98r (AALT IMG 201).

140 Allison (1958), pp. 107–8; Fryde (1999), pp. 265–6

141 Bolton (1980), pp. 151–2.

142 C. Dyer (2012 (A)), pp. 117, 128. The Celys paid a similar rate to transport their wool on the same route: Farmer (1991), p. 352.

143 In his study of transport in Cambridgeshire and Huntingdonshire in the first half of the fourteenth century, Masschaele found that overland carriage cost per ton-mile was more than double riverine cost: Masschaele (1997), pp. 208–9.

wool ready for manufacture. Suffolk cloth-makers relied upon an elaborate organisation that centred on Bury St Edmunds, but stretched from Herefordshire in the west to Lincolnshire in the north. With such a volatile commodity as wool, both grazier and woolman ran great risks. Some, such as William Copynger, could hope for modest fortunes, while others, like John Sterre, suffered losses and perhaps business failure. In the next chapter we shall consider the various stages in production that turned wool into cloth – combing, carding, spinning, weaving and fulling.

– 4 –

Cloth Production

Well away from the village centre of Norton in central Suffolk you will find a church dedicated to St Andrew. In its thirteenth-century chancel awaits the lady you have come to see. Since the 1300s she has sheltered under a stall, in the form of a misericord, quietly carding her wool ready for the spinster, weaver and fuller to turn it into cloth (Plate 4.1).[1] This chapter considers carding and all the other stages in cloth production in detail, beginning with an account of the raw material with which they worked (Figures I.1 and I.2).

The Nature of Wool

Made of the protein keratin, wool has a cellular structure well suited for the manufacture of fabric for clothing. Unbeknown to them at the time, the spinster and weaver could rely on the strength and elasticity of the wool because of millions of long spindle-shaped cells within its cortex. When stretched, rather than breaking, the links between these cells unfold, and, when released, they return to their original position. The spinster could also be confident that the twisted yarn would not unravel on account of the horny irregular scales that encase the cortex, overlap from root to tip like tiles on a roof, and make the fibres cling together. By pounding the woven cloth the fuller could make the fibres shrink, thicken and mat together because of these same scales. They give the fibres a differential frictional effect, in the sense that they move against each other more easily in one direction than another. Owing to their inherent plasticity, the fuller could then mount the woollens on a tenter frame and stretch them back to the desired length. Wool keeps the wearer warm and dry by reason of an outer sheath, or membrane, around the fibre that absorbs moisture through microscopic pores but is otherwise water repellent. The curly structure of wool, known as its crimp, re-enforces elasticity, helps hold

1 This misericord is 'exceptionally good': Bettley and Pevsner (2015), p. 442.

fibres together in spinning, and prevents them from lying flat.[2] Despite all these inherent advantages, wool 'is not the easiest fibre to handle' and it took skill to turn it into cloth.[3]

Carding, Combing and Spinning

As wool grows on sheep, its fibres exude grease, which contains lanolin, and sweat. At the very beginning of the manufacturing process, a critical decision had to be taken whether or not to scour and remove these natural oils – the so-called 'yolk'.[4] That decision marks the difference between the wet and dry draperies that Munro described as 'the true and crucial distinction' in medieval cloth-making.[5] Wet draperies involved 'wet' treatment of the wool. Generally, short staple wool was scoured in a mixture of water and urine, the ammonia of which acted as a detergent. Once dry it was beaten to remove dust and filth. Then, to facilitate spinning and protect the fine and delicate fibres during manufacture, it was re-greased with one or more of several possible lubricants. When available rape or fish oils were used and the Ipswich particular customs accounts record significant imports of both. Butter, sheep tallow, melted pig fat, or goose grease were alternatives.[6] In the wood-pasture region, in which Suffolk's industrial belt lay, butter was in plentiful supply, particularly to those cloth workers who doubled up as pastoralists. Nevertheless, it would appear that these local substitutes were a poor second best since, in 1528, the Suffolk clothiers threatened the duke of Norfolk that unless the importation of oil from Spain was resumed they would have to abandon cloth-making.[7] Dry draperies dispensed with such treatment. Longer staple wools had straighter and stronger fibres that did not need the same protection, so they normally went unscoured and relied on their own yolk for lubrication. The main product of the wet draperies was woollens, including broadcloth, kerseys and straits, and that of the dry draperies was worsteds[8] – a subject to which we will return later in this chapter.

The different production trajectories of woollens and worsteds continued at the next stage. In late-medieval wet draperies wool was normally carded, while in dry

2 Whewell (1972), p. 157; Baines (1977), p. 27; Hyde (1988), p. 556.
3 Ponting (1971), p. ix.
4 Baines (1977), p. 28.
5 J. Munro 'Distaff I – Focus on Fibre' presentation to International Medieval Congress: University of Leeds, 4 July 2013.
6 TNA E 122/52/42, 43, 45, 47, 48; Baines (1977), p. 30; Sutton (1989), p. 214; Munro (2009), p. 5.
7 TNA SP Letters and Papers, Foreign and Domestic Henry VIII, vol. IV, pt. II, no. 4239.
8 Munro (2003), pp. 184–5.

draperies it was combed.[9] Carding involved the use of two wooden boards (measuring about 30cm × 15cm), known as cards, with leather on one side impregnated with hundreds of fine, short, sharp wires.[10] These cards were of relatively low monetary value, the abbot of Sibton bought a pair for his servants for 8d.[11] They resembled wire brushes and, in the making of the wires, Spanish iron was considered best. Such iron was regularly imported, two consignments arriving in Ipswich on board a ship in December 1397.[12] In carding wool, the fibres were first separated and foreign matter removed by a process known as scribbling or teasing. Then small tufts of wool were placed on one card that was brushed across the other in a way that transferred the wool. The wool was removed as a spongy, soft, air filled roll known as a rolag. Combing involved a not dissimilar process of transferring wool from one implement to another, known as jigging, but in this case between two large upright T-shaped combs. These had rows of teeth that became increasingly longer the further they were from the handle. The combs were kept hot to provide lubrication. Wool from one was lashed onto the other with a downward sweeping movement. The wool was ultimately removed from the second comb with its longer fibres, which formed a roving, left straight and parallel. Finally, the rolag or roving was drawn out and twisted together, in a process known as slubbing, ready for spinning.[13]

In northern Europe combing pre-dated carding by many centuries. In the 1950s an Iron Age comb, made of red-deer antler, was found during the course of excavations for the foundations of the Fisons Pension Trust building in Princes Street, Ipswich.[14] The technique of carding was introduced in England at some time after 1300, possibly in response to a deterioration in the climate which would have reduced the mean length of the wool staple and so made it more difficult to comb.[15] Many European cloth-making towns resisted cards in the erroneous belief that their wires would damage the wool, or perhaps allow illicit blending of different quality wools, but no evidence of such resistance is recorded in England.[16] Each

9 A third method was known as bowing, but with little evidence of its use in western Europe and certainly not in Suffolk: Crowfoot et al (2001), p. 17; Munro (2003), p. 199.

10 Initially, straight-spiked teasels were used in cards, but by the fourteenth century were superseded by wires. The Latin term for teasel is *cardus* thus giving cards their name: Ponting (1971), p. 8.

11 Denney (1960), p. 142.

12 TNA E 122/193/33; C. Dyer (2002), p. 303.

13 Lemon (1972), p. 88; Baines (1977), pp. 31–7; Munro (2003), pp, 197–8.

14 Smedley (1960), pp. 47–9.

15 Chorley (1997), p. 7.

16 Chorley (1997), p. 16; Munro (2003), pp. 198–9; Van Bavel (2010), p. 347. A distrust of the use of wires in cloth production was, however, evident in the English authorities' attitude towards the gig mill.

method had its advantages. Yarn that had been combed was stronger than that which had been carded, a matter of particular importance in the warp. However, combing was far more wasteful than carding because shorter fibres ('noils') were discarded. It required greater upper body strength, so was often done by men who might expect higher wages than women.[17] Carding was more versatile in the sense that it allowed the blending of wools of different quality and colour in the same yarn.[18] During the fifteenth-century, in the production of woollens, combed yarn was sometimes used for the warp with carded yarn for the weft. So a single workshop might produce both. In 1418 Agnes Stubbard of Bury St Edmunds, who was probably related to the woolmen of that name, bequeathed to each of her servants Sibill Chekyneye and Margaret Tompston a pair of woolcombs, a combingstock, a wheel and a pair of cards; and to Issabell daughter of John Yoxford of Bottisham a pair of woolcombs, a wheel and a pair of cards.[19] While two of the wills copied in the Register Baldwyne (1439-74) refer to cards, no mention at all is made of combs.[20] In 1464 a statute required that 'every carder, spinster, weaver, fuller, shearman and dyer shall duly perform his duties in his occupation', but omitted any reference to combers.[21] By 1500, as a result of more extensive use of the spinning wheel, in the production of woollens, both the warp and weft yarn were generally carded.[22] That combing did not disappear altogether from East Anglia is evident from the will of Thomas Paycocke of Coggeshall who, in 1518, remembered the combers who had worked for him.[23]

Some cards were imported, perhaps from Rouen which was a known centre of card-making, and in 1491 the Ipswich particular customs accounts record the arrival of a batch.[24] Others were manufactured locally. A few men earned a living in making them, but never enjoyed great reward. The plea rolls indicate that, nationwide, more than 80 per cent of card-makers worked in urban centres, and 86 per cent of their disputes were for the recovery of debts. This suggests that they were independent operators, rather than the outworkers of rich clothiers.[25] We have the names of a thirteen card-makers in fifteenth-century Suffolk – five from Hadleigh, two from Ipswich, one each from Badwell Ash, Boxford and Sudbury and three,

17 Lipson (1921), p. 185; Bowden (1962), pp. 41–2; Baines (1977), p. 32; Chorley, (1997), p. 9, Munro (2003), p. 200.

18 Munro (2003), p. 198.

19 SROB 158 Osbern as translated by Tymms (1850), pp. 2–5.

20 SROB 388, 430 Baldwyne.

21 *Statutes*, 2, p. 406.

22 Chorley (1997), pp. 8–10, 22.

23 Power (1924), p. 156.

24 TNA E122/53/9; Kerridge (1985), p. 175.

25 Urban centres are those included in one or both of Alan Dyer's lists of top one hundred towns by taxpaying population and taxable wealth: A. Dyer (2000), pp. 761–7.

whom we know by their wills, from Bury St Edmunds. A further two card-makers were recorded in the Babergh Military Survey of 1522, one of middling wealth and the second 'poor'.[26] Seven were cited as defendants before the common bench in eight separate debt cases, six of which were for the minimum sum of 40s., one for 50s. and the eighth for £10. Robert Byrd was in business in Hadleigh for at least twenty years, defending this unusually high value claim in 1470, and another for 40s. by the town's leading clothier William Forthe in 1490.[27]

The wills of the Bury men paint a picture neither of conspicuous wealth, nor indeed abject poverty. John Candeby had done sufficiently well at one stage in his career to have employed an apprentice Roger Well, to whom he left a gift of 3s. 4d.[28] John Alpe was survived by his wife Alice and three children, Elizabeth, Nicholas and John. He gave the matrimonial home to Alice for life, and then to Nicholas. He left Elizabeth his shop, suggesting that he favoured her over her brothers to carry on the family business. His son John was betrothed to Emma the daughter of John Jugge wiredrawer who may have been a supplier of the wire Alpe used for his cards. He bequeathed John a tenement on condition that Jugge gave Emma a dowry of equal value when they married.[29] If the personal relationship between John and Emma had grown out of the business relationship of their parents, was the conditionality of this gift a hint that the relationship between Alpe and Jugge had not always been one of complete trust and confidence?

The invention of spinning, like the invention of the wheel, is lost in the mists of time.[30] We shall never know who first twisted wool fibres together and discovered that this made a yarn strong enough to weave into cloth. The greater the torque, the stronger the yarn. We do know that the discovery was momentus. The three stages of spinning were drawing out the fibres from the rolag or roving, and then twisting and winding them on to a spindle. Warp yarn was spun closer and twisted tighter than weft so as to give it the strength to withstand the stress and strain inherent in weaving. In Baines' words 'the skill of the spinner lies in drawing-out the fibres evenly by maintaining the correct balance between the speed of draw, the quantity of fibre and the amount of twist'.[31] For centuries spinning was done by hand, 'by the rock', using two wooden rods – one, a distaff, for the rolag or roving, and the other, a spindle, for the twisted yarn. The spindle was weighted by a whorl, made of stone or other heavy material (Plate 4.2). This acted as a flywheel, keeping the spindle vertical and steady, and making it revolve uniformly. The spinster drew off the rolag

26 Pound (1986), p. 133.
27 TNA CP 40/834, m. 6v (AALT IMG 874); CP 40/911, m. 5v (AALT IMG 9d).
28 SROB 187 Osbern.
29 SROB 227 Hawlee.
30 Ryder (1968), pp. 73–82.
31 Baines (1977), p. 41.

or roving from the distaff, attached it to the spindle, spun one end of the spindle with thumb and forefinger and lowered it to the ground to allow the wool to take up the twist. Once twisted the yarn was wound on to the other end of the spindle ready for weaving.[32]

Gradually, hand spinning was supplanted by the spinning wheel, first introduced to England around 1300 and illustrated in the mid-fourteenth-century Luttrell Psalter (Plate 4.4).[33] The so-called 'Great Wheel' earned its name from its size, with a diameter of well over a metre. It worked on similar principles to hand spinning, save that the spindle was in a horizontal mount and linked by a driving band to the wheel, and the twist was provided by the spinster turning that wheel. In order to draw out the rolag or roving she walked up to thirty miles each week repetitively backwards and forwards to the spindle, giving the Great Wheel the sobriquet of the Walking Wheel. Aside from saving shoe leather, hand spinning retained certain other advantages over wheel spinning. The distaff and spindle were cheap, easy to make, easy to store and highly portable. They could be carried into the fields or to market for spinning during idle moments. Some of the same European towns that frowned on carding took a similar dim view of the spinning wheel, particularly in the production of warp yarn, fearing that it would be weak, uneven and 'with too many knots'. Such fear was well founded because faster spinning made it more difficult to maintain a steady twist and so ensure an even yarn. As the spinster used one hand to turn the wheel, she only had one to control the fibres and the twist. Only with the introduction of the Saxony Wheel at the very end of the fifteenth century, with its foot treadle and U shaped flyer, were these problems addressed by freeing up both hands and enabling simultaneous drawing out, twisting and winding on.[34] The one enormous advantage that even the Great Wheel enjoyed over hand spinning was the speed with which the job could be done. In a manufacturing industry, in which spinning was so often the bottleneck of production, this was critical. Productivity increased three fold. As a result, by the end of the Middle Ages, the spinning wheel was in wide use in the production of all but the finest of yarns.[35]

---

32 Lipson (1921), pp. 132–3; Lemon (1968), pp. 84–6; Munro (2003), pp. 200–1.

33 Britnell argued for a thirteenth century introduction: Britnell (2004 (B)), p. 297; while Lipson and Baines preferred the fourteenth: Lipson (1921), p. 133 and Baines (1977), p. 54. Baines contended that the spinning wheel may have arrived with Flemish weavers, invited to England by Edward III.

34 Catling (1972), pp. 101–2; Baines (1977), pp. 42–4, 61; Munro (2003), pp. 201–2; Van Bavel (2010), pp. 347–8; Muldrew (2012), p. 504.

35 Lemon (1968), pp. 89–91; Ponting (1971), p. 27; Baines (1977), pp. 82–9; Munro (2003), pp. 202–4. Possibly, the Saxony wheel was initially used only for worsteds: see Munro's comments in 'Distaff I – Focus on Fibre'. Miller and Hatcher expressed a persuasive dissenting note in saying that the spinning wheel 'was in nothing like general use before Tudor times': Miller and Hatcher (1995), p. 94.

Evidence of the use of the spinning wheel in fifteenth-century Suffolk is surprisingly slim. We have already met Agnes Stubbard, the proud owner of at least three. In 1443 Ralph Recher claimed that his spinning wheel had been stolen when his house in Ipswich was broken into.[36] The Register Baldwyne identifies three testators who probably included them among their bequests. Anne Catour (alias Neve) of Ixworth was one such. She died surrounded by the prized belongings she had lovingly, and at length, listed in her will. She left a 'qwel' to her servant Isabel, to which she gave a value of 10d., as well as six stone of wool.[37] William Wethere of Stanton, who mentioned 'le whele' and 'le cardes', and John Barbour of Redgrave, who referred specifically to a 'spynnyngwele', were the other two.[38] In 1495 Bennett Wareyn, a prosperous coverlet weaver of Bury St Edmunds, bequeathed a wheel to each of his sons John and William.[39] As coverlets were usually made from worsted, this suggests that he used spinning wheels for combed wool – may be they were Saxony wheels.[40] A haul of just nine spinning wheels is disappointing, particularly in a county where so many spinsters were hard at work, and perhaps more will be found one day.[41] This paucity of references may be attributable to the fact that very few spinsters made wills.[42] Equally, the spinning wheels that they used may have belonged to the clothiers who employed them and who did not bother to mention such low value items in their own wills. Traditionally, historians have tended to the view that, even when artisans lost ownership of the wool or cloth with which they worked, except in the case of the fulling mill, they retained ownership of the tools of their trade.[43] Such a view may not hold true for spinning wheels. Whatever the explanation for their absence from the records, it is reasonable to assume that during the late Middle Ages much, if not most spinning in Suffolk was still done by the rock.

---

36 SROI C/2/3/1/51.

37 While 10d. may seem a modest valuation for a spinning wheel, Britnell noted that the wife of Richard Taillour of Colchester leased one for a year for the same sum in 1395/96: Britnell (1986), p. 102; while Yates cited the case of John Jones of Newbury who, in 1573, had a spinning wheel and a pair of cards valued at one shilling: Yates (2007), p. 89.

38 SROB 66, 388, 443 Baldwyne.

39 SROB 34 Pye.

40 Crowfoot et al (2001), p. 39; Oldland (2011), p. 183.

41 Jo Sear, who for her doctorate studied medieval Suffolk wills, including those of Clare and Newmarket, came across no reference to spinning wheels: personal email correspondence 16 April 2015.

42 In only one will, that of Margaret Ladde of Bury St Edmunds (SROB 32 Hawlee), proven in the courts of the archdeacons of Suffolk and Sudbury and of the sacrist of Bury St Edmunds, between 1350 and 1500, did the testatrix describe herself as a 'spinster'.

43 Lipson (1921), pp. 56–7; Carus-Wilson (1987), p. 645; Ponting (1971), p. 24.

Spinsters[44]

The tasks of carding, combing and spinning the wool were by far the most labour intensive in cloth production. Various medieval and early modern sources suggest that about half the hours needed for the production of woollens were spent in carding and spinning.[45] For example, in 1588 a Yorkshire clothier described the requisite workforce thus:

> Three score persons are thus to be devided: xii for sorting, dressing, and lyttinge the wolle; xxx[tie] for spinning and carding; xii for weaving and shearing; the odd vj persons to helpe the reste, as to goe to the myllne and tourle.[46]

So, who did all this work? Some was carried out in workshops by servants like those employed by Agnes Stubbard in Bury St Edmunds. Twenty miles away in Boxford, at about the same time, the spinster Amice Mayster may have operated on a similar scale to Agnes. She faced a claim by the Ipswich woolman Richard Dysse for the significant sum of £8 10s. which could have bought about 50 stone of wool, enough to keep a servant busy spinning for up to three and a half years.[47] Both Agnes and Amice may have operated their business from a spinning house, similar to the one in Chequer Square, Bury St Edmunds that John Baret mentioned in his will.[48] Other work was put out to the homes of capital farmers where children and servants would have provided the necessary labour.[49] Yet more was done by weavers, or at least by members of their household. Bailey cited the case of Robert Woolpit of Fornham St Genevieve who was described as 'an habitual spinner and weaver of woollen and linen cloth', and who employed an apprentice and female servant to help him.[50] Nevertheless, the lion's share of carding and spinning was done by poor women, often single or widowed, sometimes with young children, for whom it provided a meagre, yet only possible source of income.[51] In the poll tax returns of

44 In the Middle Ages the term 'spinster' meant a woman who spun cloth, not, as it later came to mean, a mature single woman. She was more likely to call herself a 'maiden' or a 'virgin': see the wills of Katherine Saunson (SROB 124 Hawlee) and Elisabeth Galion (SROB 61 Pye).

45 Miller and Hatch (1995), pp. 95–6; Muldrew (2012), pp. 505, 509.

46 Tawney and Power (1924), p. 217.

47 TNA CP 40/616, m. 284r (AALT IMG 563). In 1415 a stone of wool cost about 3s 4d. This, and later calculations, is based on the proposition that a spinster produced 1lb of woollen yarn per day working for two hundred days a year: Zell (1994), p. 166; Muldrew (2012), p. 517.

48 SROB 95 Hawlee as translated by Tymms (1850), p. 20.

49 Kerridge (1985), p. 203.

50 Bailey (1989), p. 174.

51 In Winchester spinning 'was perhaps the most characteristic means of livelihood for respectable women living alone': Keene (1985), 1, p. 300.

1381 for Suffolk none of the sixty-nine women described as spynnere is recorded as being married, although three had daughters, also spynneres, which suggests that they had once been wed.[52] The clothiers, for whom an increasing proportion worked, might in a few cases remember them in their wills, but while alive often treated them shoddily. Spinsters earned 1d. per day, 2d. if they were very fast.[53] They frequently waited a long time to be paid, so bearing a disproportionate share of the network of credit that sustained the cloth industry.[54] In Ipswich Celia Chelston, probably a spinster, had to sue the weaver John Cukhook for payment.[55] Until the first Truck Act of 1464 prohibited the practice, they might even have been passed off with 'pins, girdles and other unprofitable merchandise', rather than paid in cash.[56] Ponting described theirs as 'the most sweated of all trades'.[57] Their plight is the dark side to the story of the magnificent churches that were built on the profits of the industry.

As they had no guilds to regulate or protect them, and invariably worked from home, spinsters have left little mark on the historical record. They are 'among the most poorly documented of all occupational groups'.[58] Nationally, out of ninety-one cited in plea rolls, only thirty-seven came from towns, of whom sixteen lived in London (including Westminster and Southwark), confirming that outside the Capital spinning was predominantly a rural occupation. Whereas most cloth-workers' disputes were over debts, less than a third of those of spinsters' were. Indeed, in the countryside, discounting cases they fought on behalf of their deceased husbands, the proportion of debt disputes was less than one in six. We can assume that, in the fifteenth century, very few spinsters ever operated as businesswomen in their own right.

In Suffolk, despite the paucity of records, we can trace spinsters to most parts of the county (Map 4.1).[59] For example, in 1381, in the tiny hamlet of Ixworth Thorpe four spinsters paid the poll tax out of a total taxable population of only forty-six. In Hinderclay the figures were six out of eighty-six, in Creeting St Peter six out

---

52 Fenwick (2001), 2, pp. 504–36.

53 Britnell (1986), p. 61; Swanson (1989), p. 31. In late-thirteenth-century Laleham, Middlesex spinning cost 2d. per pound for the warp yarn and 1½d. for the woof: Lloyd (1970); p. 334. Poos cited a later example of a woman in Writtle, Essex in 1405 seeking payment of 1s 5½d for spinning at least 13 pounds of wool that equates to a rate of about 1⅓d per day: Poos (1991), p. 67.

54 Zell (1994), p. 167.

55 SROI C/2/3/1/41 (8 Nov. 1407).

56 *Rot. Parl.*, v, p. 502; *Statutes*, 2, pp. 405–6.

57 Ponting (1972), p. 25.

58 Zell (1994), p. 167.

59 The most abundant medieval records, namely manorial court rolls and wills, do not reveal much about spinsters. Courts did not routinely record occupations, and, in any event, spinsters were unlikely to appear before them. Furthermore, very few spinsters left wills.

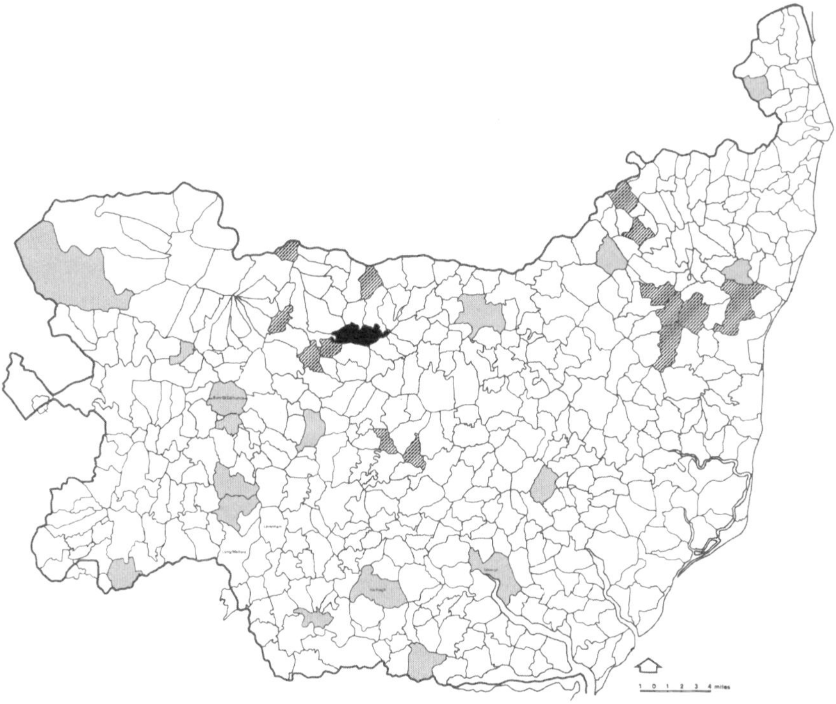

Map 4.1: *Suffolk Spinsters 1350–1500*

Notes: Grey marks a parish in which a spinster was recorded in TNA CP 40. Hatched dark grey marks a parish in which one or more spinsters was charged poll tax in 1381. Black marks a parish in which one or more spinsters was charged poll tax and a spinster was recorded in TNA CP 40.

Sources: TNA CP 40 and Fenwick (2001), 2, pp. 504–36.

Map format attribution: David Dymond and Edward Martin.

of forty-one.[60] By the late 1460s Suffolk cloth-makers were producing about five thousand broadcloths, or equivalent, each year, generating even greater demand for spinsters' labour. At that time the law required that a broadcloth weigh not less than 38 pounds, so, in a year, a single spinster might have produced enough yarn for about five such cloths. At this rate, allowing for some wastage in weaving, fulling and shearing, at least a thousand spinsters would have been needed to meet the demand – about 1 per cent of total county population.[61]

60 Fenwick (2001), 2, pp. 507–8, 522–3.

61 *Statutes*, 2, pp. 424–5. This assumes a population of about 100,000: Bailey (2007), p. 292; Broadberry et al (2015), p. 25.

With the few exceptions already mentioned, the county's spinsters appear to have been as poor as elsewhere in England. Of fifteen common bench cases, only four were disputes over debt. Another five, none later than 1420, were for infringement of the Statute of Labourers with employers complaining that the poor women had fled their service. Agnes Illans had come from Melton Mowbray in Leicestershire to work for Thomas Hethe in Mildenhall, but, finding him to be an unkind master, had transferred her services to Guy Lambholme. Joan Bory similarly left John Saxham's employ for that of the Bury St Edmunds draper John Wode. In 1475 the Sterne family were in trouble with William Aleyn, on account of foraging for firewood in his close at Nowton. Margaret was a widow, Thomas an unskilled labourer and, through her spinning, Alice may well have been the main breadwinner.[62] Sad as the Sterne case was, the most pitiable of all involved Joan Melle a spinster of Henham near Beccles. She was falsely accused and convicted of breaking and entering the house of her estranged husband William, of assaulting him, and of stealing, among other valuable items, 24 yards of woollen cloth and a pair of shears. For this felony she was sentenced to death, but, being with child, her execution was deferred until the baby was born. Happily, before the dreadful day arrived, the king showed clemency, pardoning and releasing her from gaol.[63]

Spinsters may not always have been so badly off. The Suffolk poll tax records for 1381 are fragmentary. The returns of relatively few towns and villages have survived and many of these survivals provide either no, or very sketchy occupational data. We can, nevertheless, trace seventy-seven spinsters and the tax that each of them paid (Figure 4.1). Of those spinsters for whom the tax paid is recorded, more than half paid the full 12d. or more and only five paid the minimum 4d. Married couples were generally taxed together under the husband's name, so these women paid in their own right. Their per capita contribution to the poll tax was slightly less than that of weavers and fullers, but they were far from destitute in Suffolk, or indeed elsewhere.[64]

So what went wrong? It may be that, during the fifteenth century, as the cloth industry grew in profits and prestige to become big business, women were marginalized and downgraded. As Thirsk wrote 'if a venture prospers, women fade from the scene'.[65] Bennett showed how this happened in brewing, particularly with the

62 TNA CP 40/616, m. 466v (AALT IMG 2021); CP 40/636, m. 297r (AALT IMG 600); CP 40/853, m. 286r (AALT IMG 562).

63 *CPR 1436–41*, p. 225.

64 The York alnage accounts of 1395 recorded that a quarter of cloth was presented by women, something that appears to have prompted a misogynistic backlash by envious male colleagues who expelled many of them from the weavers' guild: Heaton (1920), p. 38.

65 J. Thirsk 'Women's initiatives in early modern England', presentation to 'Women's Initiatives in early Modern England, 1500–1750' Conference, London, 4 June 1994.

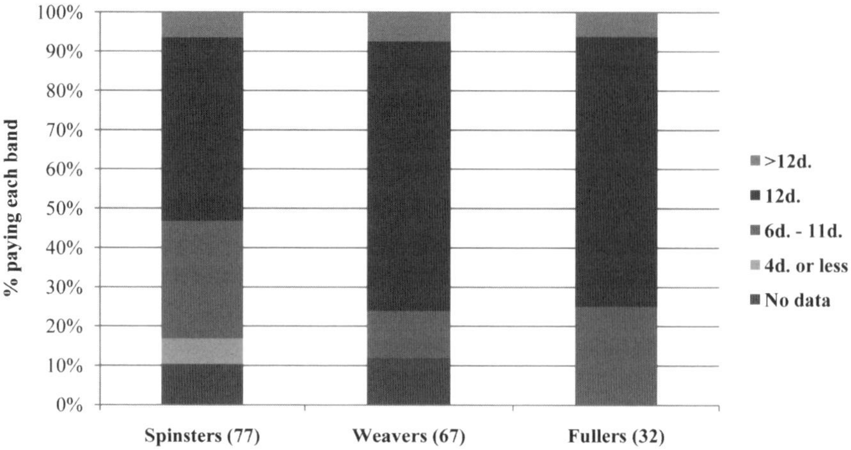

Figure 4.1: *Poll Tax Payable by Suffolk Spinsters, Weavers and Fullers in 1381*

Notes: Figure 4.1 records and compares the percentage of known cloth-workers who paid each of several different bands of poll tax, or for whom the poll tax charge is either illegible or missing. The figures in parentheses indicate the total recorded number of each type of cloth-worker.

Source: Fenwick (2001), 2, pp. 504-36; Fenwick (2005), 3, p. 570.

arrival in England of beer. 'Many fewer [women] profited from a new beerbrewing industry that required its brewers to have ready access to various forms of capital, numerous contacts in diverse and farflung markets, and firm control over a large and largely masculine labor force.'[66] Sadly, for spinsters, what was true in brewing, proved equally true in cloth production.

Weaving

Once the yarn was spun it was ready for weaving, in essence the interlacing of one series of yarn threads at right-angles with another to produce a cloth fabric. The first series, known as the warp, was stretched lengthways across the loom, while the other, the weft, was interlaced. Rev Edmund Cartwright, inventor of the power loom, identified just three movements in weaving cloth, namely opening the shed, throwing the shuttle, and beating the weft threads together.[67] A shed was opened by lifting up warp threads with the use of a heddle. A shuttle was the tool to which a weft thread was attached and was moved back and forth through the sheds and across the loom. After each pass of the shuttle, the weft threads were beaten together

66 Bennett, (1996), p. 92.
67 Lipson (1921), p. 165.

using a reed (also known as a slay) to draw them tightly towards the weaver. William Rysby of Lavenham bequeathed to his widow Joan and his servant Simond Rysby a 'wetcloth slay' and a 'settcloth slay' each.[68] The number of warp threads in a cloth was a key determinant of its quality, with the cheapest containing as few as six hundred and a fine broadcloth in excess of two thousand.[69]

By the fifteenth century most woollens had a simple plain, known as tabby, weave in which the weft thread passed over one warp thread and under the next. The resultant even surface was ideal for later finishing treatments.[70] Moreover, as the pattern of the weave was obliterated in fulling, more complicated twill weaves were redundant.[71] The 1477 Ordinances of the Bury St Edmunds weavers stress the importance of plain weaving in the town.[72] Twill weaves were, however, not unknown, particularly in worsteds in which the pattern was better preserved.[73] Woodbridge became a centre of twill weaving and remained so until the mid-seventeenth century.[74] John Albrede of that town was a relatively successful twill weaver. He numbered among his friends the prominent merchant, Peter Terry, who served twice as bailiff of Ipswich and also as Albrede's executor.[75] John died in 1450, apparently childless, but leaving a wife Agnes who survived him by eight years. Agnes was sufficiently well provided for in her old age to employ household servants whom she remembered in her will. Both John and Agnes made bequests to the parish church, as well as to Woodbridge Priory and to the churches of Bromeswell and Great Bealings. However, their most generous gift was for the erection of a rood screen in Woodbridge church, lofty and richly carved, guilded and painted.[76] The art of twill weaving did not die with them. In the subsequent seventy years John Love, William Bele and William Marriott all worked as twill weavers in

68 TNA PROB 11/15/265. The difference between the two was probably that the wetcloth slay was for woollens and the settcloth slay for worsteds.

69 Chorley (1997), p. 23; Oldland (2014), pp. 32–3.

70 Crowfoot et al (2001), p. 44.

71 Munro (2009), p. 6.

72 SROB B9/1/2.

73 Burnham (1981), p. 154; Crowfoot et al (2001), pp. 27–43, 212–3; Munro (2003), pp. 192–3. Evans stated that twill was 'a coarse linen cloth', but gave no authority for her contention: Evans (1985), p. 170. The better view is that it was a weaving pattern that could be used in woollen or linen cloth: OED http://www.oed.com/view/Entry/208062?rskey=krqGhw&result=1#eid.

74 Woodbridge was 'not a district in which many other linen weavers are known to have lived': Evans (1985), p. 58. In this light, and in view of the early dates of these twill weavers, it is submitted that they were wool weavers.

75 *Annals*, pp. 100, 103.

76 SROI IC/AA1/52, IC/AA2/20; Arnott (1897), pp. 340–1.

Woodbridge, while John Godard did so in nearby Dallinghoo, Robert Godard in Ufford and William Bobat fifteen miles further up the coast in Yoxford.[77]

The design of the loom improved significantly during the Middle Ages. In the ancient world weavers commonly used an upright loom on which the warp threads were stretched by stones or similarly heavy objects (Plate 4.3). From the eleventh century they began to use, instead, the horizontal treadle loom which enjoyed several advantages over the older model. The warp threads were spaced more evenly and stretched more tautly, while the weft threads were beaten in more firmly, so producing a uniformly better quality cloth. The heddles were operated by foot treadles, giving greater control over the opening of the sheds. Henry Borham of Clare left his son William a pair of looms with a treadle.[78] The horizontal loom could produce much longer cloths and could do so three times faster (see Plate 4.5). Initially designed for use by a single weaver and so limited to the production of narrow cloth, in time it evolved into the broadloom that was operated by two weavers and produced broadcloth. The broadloom, being about two yards wide, enabled the master weaver to work with an apprentice, so increasing productivity, reducing labour costs and enabling him to keep an eye on the work of his youthful companion.[79] The narrow loom never disappeared altogether. It cost less to build, operate and maintain and was preferred by those artisans who produced worsteds or just chose to work alone.[80]

Narrow and Broad Cloths

In describing English cloth, the late thirteenth century Ipswich Domesday charmingly differentiated between 'doubele werk that they call tomennesette', that is cloth produced on a broadloom with two men's seats; and 'cloth of longe webbe, that they call omannessete', that is cloth produced on a narrow loom with one man's seat.[81] Narrow cloths might be called, cogware, kerseys, matsales, narrow dozens, or straits; while broad cloths might be called broad dozens, cloths of assize, vesses, whole cloths or, indeed, just broadcloths.[82] In 1414 John Sutton shipped from Ipswich six hundred straits worth 6s. 8d. each and another five hundred worth 5s. 6d. each; two hundred cogware worth 4s. 10d. each and two hundred matsales

77 TNA CP 40/796, m. 195v (AALT IMG 1232); CP 40/951, m. 166r (AALT IMG 341); CP 40/807, m. 357v (AALT IMG 1510); CP 40/807, m. 357v (AALT IMG 1510); CP 40/826, m. 97r (AALT IMG 194); SROI IC/AA7/146.

78 SROB 59 Newton.

79 Kerridge (1985), p. 11.

80 Munro (2003), pp. 194–7.

81 Twiss (1873), p. 197.

82 The term 'broadcloth' is used as a general descriptor in this volume.

(cloths of Mattishall, Norfolk) worth 4s. each; as well as twenty broadcloths worth 30s. each and two hundred broad dozens worth 14s. each.[83] Most of these cloths were woollens, but, because worsteds did not figure in alnage accounts and very rarely in customs accounts, it is easy to forget that they comprised a significant proportion of total output. Salzman remarked on Suffolk's 'considerable trade in a cheap, coarse variety of cloth known as "Vesses or set cloths"' that may well have been worsteds.[84] Roger Man, a London mercer, sought a court order for delivery by Thomas Parker, clothier of Lavenham, of twenty woollen cloths called vesses, while John Melton of Ipswich pledged a broadcloth called a vesse measuring 24 yards long for £1 12d.[85] Nicholas Gosselyn of Lavenham bequeathed 43 vesses to various members of his family, each worth 36s.[86]

As Bailey rightly contended, Suffolk cloths 'were produced in an eclectic variety of styles and colours'.[87] The dimensions of these various cloths were fixed and unfixed by a bewildering series of statutes, subsequent repeals and exemptions as the government sought out the best policy: whether regulation to give overseas buyers confidence in English cloth, or laissez faire to promote the domestic textile industry. In 1376 the cloth-makers of Essex and Suffolk successfully petitioned Parliament to exclude cogware, kerseys and other narrow cloths from statutory control.[88] In another petition they explained that, traditionally, their cloths measured 12 by 1 yards (Appendix A).[89] Nearly a hundred years later, in 1468, Parliament enacted that in Essex, Suffolk and Norfolk broadcloth should be 28 yards 28 inches long, by 7 quarters of a yard wide, while straits should be half as long, half as wide and one quarter the weight.[90] Even after 1468 cloth-makers did not always adhere to these statutory criteria. John Fycches of Sudbury sold four unfinished cloths each 24 yards long, while John Marten of Hadleigh sold to the prior of Christchurch, Canterbury broadcloths of various quality measuring between 22¾ and 32¼ yards and John Reynham kept in his warehouse broadcloths 33 yards long.[91] The proscribed length of the broadcloth elsewhere in England was, in fact, 24 yards.[92]

83 TNA E 122/51/39.

84 Salzmann (1913), p. 168. OED http://www.oed.com/view/Entry/222876?redirectedFrom=vesse#eid. A statute of 1487 excluded vesses from the prohibition on export of cloths that were not first 'barbed rowed and shorne' ie finished: *Statutes*, 2, p. 520.

85 TNA E 122/51/39; TNA CP 40/883, m. 217v (AALT IMG 396d); SROI C/2/10/3/3 f. 83.

86 TNA PROB 11/8/360.

87 Bailey (2007), p. 271. See also Betterton and Dymond (1989), p. 37.

88 *Rot. Parl.*, ii, p. 347; *CPR 1374–77*, p. 396. An exclusion that was re-confirmed in 1381: *CPR 1377–81*, p. 615.

89 *Rot. Parl.*, iii, p. 320.

90 *Statutes*, 2, pp. 424–5.

91 TNA CP 40/911, m. 179v (AALT IMG 351d); CCA-DCc-ChAnt/C/244; TNA PROB 2/87.

92 *Statutes*, 2, p. 154.

So, maybe even contemporary Suffolk cloth-makers were confused as to the rules they should follow. In penalizing breach of regulations little evidence survives of rigor on the part of the government or of the alnager who acted on its behalf locally. In 1434/35 eight pieces of offending cloth were forfeited: three in Ipswich, two belonging to Robert Stanesby in Bildeston and one each in Chelsworth and Lavenham. Two years later, three Hanseatic merchants in Ipswich were made to surrender five broad and three narrow dozens, while John Derby of Sudbury was the only denizen to suffer, being deprived of 12 yards of unfinished narrow cloth.[93]

The quality of cloth was determined by factors that included the origin of the wool, the number of warp threads and the skill of the craftsman. The value of John Sutton's exports suggest that in the early-fifteenth century much of Suffolk's cloth was narrow and relatively coarse. Worsteds were generally lighter than woollens. As the county's industry began to suck in finer wool from the Cotswolds and the Welsh Marches, and became more geared towards the export trade in better quality woollens, its product improved. Weight is a crude, but not wholly unreliable measure of quality. The 1468 statute that fixed the size of East Anglian broadcloth, also stipulated a minimum weight of 38 pounds. In 1552 another statute raised the bar significantly to 80 pounds.[94] Both weights may be reasonably assumed to have reflected best practice at the time. The standard weight of Suffolk broadcloth had probably slowly increased between the two dates. This increase must have given impetus to the transition from foot to mill fulling.

At the end of the fourteenth century the output of narrow cloths, as measured by the alnage accounts of Suffolk for 1394/95 and 1396/97 and by the particular customs accounts of Ipswich for 1396/98, easily exceeded that of broadcloth. By the late 1460s, when we next have full and reliable sets of alnage accounts, the position had reversed.[95] And as the pull of London and overseas markets, and the demand for premium broadcloth grew, this trend was to continue.[96] Some textile towns, notably East Bergholt, Lavenham, Long Melford and Sudbury, consistently favoured broad over narrow cloth. John Risby of Lavenham and John Amyot the Elder of Long Melford were among many who made a living from their broadlooms.[97] Nevertheless, even in these towns, some weavers produced narrow cloth. William

93 TNA E 101/342/21.

94 *Statutes*, 4, p. 136. The prescribed dimensions of Suffolk broadcloth in 1468 and 1552 were similar. By way of comparison the 1552 statute stipulated that a Kentish broadcloth should weigh not less than 90 pounds.

95 TNA E 101/342/8, E 101/342/10, E 101/342/25, E 101/343/2, E 101/343/4, E 101/343/5, E 122/193/33.

96 Oldland (2014), p. 36.

97 TNA PROB 11/9/365; SROB 79 Hervye.

Peverell bequeathed a narrow loom in his will of 1488, though all his sealed cloths in the late 1460s had been broad.[98]

The fourteenth-century Minister's Accounts for Sudbury sporadically recorded the big and little woollen weavers, among others, who paid burghal dues. The number of those working with broadlooms, as many as thirty in 1371, always heavily outnumbered those working with narrow looms.[99] In the late 1460s more than 90 per cent of the cloth produced by Sudbury cloth-makers was broadcloth.[100] The Shorthose family were prominent weavers in the town for nearly sixty years. As early as 1400 Richard Shorthose faced a claim for £10 compensation for allegedly failing to weave, well and competently, three woollen cloths. The value of this claim suggests that the original order had been for good quality broadcloth. For twenty years from 1420 Richard traded regularly with the Sudbury draper Richard Othelso, several of their deals again being of a relatively high value – £10 10s., £7 and £20.[101] The name of Richard's son Thomas first appeared in the records in 1440 and he died in 1459, so his career was probably a much shorter one. Among just nine testators, whose wills are copied in the Register Baldwyne and who made a bequest of looms, only Thomas provided any detail. His apprentice John Kent received not only the 'broadloom on the east side of my shop', but also a reed (described as a '*broadslay*') and a roller on which the cloth was wound (described as a '*wheel-trendle*').[102] He left his house, le Stonhalle, to his wife for life (Plate 4.6).[103] Although Thomas paid for one of the friars of Sudbury to pray for the souls of his children, he made no reference to any living offspring, so one may sadly assume that they had all predeceased him, leaving only the apprentice to continue in business with the family broadloom. A John Kent made cloth in Great Waldingfield in 1466/67, perhaps the same man and someone to whom we shall return below.[104]

In Bury St Edmunds in the late 1460s cloth production was much more evenly divided between broad and narrow cloths and this is reflected in testamentary references to looms. They were mentioned in sixteen wills of the town's weavers

---

98 SROI IC/AA3/83v.

99 Redstone (1910), p. 340. Britnell detected 'obvious weaknesses in the figures as a reliable guide to the fortunes of the cloth industry', but his concerns do not detract from the point made above: Britnell (1986), p. 83.

100 TNA E 101/342/25, E 101/343/2, E 101/343/4, E 101/343/5.

101 TNA CP 40/556, m. 224r (AALT IMG 445); CP 40/636, m. 175r (AALT IMG 355); CP 40/656, m. 370v (AALT IMG 1707); CP 40/716, m. 54v (AALT IMG 1144).

102 SROB 235 Baldwyne.

103 A house called *The Stone* stands on the corner of Stour Street and School Street and may possibly have been the Shorthose home: Grimwood and Kay (1952), p. 121; Northeast (2001), p. 405n; Bettley and Pevsner (2015), p. 525; http://list.historicengland.org.uk/resultsingle.aspx?uid=1037434

104 TNA E 101/343/2.

and clothiers who died between 1426 and 1518 – six were broadlooms, five narrow looms, two bedlooms and thirteen unspecified.[105] The bedlooms may have been the broadest of all, up to four yards wide.[106] The coverlet weaver Wareyn Bennet spelt out that his son John should receive a 'loom three yards broad' and his other son William a 'loom of 9 quarters broad'.[107] Roger Shaldeford bequeathed both a broad and a narrow loom to his widow Amy, while John Gardener bequeathed a broadloom and two narrow looms to his son William.[108] In some towns, such as Winchester, the dual use of narrow and broadlooms by the same weaver or clothier was prohibited, evidently not so in Bury.[109]

An Industry in the Valleys

Weavers are recorded, in the plea rolls, in ninety-nine, and fullers in eighty different parishes, nearly all of them in river valleys (Maps 4.2 and 4.4). The towns and countryside of the valleys of the rivers Stour and Brett are among the most picturesque in East Anglia (Plates 4.8 and 4.9). They have long been a favourite haunt of Suffolk artists, among them John Constable and Thomas Gainsborough. As Barbara McClenaghan wrote in the early 1920s in opening her volume on *The Springs of Lavenham*:

> There are many villages in West Suffolk far away from the sound of a train, where the only noise that breaks the drowsy stillness is the rattling tune of the haymaking machine, or the whir of the reaper and the hum of the thresher. There are few manufactures in West Suffolk to-day, and it is difficult to imagine these same sleepy villages as the centre of a thriving cloth-making industry.[110]

But in the fifteenth century they were very busy. You could walk through twenty-three parishes, from Haverhill in the west to Shotley at the mouth of the Stour in the east, more than thirty-five miles as the crow flies and many more as the duck paddles, and in all but four you would have found a weaver or fuller hard at work. In some, such as Long Melford, Sudbury, Nayland and Stoke-by-Nayland, you would have found dozens. At Higham the river Stour meets the river Brett that has, by then, flowed through Lavenham, Monks Eleigh, Chelsworth and Hadleigh, all of

---

105 Several wills refer to a '*pair of looms*', a phrase that meant one loom, rather than two: Betterton and Dymond (1989), p. 31.

106 Kerridge (1985), p. 11; Sutton (1989), p. 213; Swanson (1989), p. 38.

107 SROB 34 Pye.

108 SROB 206 Osbern; 196 Pye.

109 Carus Wilson (1987), p. 638. In the Weald dual use was permitted: Zell (1994), p. 171.

110 McClenaghan (1924), p. 1.

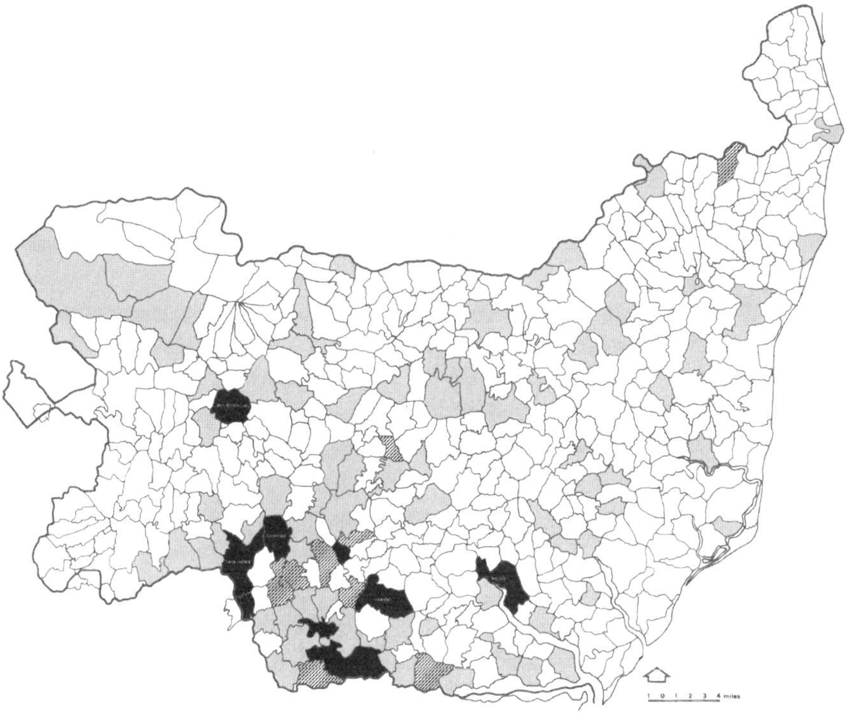

Map 4.2: *Suffolk Weavers 1350–1500*

Notes: Grey marks a parish in which between one and four weavers were recorded. Hatched dark grey marks a parish in which between four and nine weavers were recorded. Black marks a parish in which not less than ten weavers were recorded.

Source: TNA CP 40.

Map format attribution: David Dymond and Edward Martin.

them major centres of the woollen cloth industry. These towns and villages lay in the south Suffolk Hundreds of Babergh and Cosford.[111] Of the 381 weavers cited in the plea rolls 168 (44.1 per cent) resided in twenty-three parishes in Babergh and forty-eight (12.6 per cent) in six parishes in Cosford. From Lavenham alone we have the names of forty-three different weavers. By the late 1400s this 160 square miles was the most heavily industrialised area in the whole of England.[112]

111 E. Martin (1999), pp. 26–7.

112 See Appendix B. In their classic study of the changing distribution of wealth in the Middle Ages Darby and his colleagues referred to 'another source of wealth near the Essex-Suffolk border – the making of cloth in settlements along the south-eastward flowing rivers': Darby et al (1979), p. 258.

Suffolk is blessed with several other rivers that reach the sea along its coast, the Gipping/Orwell, the Deben, the Alde and the Waveney. Cloth-making was a component of the local economy along each of them. Nine weavers worked in Stowmarket and three in Needham Market. Both grew in importance as textile towns in the second half of the fifteenth century.[113] We have already met the prosperous twill weavers of Woodbridge and, further upstream near the source of the Deben, a group of central Suffolk villages around Debenham had a vibrant textile industry.[114] In Beccles on the Waveney in the far north-east of the county Isebraven Ducheman worked as a master weaver, as did Henry Horn a generation later.[115]

River valleys were well suited for cloth-making. Rivers provided a ready source of water that was of particular importance in fulling and dyeing. Even in the absence of a natural water supply, one could be constructed. An archaeological survey and inspection in Lavenham investigated the brick culvert running below the modern ground surface of Water Street. This probably channelled water for the late medieval textile industry.[116] The Suffolk rivers also provided cloth-makers with access to the ports of Colchester and Ipswich, and, once London began to dominate the export trade, to the coastal routes that led to the Capital.[117] The waterman Thomas Lukat was just one who plied his trade on the Stour.[118] Every rule has an exception. The only important textile town without a riverine link to the east coast was Bury St Edmunds where more weavers were to be found than anywhere else in the county. The reasons for its extraordinary success are considered next.

113 Amor (2004), pp. 419–21; Amor (2006), pp. 189–90.

114 The medieval Gough Map clearly marked the river Deben and the town of Debenham near its source: Millea (2007), pp. 70–1, 81.

115 TNA CP 40/736, m. 300r (AALT IMG 587); CP 40/895, m. 268r (AALT IMG 494). The Waveney valley became an important area for linen manufacture only after 1500 and, before 1700, there were 'very few' linen weavers in Beccles: Evans (1985), pp. ix, x, 41, 46. So those weavers named in TNA CP 40 would very probably have worked with wool.

116 Fradley (2007), pp. 10–11.

117 For a discussion of riverine traffic down the Gipping/Orwell: Amor (2006), p. 190; Amor (2011), p. 96. For a discussion of riverine traffic down the Waveney; Langdon (2007), pp. 114, 120; and Gransden (2015), p. 183. Admittedly, neither Langdon, nor, before him, Edwards and Hindle, identified the rivers Stour, Brett or Gipping as major navigable rivers, although the latter recognised the Orwell/Gipping and the Stour as 'minor navigable rivers': Hindle and Edwards (1991), pp. 130–1. Both papers were, however, based on the study of records of a nature and date that would not have identified the navigation of rivers for the purposes of the Suffolk cloth trade.

118 Lukat was a witness to the will of the Nayland clothmaker Stephen Raynham: TNA PROB 11/18/55.

The Weavers of Bury St Edmunds

As a centre of cloth-making Bury St Edmunds enjoyed many advantages, being the focal point of the Suffolk wool trade and the hub of many of the county's busiest roads. Any merchant or chapman who rode due south soon arrived in the industrial belt. With a population of over three thousand souls, local demand for cloth was high and was enhanced by the presence of the abbey and also of six hospitals.[119] The abbot and his brothers had long abandoned the simple life proscribed by St Benedict and would have been important customers of the town's cloth-makers. Furthermore, the shrine to St Edmund that they carefully tended was one of Europe's most popular pilgrimage sites, attracting people in holiday mood and with open purses.[120] Hospital care was not primarily targeted at the sick or destitute, but 'often sold to the highest bidder or reserved for those [...] aptly described as a "begging aristocracy"'.[121] Residents could afford to buy nice clothes.

Despite these advantages, an analysis of the worldly possessions that they bequeathed in their wills, tells us that, during the Great Slump, Bury's weavers had suffered a prolonged period of economic hardship (Table 4.1). In the first half of the fifteenth century most weavers mentioned something precious to them. Bedding, brass and latten were the most widely owned consumer items, but two weavers left pewter and two silver. Several bequeathed their own looms, and others may have sold or passed them on when they retired. John Mildenhale was perhaps the most affluent, bequeathing to his daughter Margaret a coverlet with tester, two blankets, two small sheets, a mazer set with silver and adornment, six silver spoons, two basins with two lavers, a laver for washing, a large brass pot, another small pot, a brass pan, three candlesticks, a spit, an andiron, an iron skewer, a tripod, and half a garnish of pewter vessel.[122] In the third quarter of the century references to such goods and chattels became few and far between. In their final hours many a weaver was preoccupied with concern that his wife would be able to 'pay my dettes and bryng me honestly to the erthe'.[123]

At the same time, the more fortunate cloth-makers of Suffolk's industrial belt continued to thrive. They enjoyed the benefit of an export led demand for finished cloth that was shipped from Colchester and Ipswich by merchants of the Hanseatic League, particularly those of Cologne.[124] Bury St Edmunds may well have missed out on this trade. Furthermore, the town's independent cloth-workers faced growing

119 A. Dyer (2000), p. 761.
120 Bailey (2007), p. 129; Gransden (2015), p. 179.
121 Rawcliffe (2013), p. 316. See also McIntosh (2011), p. 76.
122 SROB 194 Osbern.
123 SROB 68 Pye.
124 Britnell (1986), pp. 171–6; Amor (2011), pp. 114–38.

TABLE 4.1: TESTAMENTARY BEQUESTS OF BURY ST EDMUNDS WEAVERS

Date of Will	*Bedding*	*Brass/ latten*	*Clothing*	*Furniture*	*Loom*	*Pewter*	*Silver/ mazer*
1415		■					
1421							
1422							
1426			■		■		■
1430	■	■				■	■
1434							
1434			■		■		
1436		■				■	
1438							
1440							
1442	■	■		■			
1447							
1449					■		
1452	■		■		■		
1452							
1452							
1452							
1452							
1454							
1456							
1457	■	■					
1459							
1462	■	■			■		
1464							
1466					■		
1466							
1468							
1469							
1471					■		
1472							
1476							■
1479							
1479							
1479	■		■				
1481	■						
1495					■		
1496							

Notes: Each line reflects the will of a Bury St Edmunds testator who describes himself as a weaver, giving his date of death and the goods and chattels bequeathed in his will. It includes only specific references to goods and chattels, not the generic reference to household goods found in most wills.

Sources: SROB Osbern, Hawlee, Pye, Mason and Hoode.

competition from clothiers who had begun to organise and exploit a rural work-force that could undercut them.

Something had to be done. In 1477 the weavers persuaded the abbot to grant them new ordinances. Twenty-nine of them signed up to the terms of an indenture which recognized that 'many persones and greate multytude of people as weell of menne as womenne and childrenne withinne the toune of Bury Seynt Edmund bee dayly gretly occupyed be the mene of the crafte of wevers' (Plate 4.10). The ordinances provided for the holding of an annual assembly to elect four wardens, with powers of search, to police their craft; permitted only those admitted to the craft, following a seven year apprenticeship, to work in the town as weavers; limited each weaver to no more than four looms; required both masters and journeymen to honour the terms of any contract of service between them; and prohibited any weaver from poaching another's journeyman or apprentice.[125] With one Nelsonian

125 SROB B9/1/2.

blind eye on the other ordinances and the other eye firmly on guild finances, the wardens allowed outsiders ('foreigners') to set themselves up in the town as weavers so long as they paid a hefty fee. Several of the ordinances, such as the limit on the number of looms, were designed to prevent clothiers from hijacking cloth-making in Bury. Were these ordinances a sign of strength, or another example of the restrictive practices that bedeviled late medieval towns, and of the axiom that 'craft ordinances tended to be revised only in times of upheaval'?[126] In a town where formal institutions of commercial organisation were actively discouraged, the indulgence given to the weavers by these ordinances is intriguing. Happily, in the final twenty-five years of the fifteenth century, they enjoyed better times. Any upturn was, however, almost certainly due more to demand for bedding, than to the ordinances.

The contemporary commentator, John Fortescue Kt, writing nostalgically from exile in France, stressed that English people enjoyed 'abundant bedding'. According to Hatcher, the bed became an 'object of conspicuous expenditure' which demonstrated both economic and social success.[127] Weaving bedding and coverlets required a degree of skill beyond most rustic artisans, a broader than usual loom and a ready supply of worsted yarn.[128] For these reasons the craft was more securely protected from the competition of the countryside. Although, in general, bed and coverlet weavers did not engage in appreciably more valuable trade than woollen weavers, they were better able to carry on as independent operators (Table 4.2). Theirs was predominantly an occupation of East Anglian townsfolk, providing a good living for many not only in Bury St Edmunds, but in King's Lynn and Norwich too. In Bury, in the period 1480 to 1500, very nearly half the twenty-one weavers cited in the plea rolls were making bedding and coverlets. They had established a presence much earlier, but in these later years they flourished. When he died John Hermer was able to bequeath to his son his best doublet, gown and coat; while John Stynge could instruct his executors to sell his two feather beds, a mattress with transoms and pillows, a third bed with a coverlet and a pair of blankets.[129] John Roket was the most active litigant making two claims and defending eleven between 1470 and 1492, with a mean value of over £8. Bed and coverlet weavers benefited from strong local demand. Nearly half their debt disputes were with fellow townsfolk, including a carpenter, a dyer, a goldsmith, a mercer, a scrivener and a shoemaker (Table 4.3).[130] They also engaged in more long distance business

126 Swanson (1989), p. 39.

127 Fortescue (1942), p. 89; Hatcher and Barker (1974), p. 47.

128 Kerridge (1972), p. 33; Sutton (1989), p. 203; Oldland (2011), p. 184.

129 SROB Pye 186, 193.

130 TNA CP 40/736, m. 168v (AALT IMG 1367); CP 40/885A, m. 5v (AALT IMG 116); CP 40/888, m. 52r (AALT IMG 107); CP 40/943, m. 38r (AALT IMG 79).

I.1 *Alnage seal matrix*
(Reproduced by kind permission of Colchester and Ipswich Museums)

2.1 *Bird's eye view of Lavenham*

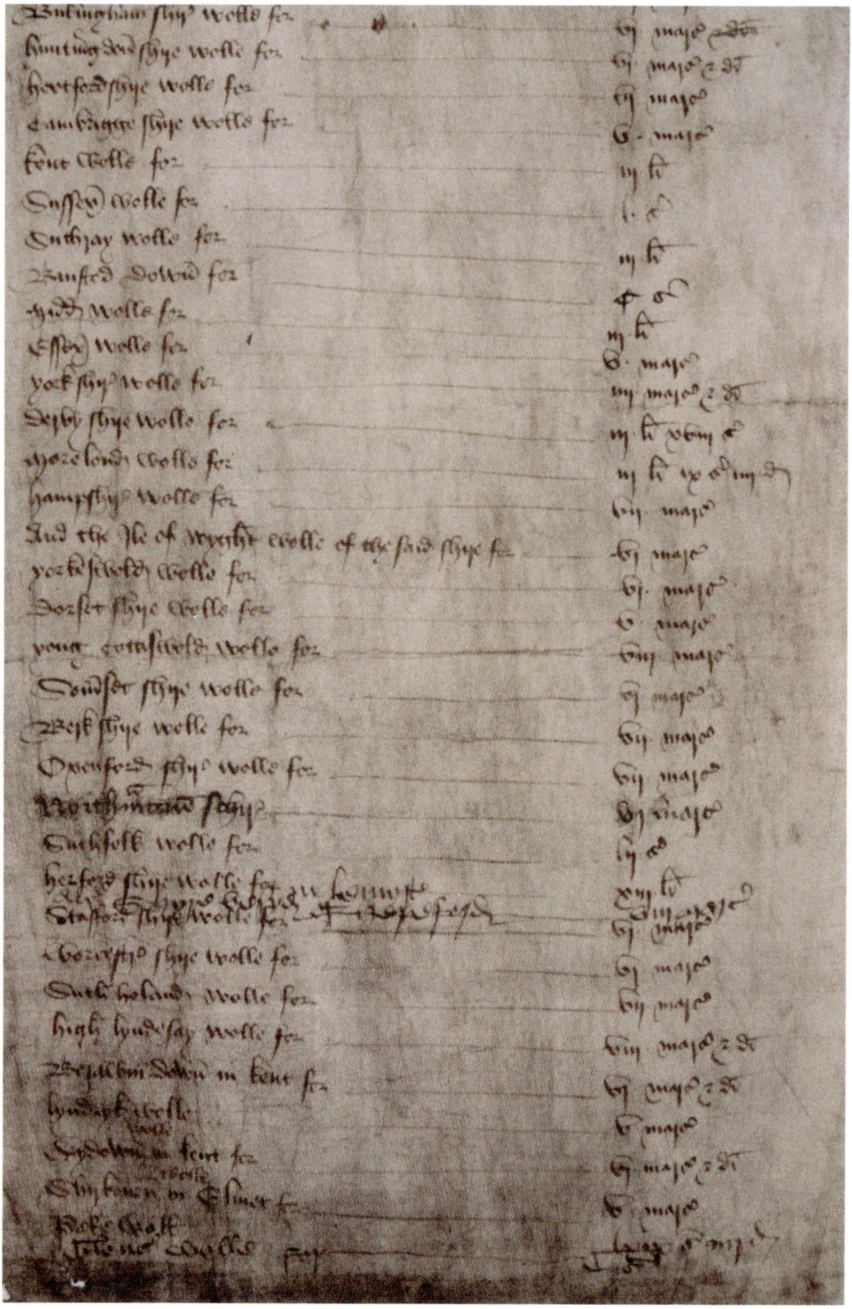

3.1 *Extract from the Parliamentary schedule of wool prices 1454*
(Reproduced by kind permission of The National Archives)

3.2 *Sibton Abbey refectory ruins*

3.3 *Sibton Abbey arcade (St Peter's church, Sibton)*

4.1 *Spinster carding wool (misericord St Andrew's church, Norton)*

4.2 *Spinning whorls (Lavenham Guildhall)*

4.3 *Loom weights (Lavenham Guildhall)*

4.4 *Woman with spinning wheel (Luttrell Psalter, British Library)*
(© The British Library Board)

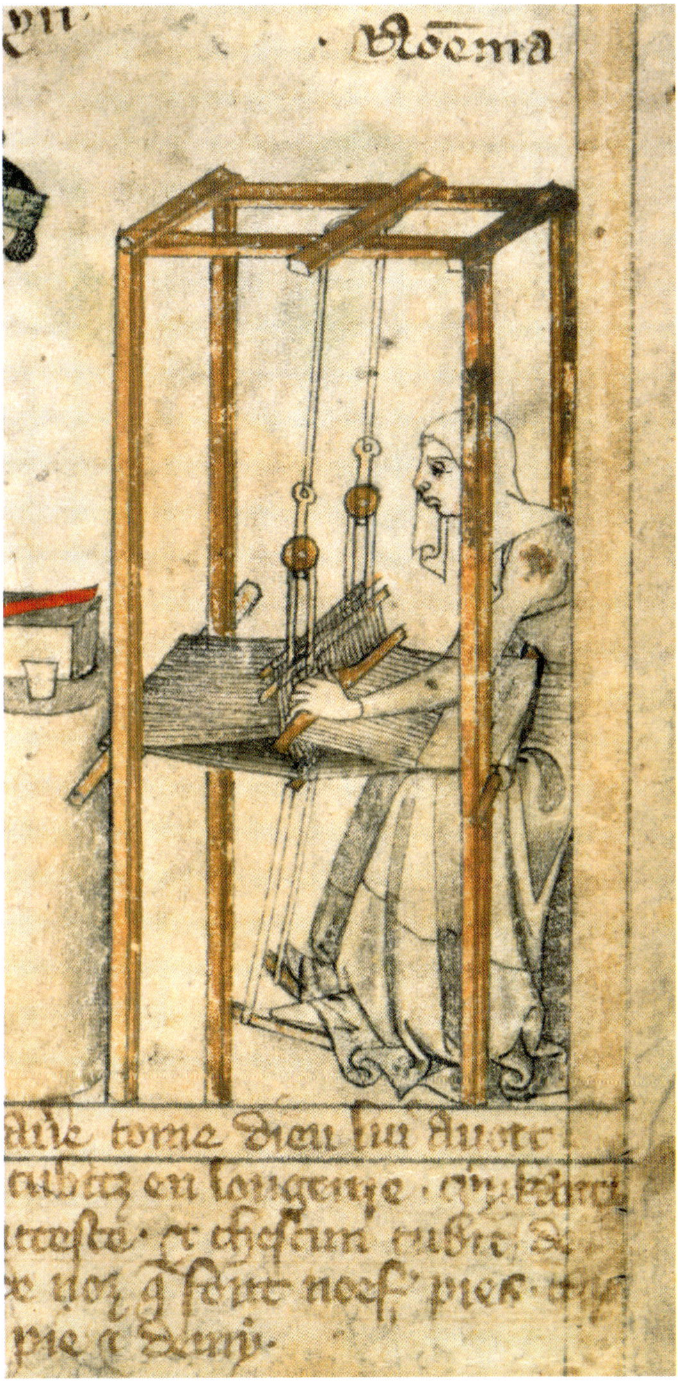

4.5 *Woman using narrow loom (Egerton Genesis, British Library)* *(© The British Library Board)*

4.6 *'le Stonhalle', Sudbury (home of Thomas Shorthose?)*

4.7 *Hall Street, Long Melford*

4.8 *River Stour, Fullingpit Meadows, Sudbury*

4.9 *River Brett, Hadleigh*

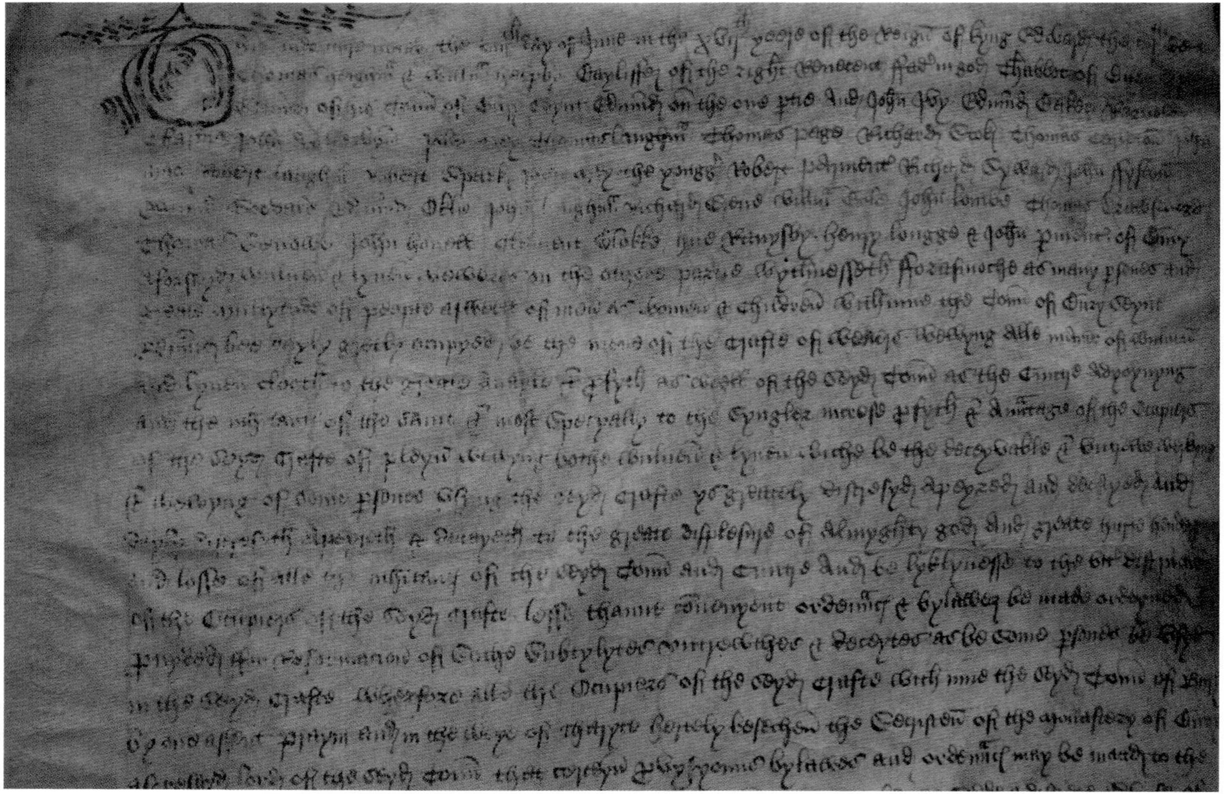

4.10 *Opening section of the Bury St Edmunds weavers' ordinances 1477*
(Reproduced by kind permission of Suffolk Record Office, Bury St Edmunds branch)

4.11 *St James the Less with fulling club (stained glass, St Ethelbert's church, Hesset)*

4.12 *St James the Less with fulling club (rood screen, St Edmund's church, Southwold)*

4.13 *Teasels (Lavenham Guildhall)*

5.1 *Woad (Lavenham Guildhall)*

5.2 *Dyers at work*
(© The British Library Board)

5.3 *Shearman's shears (Christ of the Trades painting, St Ethelbert's church, Hessett)*

6.1 *Merchant clothier (church of St Peter and St Paul, Clare)*

6.2 *Edge Hill, Hadleigh (home of William Forthe?)*

6.3 *Merchant mark of Thomas Spryng II (exterior of tower, church of St Peter and St Paul, Lavenham)*

than woollen weavers or fullers. Over 25 per cent of their debt disputes were with Londoners. Among those seeking payment from Roket were four men from the City, including a draper, grocer and shearman.[131]

How Many Weavers?

In the 1460s Suffolk cloth-makers produced five thousand sealed broadcloths (or equivalent in narrow cloths) each year. What is the minimum number of weavers who could reasonably have been expected to achieve such an output? Munro estimated an annual output per loom of twenty broadcloths in Flanders, while Zell more cautiously suggested eighteen and a half in the Weald.[132] In both those regions the quality and weight of the cloth was much greater than in mid-fifteenth-century Suffolk. If, instead, we plump for a round figure of forty Suffolk broadcloths per loom per annum, then the county's industry would have needed the output of 125 looms, each operated by two artisans, one perhaps a master and the other a journeyman or apprentice.[133]

In addition to woollens, Suffolk's cloth-makers also produced worsteds. In the Middle Ages these were not subject to alnage and so were not counted in the five thousand sealed cloths. The bed and coverlet weavers of Bury St Edmunds were the main users of worsted, although, as discussed below, there were probably some in other Suffolk towns such as Lavenham.[134] By the end of the fifteenth century at least ten such weavers carried on their craft in Bury and some, like John Roket, had built substantial businesses. It would be fair to assume that between them they owned twenty looms, and employed the same number of journeymen and apprentices. In the 1460s, however, their output was much less, so for present purposes let us discount them.

Turning to the plea rolls, the number of recorded Suffolk weavers is shown in Figure 4.2.[135] Each column covers a twenty-year period and so represents a generation who might have been in business at the same time. That for 1460 represents 104 weavers in the period 1450 to 1470, and that for 1470 ninety-four in the period 1460 to 1480. Nearly all the cases in which they were cited were writs of debt, so it is fair to assume that most of them were independent master weavers, rather than

131 TNA CP 40/837, m. 63v (AALT IMG 410); CP 40/888, m. 190v (AALT IMG 990); CP 40/895, m. 57r (AALT IMG 111).

132 Munro (2003), pp. 196–7; Zell (1994), p. 179.

133 Swanson (1989), p. 34.

134 Sutton (1989), p. 203; Swanson (1989), p. 38; Crowfoot et al (2001), p. 39.

135 Even a search of all the plea rolls would probably not reveal every Suffolk weaver, so at best we have samples rather than comprehensive lists.

journeymen, apprentices or even the outworkers of clothiers. Neither of these two samples is far short of the 125 required to weave five thousand broadcloths.

Let us try to reconcile this figure of 125 weavers, with the figure of 577 clothmakers who presented cloth to the alnager in four years in the late 1460s. Of these 577, only sixteen presented, on average, more than forty broadcloths (or equivalent in straits) each year. More than three hundred of them presented less than sixteen broadcloths over the period, or, on average, less than four each year. So, while the plea rolls mostly recorded independent master weavers, the alnage accounts also identified 'Jack of all Trades' who dabbled in weaving in winter when husbandry was slack, would probably never have been sued before the common bench and, if he was, would have been described as a yeoman or husbandman rather than as a weaver. He was to have a continuing role in cloth production for many years to come.

The number of Suffolk weavers almost certainly rose during the first half of the fifteenth century and declined in the second (Figure 4.2).[136] This matched very closely the rise and fall in Hanseatic cloth exports from Colchester and Ipswich (Figure 2.3). The downward trend would have been even sharper but for

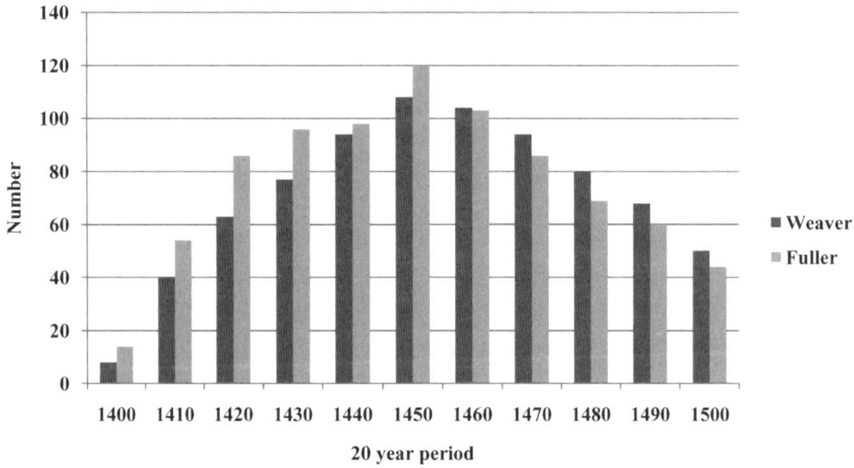

Figure 4.2: *Numbers of Suffolk Weavers and Fullers*

Notes: Each dark column records the number of Suffolk weavers, and each light column the number of fullers in the twenty years centred on the given date.

Source: TNA CP 40A.

136 The first two columns are of limited statistical value because all or most of the rolls predated the Statute of Additions 1413, by which the provision of details of a defendant's residence and occupation became compulsory: *Statutes*, 2, p. 171.

the growth in the number of Bury's bed and coverlet weavers. For this decline there are several possible explanations. Cloth production may have contracted in the late 1460s as the Hanseatic merchants departed and exports from Colchester and Ipswich fell. While this cannot be entirely discounted, the nationwide survey in chapter 1 suggests that it is most unlikely. A better explanation is that the Hanseatic exodus triggered a major change in the organisation of cloth production. The more substantial weavers increasingly titled themselves and operated as 'clothiers', while many of the others became their outworkers. As early as 1450 Henry Galyot of Lavenham was described as a weaver alias '*wesser and clothmaker*' and within thirty years the number of recorded clothiers easily exceeded weavers (Figure 6.1).[137] Those who continued to call themselves weavers were less visible in the plea rolls, as the value of their transactions fell below the 40s. threshold, although as many as 122 of them were cited in twelve rolls between 1480 and 1500.

The Babergh Military Survey of 1522 recorded 112 weavers, of whom fifty-seven had total wealth not exceeding £1 – barely enough to buy a loom and a pair of shuttles.[138] These poorer men, like Jack of all Trades, were almost certainly not independent master weavers, made only a limited contribution to the total output of cloths and would probably not have been litigants in the common bench. Assuming that the fifteenth-century ratios continued into the sixteenth, then just under half of Suffolk's woollen weavers lived in Babergh. So, if the number of independent master woollen weavers in 1522 in the hundred was fifty-five, then the total number in the county might still have been about 125. Arguably, any increase in output since the late 1460s was largely attributable to outworkers and to worsted weavers making bedding and coverlets in Bury St Edmunds and elsewhere.

From Apprentice to Merchant Weaver

When the apprentice John Kent inherited a broadloom from Thomas Shorthose of Sudbury in 1459 he may have harboured ambitions one day to be as successful as his master. He too would own a spacious town house in a pleasant location from which to carry on a thriving business. Within his craft a career hierarchy stretched before him that one day might put him on par with merchant weavers who made fortunes trading in London and even overseas.

---

137 TNA CP 40/756, m. 244v (AALT IMG 1399). The term 'wesser' probably meant maker of vesses.

138 Pound (1986), p. 133. The difficulty in identifying Jack of all Trades weavers is illustrated by a list of people indicted for their participation in the uprising of 1525 which included fifty-six otherwise unrecorded weavers. Of these fifty-six, thirty-six were also listed in the 1522 survey. In 1522 nineteen of the thirty-six had been described as weavers, eight as labourers, six had been given no occupation and one had been each of a baker, fuller and painter: Pound (1999), pp. 319, 329.

First, he may have had to find another master and complete his training. Had he been working in Bury St Edmunds then he would have had to serve as an apprentice for seven years. In Ipswich he might have struggled to find a master at all, and, even had he done so, he would probably never have been admitted as a freeman of that borough.[139] The relationship between master and apprentice could be close. Like Kent, John Hosteler of Bury St Edmunds was a beneficiary under the will of his former master weaver John Couper.[140] If Kent was unsuccessful in completing his training then he could take his loom into the countryside and carry on his craft as the outworker of a clothier. Perhaps that is why we find him later in Great Waldingfield. Combining weaving with husbandry might make all the difference between a secure living and subsistence.[141] John Laryot of Stoke-by-Nayland was one who did so. After assaulting Laryot, John Holand Esquire resisted his claim for compensation on the basis that he was his bondsman.[142] Conversely, if Kent was successful, he would have moved on to become a journeyman working in another man's premises.[143] As such he might have earned 6½–8d. per day, more than a skilled building worker.[144] Eleven alien cloth-makers sweated in John Stanesby's workshop in Bildeston, but such large-scale operations were rare.[145] Normally, a master would employ one or two journeymen whom he could ill afford to lose. For that reason the Bury St Edmunds ordinances tied master and journeyman together and prohibited poaching. Elsewhere, a deserted master could call on the Statute of Labourers to bring his helpmate to heel. So, we find such claims against Nicholas Spar at Ipswich, John Roundes at Boxford, and, as late as 1489, John Crosse at Lavenham.[146]

Once he had acquired more experience and made some contacts, John Kent could move on from journeyman to life as an independent weaver.[147] Had he not already owned a loom, he could have acquired one at modest cost of between 16–18s. together with a pair of shuttles for between 12–22d., and needed very little other

---

139 SROB B9/1/2. No weaver apprentice was recorded as being admitted as a freeman of Ipswich during the fifteenth century: Amor (2012), p. 418. Indeed, weavers enjoyed very little status in the town and were lumped together with spinners, seamstresses and other poor people of low degree, whose goods might not belong to them and were therefore not to be taken as security in legal disputes without the necessary precautions: Twiss (1873), pp. 133–5.

140 SROB 217 Hawlee.

141 Braudel (1982), pp. 306–7; Carus-Wilson (1987), p. 675; Zell (1994), pp. 174–6.

142 TNA CP 40/616, m. 324r (AALT IMG 645).

143 Bridbury (1982), p. 31.

144 Swanson (1989), p. 34.

145 TNA E 179/180/111; Zell (1994), p. 207.

146 TNA CP 40/477, m. 161v (AALT IMG 328); CP 40/616, m. 407r (AALT IMG 813); CP 40/907, m. 241r (AALT IMG 454).

147 C. Dyer (2002), p. 325.

plant or capital equipment. 'The loom was not an expensive piece of equipment compared, for instance, with the seed and stock of a husbandman.'[148] Even so, Table 4.1 suggests that not all weavers owned their own loom, or, at least, still had one to bequeath to family or friends when they died. Unless he was extraordinarily successful, Kent would have made only a modest living. When faced with a demand for poll tax in 1381, few weavers paid more than the standard 12d. per head (Figure 4.1). Most of their business was less than £5 in value and within a radius of ten miles from home (Tables 4.2 and 4.3). A sum of £5 would barely have bought two second-quality broadcloths.[149]

Hadleigh provides some good examples of weavers operating on this relatively modest scale. Twelve were recorded in the incomplete poll tax return of 1381, none of them paying more than the standard 12d. per head.[150] At least one, John atte Mille, was probably still in business a decade later, presenting ten broadcloths to the alnager in 1394/95. His contemporary, Walter Dering, presented six broadcloths, while John Dering, perhaps Walter's kinsman, presented sixteen narrow

TABLE 4.2: VALUE OF DEBT LITIGATION OF SUFFOLK WEAVERS, BED/COVERLET WEAVERS AND FULLERS 1350–1500

	All weavers		*Bed/coverlet weavers*		*Fullers*	
Value of claim	*No. of claims*	*Percentage of claims*	*No. of claims*	*Percentage of claims*	*No. of claims*	*Percentage of claims*
£20 or more	12	3.1	4	9.1	15	3.3
£10 or more but less than £20	40	10.3	4	9.1	40	8.9
£5 or more but less than £10	90	23.2	11	25	91	20.2
Less than £5 but more than 40s.	86	22.2	6	13.6	81	18.0
40s. or less	159	41.1	19	43.2	215	47.8
Unknown	0	0	0	0	8	1.8
Total	387		44		450	

Source: TNA CP 40

148 Britnell (1986), pp. 75–6; Swanson (1989), p. 33; Zell (1994), p. 172. A Westminster Abbey account roll for 1295 recorded the cost of a beam or balance as 7½d., twelve shuttles as 9d., a wheel as 9d. and slays as 1s. 6d: Lloyd (1970), p. 333.

149 Munro (2009), p. 40.

150 Fenwick (2001), 2, pp. 512–4.

TABLE 4.3: DISTANCE BETWEEN PARTIES TO DEBT LITIGATION WITH SUFFOLK WEAVERS, BED/COVERLET WEAVERS AND FULLERS 1350–1500

	All weavers		*Bed/coverlet weavers*		*Fullers*	
Distance	*No. of claims*	*Percentage of claims*	*No. of claims*	*Percentage of claims*	*No. of claims*	*Percentage of claims*
Same town or village	44	29.5	9	47.4	30	19.5
Less than 6 miles	21	14.1	0	0	36	23.4
More than 6, but less than 10 miles	18	12.1	0	0	25	16.2
More than 10, but less than 20 miles	24	16.1	1	5.3	39	25.3
London	18	12.1	5	26.3	10	6.5
Norwich	5	3.4	1	5.2	1	0.6
Overseas	2	1.3	0	0	2	1.3
Other more than 20 miles	17	11.4	3	15.7	11	7.1
Total	149		19		154	

Source: TNA CP 40.

dozens that same year, and thirty-one in 1396/97. In 1415 Walter supplied the local chapman William Armesby with six pieces of woollen cloth for 40s., perhaps to re-sell door to door or at a local market.[151]

Some families were able to build successful weaving businesses over the generations. So, if blessed with children, John Kent might hope to see his progeny better themselves. In Long Melford several members of the Talbon clan were cloth-workers. In 1441/42 a Robert Talbon leased a small parcel of land in Hall Street.[152] Melford men of the same name were described as a weaver in 1475 and a cloth-maker in 1480. A Roger Talbon (alias Beste) was cited as a weaver in 1460 and a John Talbon the Younger as another in 1480 and 1483, defending claims for 40s, £7 6s. and £12.[153] Both Robert and John Talbon produced broadcloth in the late 1460s, although Robert did so more regularly and in larger quantities – 106 broadcloths in total, putting him in thirty-fifth position in a league table of the county's top cloth-makers.

151 TNA E/101/342/8, 10; CP 40/616, m. 181v (AALT IMG 1445).

152 SROB J/523 Extent of the manor of Melford 1441–2 (transcribed by David Dymond); E. Wigmore *Long Melford in 1441/2*.

153 TNA CP 40/871, mm. 25r, 441r (AALT IMG 53, 824); CP 40/883, m. 250r (AALT IMG 448).

While none of the recorded Talbon weavers was a woman, a widow could continue the business of her late husband, having acquired weaving skills during his lifetime. Several wills included a bequest of a loom to a widow. Edmund Goodale of Bury St Edmunds left more specific instructions, providing that his son should inherit his looms once his wife Margaret had ceased to use them.[154]

A fortunate few, by birth or ability, had the resources and contacts necessary to undertake high value, long distance trade and so amass riches as merchant weavers. One such was Roger Crytott, the foremost Lavenham weaver of his generation. Together with his kinsman Thomas, he presented more broadcloth to the alnager during the 1460s than any other of the town's cloth-makers. He became a freeman of Ipswich and was among a small handful of men who not only produced cloth, but also exported it from that port. He traded with merchants of King's Lynn in Norfolk, pursuing John Thorysby for payment of the enormous sum of £105 5s, and remembered the Greyfriars of that town with a bequest when he died. He owned two dye-houses, one located 'Chercheward' and the other 'Buryward', as well as tenters for stretching cloth. He was survived by his wife and three children to each of whom he left, among other property and goods, a fine piece of silverware. At a more mundane level, he consented, and was trusted to act as executor by his less-successful neighbours such as John Cobold, a fellow weaver, John Stoke and Andrew Grygges.[155] In 1522 in Babergh only two weavers had wealth exceeding £20, none exceeding £40, so however hard he worked John Kent was unlikely to ever become another Roger Crytott.[156]

Fulling and Tentering

On a warm spring afternoon the Fullingpit Meadows of Sudbury are delightful. At their northern end, by the weir, they provide an excellent vantage point upstream on the river Stour (Plate 4.8). Six hundred years ago they would have been far less pleasant. Fullers inherited their name from the 'fulliones' who, in Roman times, like them, collected urine.[157] Fullers' pits would have been full of it, slowly going stale to produce the ammonia with which cloth could be scoured. The stink was dreadful.[158] And it emanated not only from the pits themselves, but from the naked men who

154 SROB 130 Pye.

155 BL, Ms Add. 30,158, f. 13r; TNA E 101/342/21, 25; TNA E 101/343/2, 4, 5;TNA CP 40/756, m. 5v (AALT IMG 922); TNA CP 40/796 m. 68v (AALT IMG 978); TNA E 101/343/5; SROB 52 Hervye; SROB 132, 310 Baldwyne.

156 Pound (1986), p. 133.

157 Ponting (1972), p. 9.

158 In nearby Nayland the clothier John Ringbell bequeathed 13s. 4d. to 'amend the fowle slowgh' at Peyton Field – had this been the site of his fulling operations?: TNA PROB 11/9/295.

immersed themselves in the foul liquid.[159] Only these hardened artisans would have wanted to hang around and they were not there for the view. Their task was to clean, shrink and felt the woven cloth. In doing so, they increased its density and weight and gave it not only a smooth and soft surface, but also much greater resistance to weather and wear. Clothes were frequently passed down from one generation to the next.

As explained by Langland in the late 1300s in *The Vision of Piers Ploughman*:

> You've seen the cloth that comes straight from the weaver. It isn't fit to wear till it's been scoured and beaten under foot or in fulling stocks. It has to be thoroughly washed, and carded with teazles, tucked and stretched, and only then can it come under the tailor's hands.[160]

For centuries cloth was fulled in pits or tubs that contained a mixture of warm water, urine and fuller's earth.[161] Not found in Suffolk, in the Middle Ages, fuller's earth was sourced from Kent.[162] This may explain why John Feerby Esquire of that county made three separate claims for payment of debts due from Suffolk fullers, namely John Laurence of Stoke, John atte Melle of Stoke-by-Nayland and Thomas Lylye of Hadleigh.[163] While in the pits the cloth was trampled under foot and/or beaten with a club. Images of the fuller's club can be seen in the hands of their patron saint, St James the Less, in the churches of Hesset and Southwold (Plates 4.11 and 4.12). The cloth was generally immersed and pummelled twice: once to scour and cleanse it; and again, more vigorously, to shrink it so that the wool fibres clung together ever more tightly and formed the felted finish. In between, the cloth was hung over a perch and burled to remove knots and other imperfections. Finally, it was thoroughly washed clean, often in a local river. Sometimes, the riverbank was so congested with fullers washing cloth that their neighbours complained that they could not draw water.[164] Working together a master and two journeymen might hope to complete this process in three to five days, and full between sixty and seventy broadcloths each year.[165]

Having emerged from the second immersion between a half and a third of the size it had started, the cloth had to be stretched back into the correct size and shape.

---

159 The medieval Draper's Window in the church of Notre-Dame at Semour-en-Auxois includes an image of a fuller in a tub. In the nineteenth century this image was 'restored' to add clothes to the previously naked figure.

160 Langland (1992), p. 182.

161 http://en.wikipedia.org/wiki/Fuller's_earth

162 Thirsk (1984), p. 218; Zell (1994), p. 236.

163 TNA CP 40/629, m. 44r (AALT IMG 90); CP 40/636, m. 343v (AALT IMG 1719); CP 40/647, m. 78r (AALT IMG 159).

164 Rawcliffe (2013), p. 205.

165 Munro (2003), pp. 204–5.

The fuller achieved this by attaching it, using tenterhooks, to a tenter-frame. Such stretching required experience, skill and honesty. Overstretching could leave the cloth with fatal flaws, rendering fabrics liable to shrink in the wet, so was prohibited by statute. To minimize the risk of subterfuge tenters had to be erected in 'open places' and not behind closed doors.[166] In Ipswich a plot of land on the great embankment of the town was set aside for a tenter-yard measuring in length 40 ells (about 45 metres).[167] Another in Long Melford, to the west of Hall Street and close to the Chad Brook, covered an area of 33 perches (about 0.08 hectares).[168] Tenters' and tenter-yards were occasional bequests in wills. Stephen Fushok of Bury St Edmunds leased two and owned a third in Looms Lane, while Stephen Gardiner had one in nearby Eastgate Street.[169] John Reyneham owned land in Nayland called 'Teyntour Hille'.[170] Two testators from Glemsford, namely John Petywate and Thomas Heggeman, both mentioned tenters. Petywate spelt out that his wide tenter in Egremont Street was 'for the hanging and working' of cloth.[171] John Harry of Lavenham evidently had faith in his daughter Elizabeth, bequeathing his tenter-yard to her rather than to her brother.[172] In the early 1980s a tenter-frame was found, during the course of restoration work, in the fabric of 80 Church Street, Lavenham that Betterton described as follows:

> The frame was constructed of seven vertical 7ft (2.12m) timbers spaced at slightly varying intervals over a distance of 10ft (3.05m). Two diagonal strengthening members intersected six of the verticals. From this eleven lengths of oak were salvaged bearing rows of tenterhooks fixed at an average interval of 2½ in (6.3cm) along one face. With one slightly larger exception the timbers measured 3ins x 3½ ins (7.8cm × 8.8cm) in section. Four of the timbers had mortices cut into them, presumably to receive the tenter frame's verticals, and these faced in the opposite direction to the points of the hooks. It could therefore be argued that an upper rail of a tenter had been re-used here since a lower one would have been moveable and not permanently fixed to a vertical. More advanced weathering on what would then have been an upper face of most of the timbers seemed to confirm this (Figure 4.3).[173]

166 *Statutes*, 2, pp. 484–6.
167 SROI C/2/1/1/3.
168 SROB J/523.
169 SROB 259 Osbern; 125 Hawlee.
170 TNA PROB 11/11/743.
171 SROB 474 Baldwyne; 156 Hervye.
172 SROB 536 Baldwyne. The hanging of cloth on the tenter-frame was, in fact, a task often entrusted to women: Carus-Wilson (1987), p. 638.
173 Betterton (1984), pp. 317–9.

SUFFOLK INSTITUTE OF ARCHAEOLOGY AND HISTORY

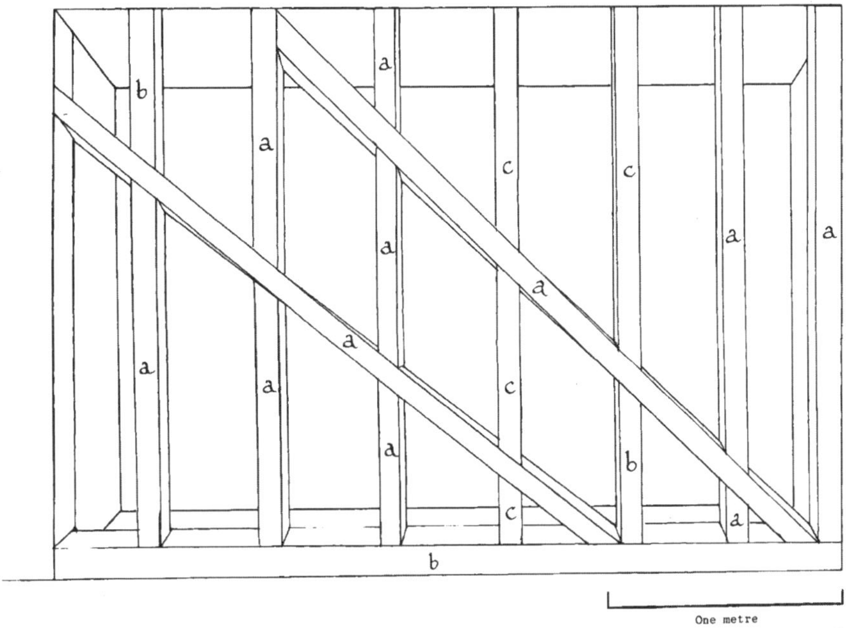

FIG. 77 — Internal wall (now removed) at No. 80 Church Street, Lavenham: wall frame with laths and plaster removed, showing *a* — timbers bearing complete rows of tenter-hooks; *b* — timbers with no tenter-hooks; *c* — missing timbers.

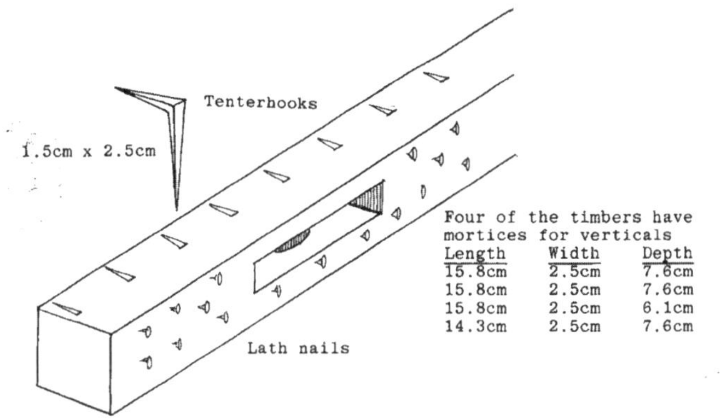

FIG. 78 — Detail of tenter rail.

318

Figure 4.3: *Lavenham Tenter-Frame*

Notes: Courtesy Suffolk Institute of Archaeology and History and the late Alec Betterton.

While the cloth hung drying on the tenter-frame, fullers subjected it to preliminary wet napping and shearing. Rowland Whitefeld of Bury left tools for napping. His servant James Parmenter received two dozen pairs of handles, a dozen wet and a dozen dry, and his son John all his other handles.[174] A handle was a rectangular wooden frame with a long haft, by which to hold it, in all about a foot long, and was drawn across the cloth.[175] The wet handles were used for preliminary napping and the dry for cloth finishing by shearmen. Teasels, a flowering plant of the genus *dipsacus*, usually the variety with hooked rather than straight spikes, were fitted tightly within the handle (Plate 4.13). These spikes picked up the loose or protruding fibres, in a process known as raising the nap, which could then be cut off with fuller's shears such as those John Chyldriston of Mildenhall owned when he died.[176] The secret was to form a perfectly proportioned raiser by the correct combination of teasels that were small and large, new and old.[177] Some teasels were grown locally and are recorded at Clare, Glemsford and Long Melford. The account of the reve of Clare for 1336/37 records expenditure of 3d. on 2 quarters of teasel seed, but from the sale of 'fullers' teasels, nothing because they remain to be sold'.[178] Others were imported from overseas and baskets of Flemish teasels were a small, but regular item in many Ipswich particular customs accounts.[179]

Fulling Mills

In her well-known paper on medieval fulling mills Carus-Wilson included a list of those recorded in which only one, that of the Hadleigh, was in Suffolk.[180] Although Kerridge was sure that 'the East Anglian plain hummed and clacked with fulling mills', a diligent search by other historians over several decades has identified only thirty-one in the county (Map 4.3).[181] Poos found the same number in Essex.[182] Fulling mills were relatively expensive to build and maintain, entailing 'high capital costs'. The labour costs involved in the construction of a fulling mill were

174 SROB 26 Mason. The Nayland shearman Richard Davy referred in his will to 'my sheryng sheres and handulles with all my crafte that I occupied': TNA PROB 11/18/50. Sixteenth-century Wealden shearmen included 'handles' in their inventories: Zell (1994), p. 185. See Munro (2003), p. 209.

175 Carus-Wilson (1967), p. 231.

176 SROB 179v Baldwyne.

177 Ponting (1971), pp. 171, 176–7; Munro (2003), p. 209; Oldland (2007), p. 98.

178 *Nonae* (1807), pp. 103–4; Ward (2014), pp. 106, 109.

179 TNA, E122/193/33, E122/51/39, E122/52/43, E122/53/1. Kerling (1954), p. 125.

180 Carus-Wilson (1967), p. 197.

181 Kerridge (1972), p. 23. Thirty-two, if William Alysaunder's mill at Monks Eleigh was a fulling mill.

182 Poos (1991), pp. 60–1.

significantly greater than in the construction of other types of mill – more than £17 in pre-Black Death Marlborough, much more in post-Black Death Suffolk.[183] In regions where rivers were slow and lazy, rather than swift and clear, overshot mills worked best, in which the water flowed over rather than under the wheel. They were more efficient, but also more expensive to build and maintain.

From time to time, different parts of the overshot mill at Priestbridge in Hadleigh required replacement or repair. Christchurch Canterbury priory's accounts tell us what needed to be done and give an impression of how the mill worked.[184] The river Brett was dammed by a weir and the waterway diverted to make a mill-race that would power the wheel. Floodgates (three are mentioned), that were fixed in the weir using stakes, controlled the flow of water so as to ensure that the wheel turned at a steady speed. Boards were used to create a wall that protected the mill from being undermined by the flow of water. Concave troughs, made from oak (100 feet in length is mentioned), together with gutters directed the water to the wheel. Rather than the water driving the wheel directly, it flowed into tubs (eight are mentioned) that were attached to the wheel and so provided the gravitational force to keep it turning. The wheel was held in place by an axle-tree and around its circumference were cogs or tappets which tripped the stocks that hammered the cloth. Repairs, as well as drought and heavy rain, intermittently interrupted operations for days or weeks on end. In the mid fourteenth-century the mill was rebuilt twice with an upper floor and tiled roof. In 1415 the fuller Thomas Benton of Hadleigh was probably keen to maintain the mill, and to secure the best man for the job, when he infringed the Statute of Labourers by enticing away the carpenter Robert Bacon to work for him.[185] In 1449/50 the work of four carpenters, five labourers and materials came to £10 2s. 9d., while fourteen years later the bill for three carpenters, two labourers and materials was £6 12s. 11d.

Before the Black Death many manorial lords, both lay and ecclesiastical, built or acquired fulling mills or converted grain mills to fulling.[186] These could contribute to their social and economic standing.[187] By insisting on their tenants using them, they hoped for a reasonable return on their capital investment.[188] They may have catered for an existing community of cloth-makers, or hoped that building a mill

183 Langdon (2004), pp. 98–103, 249.

184 CCA-DCc-BR/Hadleigh; CCA-DCc-MA (courtesy of Margaret Woods).

185 TNA CP 40/616, m. 342r (AALT IMG 682).

186 In the West-Country too 'almost all early mills were the property of the Lords of the Manor': Ponting (1971), p. 16. Munro suggested that the overshot mill was not introduced into England before the 1330s, so many of these early seigniorial mills may have been of the less expensive, but less efficient undershot type: Munro (1999), pp. 126–7.

187 Lucas (2014), p. 83.

188 Carus-Wilson (1967), pp. 198, 201; Ponting (1971), p. 16; C. Dyer (2005), pp. 92–3.

would attract some to the locality. For lords, investing in a fulling mill was a similarly speculative venture to that of a previous generation who had invested in a windmill or the purchase of a market charter. The investment proved to be a wiser one in some places than others. Already standing in 1273, the fulling mill in Hadleigh generated an annual rent which rose fairly steadily over the following hundred years from 66s. 8d. to £5, before falling back again in the fifteenth century.[189] The location of many of the early fulling mills, outside the principal centres of textile production in the late Middle Ages, suggests that these ventures were frequently unsuccessful, or at best short-lived. The priory of Blakenham was a daughter house of the great Benedictine abbey of Bec-Hellouin in France. In the late-thirteenth century the monks erected a fulling mill on the banks of the Gipping that generated an annual rental income of 17s., significantly less than that from their nearby grain mill. A doomed enterprise, no more was heard of it after 1489.[190] No doubt similar stories from similar dates can be told of Chattisham (1303), Hintlesham (1325), and Hollesley (1295) where an early fulling mill was recorded, but not one cloth-maker is mentioned in the plea rolls.[191] Even if the lord attracted cloth-makers he had no guarantee that they would use his fulling mill, as the notorious battle between St Albans Abbey and its tenants proves.[192] With the Black Death, the rapid decline of serfdom and the loss of the seigniorial right of millsuit, many of these mills would have fallen into disuse and ultimately been abandoned or converted back to grain.[193]

Thereafter, it took financial courage or recklessness to invest in a fulling mill. In Holt's words, 'milling was to be a more hazardous venture, with profits increasingly uncertain'.[194] Like many Benedictine houses, the abbey of Bury St Edmunds owned and leased out several fulling mills, in this case in Suffolk Breckland.[195] That at Lackford retained a significant rental value throughout the later Middle Ages, and, in 1442, was leased by the fuller William Cobald of Bury St Edmunds.[196] The lord of the manor of Glevering at Hatcheson granted a lease of his fulling mill for five years from Michaelmas 1402. As an inducement he agreed with his tenants to

189 CCA-DCc-BR/Had/16, 21, 22, 25 (courtesy of Margaret Woods).

190 Chibnall (1951), pp. xiii, 132. Lucas contended that 17s. was a rental payment for six months, so the annual rent may have been 34s., still much less than the grain mill: Lucas (2014), p. 345.

191 Anon (1899), pp. 292–3, 312; Holt (1988), p. 156.

192 Carus-Wilson (1967), pp. 201–3; Miller (1965), pp. 72–3.

193 Bailey (2014), p. 288. The battle in St Albans suggested that some lords at least considered millsuit to include fulling.

194 Holt (1988), p. 159. Fulling mills 'all but disappeared from urban cloth production' in the medieval Low Countries after the early fourteenth century: Van Bavel (2010), p. 347.

195 Lucas (2014), p. 85.

196 Bailey (1989), p. 308; SROB 262 Osbern.

Map 4.3: *Medieval Suffolk Fulling Mills*

Notes: Black marks a parish with a recorded fulling mill.

Sources: Dymond (1994), p. 123; Crawford (1992), p. xiv; Bailey (2009), p. 15; Peter Northeast archive, SROI HD 2448/3/351.

Map format attribution: David Dymond and Edward Martin.

supply them with fuller's earth and any board needed for repairs.[197] Hacheston still had its own fuller at the end of the century so the inducement may have worked.[198] Even the Hadleigh fulling mill was no cash cow. Its annual rental value fell from £5 in 1376/77 to £4 13s. 4d. in 1482/83, before recovering to £7 6s. 8d. a decade later.[199] Sometimes the cost to Christchurch priory of repairs exceeded their income.

The experience of the Ipswich borough authorities was even less happy. They owned two fulling mills in the town, one at Stoke bridge and the other, better

197 Andrews and Redstone (1929), p. 204.

198 TNA CP 40/943, m. 33r (AALT IMG 69f).

199 CCA-DCc-Br/Hadleigh/25; CCA-DCc-MA/168; CCA-DCc-MA/8 (courtesy of Margaret Woods).

documented, known as Horsewade mill at Greyfriars bridge.[200] The latter is first mentioned in the extant records in 1345 and is referred to as a fulling mill in a recognizance roll of 1349. In the mid-fifteenth century a series of prominent burgesses leased it out, none of whom, it would appear, could make it pay. In 1448 the general court ordered Peter Terry to pay arrears of £5 on his lease. Seven years later William Ridout and William Heede were in default. Ridout had taken a lease in October 1447, in succession to Terry, for a term of thirteen years at a rent of £20 per annum, only two-thirds of the amount that had been payable in 1345. On 26 September 1455 the general court first ordered him and Heede to give security for the settlement of arrears of £30 within the week. The deadline passed without payment, and was extended to 1 November. On 9 April 1456 a fresh one was set for the surrender of £20 of the arrears within the fortnight. A day before this deadline expired Ridout told the court that he refused to pay anything. The authorities appear to have accepted defeat and in the following January the mill was re-let to John Lackford for the same rent. By 1463/64 it had been re-let again, this time to Robert Deye, at an even lower rent of £10 per annum.[201] At a time when most of England was in the grip of the Great Slump and cloth-making in Ipswich was in decline, this sorry tale is unsurprising.

In Essex all known medieval fulling mills were in the north-east of that county, within its cloth-making area.[202] Not so in Suffolk. Only ten, less than a third of the total so far identified, were situated in the main industrial belt alongside the rivers Brett and Stour. Why so? Cloth-makers used the more mature sections of these rivers as transport routes. The weirs and dams that were essential to the operation of overshot mills were an obstacle to river traffic.[203] Even if the occasional obstacle was accepted as a necessary evil, the tolerance of the boatmen and their customers was limited. Further upstream, there were other issues. Despite intense study, no record of any fulling mill in Lavenham or Long Melford has been found.[204] How can this be explained? It may just be that the records have been lost or destroyed, or remain to be found, but there may be other reasons. In Lavenham the available running water may have been insufficient to drive a mill. In Long Melford, with perhaps a score of fullers working at any one time and all competing against each other for business, none of them could establish sufficient market dominance

200 The fulling mill at Stoke bridge was mentioned in the chamberlains' accounts for 1446–7: SROI C/3/3/1/1.

201 BL, Ms Add. 30,158, ff. 10v, 11r; SROI C/2/4/1/38; SROI C/3/3/1/2; *Annals* (1884), pp. 71, 114–5.

202 Poos (1991), pp. 60–2.

203 Miller (1965), p. 72.

204 The water mill to the north of Lavenham may possibly have been used for fulling: Betterton and Dymond (1989), p. 35.

to afford to build, and thereafter operate a fulling mill. As Zell stated 'the mill itself represented a major investment, well beyond the resources of the typical fuller'.[205] Even if one of them could have raised the necessary funds, the others would have resisted his taking away their business by doing so.[206] However, the wealthy clothiers who emerged in the late fifteenth century were in a position to build and operate fulling mills profitably. They had the necessary capital and controlled local fullers by employing them as outworkers.[207] John Horrold of Clare's estate included a fulling mill in 1478, as did Hugh Turnour's of Stoke-by-Clare in 1505, while Richard Davy of Nayland referred to a 'fulling mill river' in his will of 1514.[208]

The hypothesis that the number of Suffolk fulling mills began to increase at the end of the fifteenth century finds further support in an analysis of the changing number of fullers (Figure 4.2). The trend is similar to that of the weavers, for the same reasons, with numbers rising up to 1450 and falling thereafter. Until 1450 fullers invariably outnumbered weavers, thereafter the reverse was true. Twelve rolls for the period 1480 to 1500 recorded 122 weavers, but only 96 fullers. So, even if you look hard, their number was in decline. The Babergh Military Survey of 1522 recorded forty-one fullers, in contrast to 112 weavers. This equated to about ninety-three fullers across the county as a whole, even less than the number in the twelve rolls, suggesting that the decline in their numbers continued into the sixteenth-century.[209] As there is no evidence of a fall in cloth output in the county, and as cloths may well have been more heavily fulled at the end of the fifteenth century than at the beginning, this trend suggests that more fulling was being done by fewer people, which can only mean greater use of fulling mills. Nevertheless until that time one can only agree with Bailey's conclusion that 'fulling mills were not established in large numbers: many important textile centres functioned without them and most mills generated modest profits'.[210]

In the fifteenth century fullers were to be found in most parishes along the southern bank of the Orwell – at Wherstead, Freston, Chelmondiston and Shotley

205 Zell (1994), p. 180.

206 In 1485 the owner of a fulling mill in the Stroud Valley was attacked by a crowd of malefactors among whom, one may assume, were fullers: Lipson (1921), p. 121. See also Carus-Wilson (1987), p. 670. In the Weald in the sixteenth century very few fulling mills were required to process thousands of broadcloths: Zell (1994), pp. 180, 236.

207 Ponting (1971), p. 16; Britnell (1986), p. 185; Carus-Wilson (1987), p. 686; C. Dyer (2002), pp. 325–6. Langdon noted a 'second increase' nationwide in fulling mills at the end of the fifteenth century, although he acknowledged that his database for the East Anglia region was too small to be reliable: Langdon (2004), pp. 40–7.

208 TNA PROB 11/6/472; PROB 11/14/728; PROB 11/18/50.

209 Pound (1986), p. 133.

210 Bailey (2009), p. 17.

(Map 4.4). As the river here is tidal, it raises the intriguing possibility that tidal fulling mills were in use. Although these were rare and no direct evidence of their existence here survives, Holt's finding of two medieval tidal mills at nearby Walton, and the known existence of a third at Woodbridge, shows that the technology was exploited in east Suffolk.[211] No gig mill of the type used from the mid-fifteenth century at Castle Combe, Wiltshire is known in medieval Suffolk.[212] This mechanized the raising of the cloth, by using teasels (known colloquially as 'gigs') mounted on rollers, for one or other, or perhaps both, of two processes, namely loosening and removing detritus before fulling, and wet napping afterwards. The gig mill was viewed with suspicion by the authorities who were concerned that it might damage the cloth, particularly if teasels were replaced by wired cards. In 1464 fullers were prohibited from using cards for this purpose, to avoid the risk of their 'deceitfully impairing the cloth', and a century later gig mills were banned altogether.[213]

Fuller Free Zone

The geographical extent of fullers in Suffolk was similar to that of weavers, with 388 spread over eighty parishes (Map 4.4). One hundred and seventy worked in Babergh (43.8 per cent) and seventy-seven (19.8 per cent) in Cosford. All the major cloth-making towns had significant numbers of both fullers and weavers. Nevertheless, there were some important differences. In particular, there were no recorded fullers in the parishes close to Bury St Edmunds. Just as the abbot had established a market-free zone around his town, so he appears to have established a fuller-free zone too.[214] Icklingham, with four, West Stow with three, Tostock with two and Lackford and Norton with one each, were the closest villages with recorded fullers. Should it have been the case, how and why was this fuller-free zone established? One can be sure that it helped fill abbey coffers. The fulling mills beside the river Lark sucked in much of the work that would otherwise have been done locally by hand or foot. One may also speculate that, as at Bristol in 1381, Winchester in 1402, and Coventry in 1518, cloth-makers were perhaps forbidden or restricted from receiving cloth that had been fulled outside Bury, or sending their cloth out

211 Holt (1988), pp. 135–6; http://list.historicengland.org.uk/resultsingle.aspx?uid=1300451. See also Gransden's reference to a possible twelfth-century tidal mill at Southwold: Gransden (2007), p. 294.

212 Beresford and St Joseph (1979), pp. 267–9. Very few records of medieval gig mills survive, although in the early-sixteenth century another was situated at Bishopstrow, near Warminster in Wiltshire: Lucas (2014), p. 159.

213 *Statutes*, 2, p. 406; Ponting (1971), p. 28; Munro (2003), p. 210; Oldland (2007), p. 102.

214 Dymond and Northeast (1995), p. 46; Scarfe (1999), pp. 76–7.

of town for fulling. This would also have greatly enhanced the bargaining power of Bury fullers, as it undoubtedly did in Winchester.[215] It is, therefore surprising to find relatively few of them in Bury, less than one for every two weavers, and furthermore no recorded fulling mill. A study of debt claims against Bury fullers suggests that they engaged in higher value transactions than their counterparts elsewhere, but the small number of recorded cases rules out any definitive conclusion about their financial wellbeing. William Alysaunder was the most successful of them. He started his career as a fuller in Monks Eleigh where he was first recorded in 1440 defending a claim for just 40s. While ten years later he was still there, by that time he faced claims for much larger amounts, suggesting that his business had grown in the interim. The prominent London grocer Thomas Canning sued him for £39 11s. 4d. By the late 1450s he had moved his operation to Bury St Edmunds, although he may not have severed all ties with Monks Eleigh. In the 1460s a William Alysaunder acquired there a lease of a water mill, perhaps a fulling mill, known as *Old Northes* to the north of the Bildeston Road.[216] In Bury, assuming that all references were to the same man, he was described variously not only as a fuller, but as a clothman, merchant and yeoman. Among those with whom he dealt were two Londoners, one a draper and the other a mercer, as well as Henry Bermycham, a merchant of King's Lynn, who sued him for the princely sum of £80. In 1470 he defended a claim by Robert Cake of Stowmarket for £6.[217]

The Status of Fullers

Just as a wide spectrum of weavers extended from apprentice to merchant, so too fullers enjoyed varying fortunes. William Alysaundre was not alone in pursuing a successful career and conducting high value business. Also, many fullers, just like many weavers, pursued other occupations. Richard Cavenham of Clare, Gregory Belcham of Hadleigh and Robert Holton of Kersey were all fuller alias husbandman.[218] John Debenham of Stoke-by-Nayland was merchant alias fuller alias

---

215 Carus-Wilson (1967), p. 208; Ponting (1971), p. 237; Phythian-Adams (1979), p. 48; Swanson (1989), pp. 41, 160, 170; Davis (2012), p. 217.

216 CCA-DCc-MA/33 f. 47r. Although not described as such, the suspicion that this may have been a fulling mill, is reinforced by the fact that one of its subsequent owners was William Forthe the great clothier of Hadleigh.

217 TNA CP 40/716, m. 426r (AALT IMG 859); CP 40/756, m. 208v (AALT IMG 1325); CP 40/758, m. 164r (AALT IMG 337); CP 40/776, mm. 198r, 183v, 198v, 336v (AALT IMG 393, 1476, 1504, 1773); CP 40/834, m. 214r (AALT IMG 381).

218 TNA CP 40/656, m. 341v (AALT IMG 1648); CP 40/853, m. 52v (AALT IMG 1026); CP 40/907, m. 239v (AALT IMG 1299d).

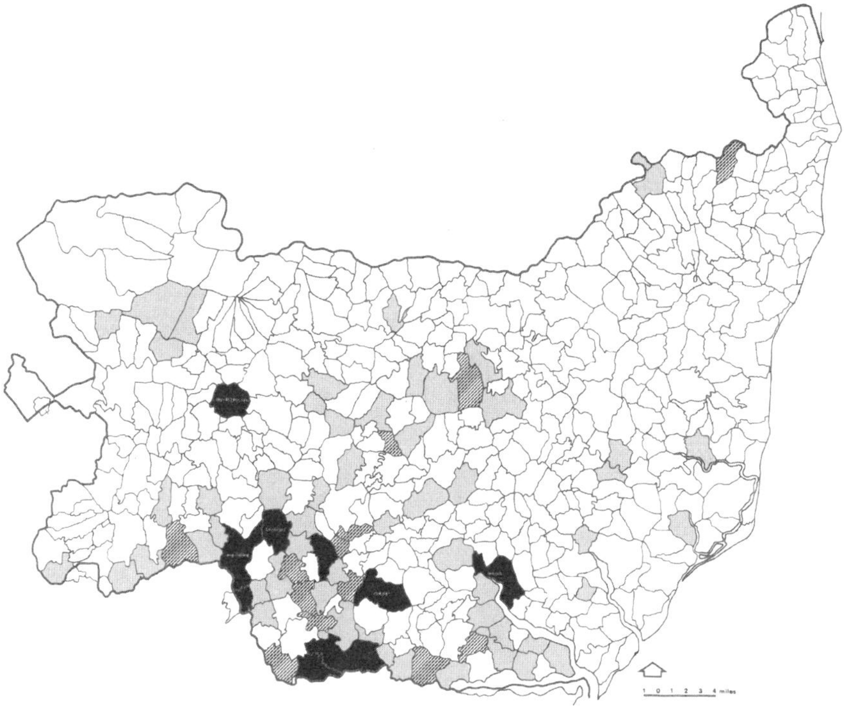

Map 4.4: *Suffolk Fullers 1350–1500*

Notes: Grey marks a parish in which between one and four fullers were recorded. Hatched dark grey marks a parish in which between four and nine fullers were recorded. Black marks a parish in which not less than ten fullers were recorded.

Source: TNA CP 40.

Map format attribution: David Dymond and Edward Martin.

tanner.[219] When he was not working ashore as a fuller, John Quylter of Ipswich was at sea as a mariner.[220]

The fullers of Hadleigh provide some archetypal examples. Several presented cloth of their own to the alnager both in the 1390s and the 1460s. Foremost among them were John Apulton who presented twelve dozens in 1394/95, and William Giffard who presented five straits in 1466 and thirty broadcloths in 1467/68. Another John Apulton, perhaps a son, was working as a draper in Little Waldingfield in 1440, and a century later a Walter Appulton, also of Little Waldingfield, styled

219 CP 40/895, m. 24r (AALT IMG 47f).

220 CP 40/943, m. 35v (AALT IMG 902d).

himself as a gentleman.[221] A step down from the Apultons, John Warde was busy in 1445, seeking payment from customers in Elmsett, Layham, Shelley and, slightly further afield, Woolpit. All his claims were for just 40s. so he was clearly not engaged in big business.[222] At the bottom of the ladder of fortune were those four Hadleigh fullers implicated in Jack Cade's rebellion of 1450, and subsequently included in the general pardon – John Raymes, Thomas Dexter, John a Bury and Henry Wodecok.[223] As none of them is included in the list of twenty Hadleigh fullers recorded in the plea rolls for the period 1440 to 1460, we can assume that they were of limited economic consequence. If they were the poor souls who earned a meagre living by stomping up and down in pits of stale urine, then it is hardly surprising that they turned to violence to try and better themselves.

In London in 1538, on the back of their role in cloth finishing, fullers joined with the shearmen to form the twelfth and last merchant company of the City.[224] In a few other places, such as Stroud, they assumed the profitable role of entrepreneur. Not so in Suffolk. Even though they possessed the cloth at the stage it became saleable or ready for finishing, evidence of their involvement in finishing or marketing is limited.[225] Men such as William Alysaundre, John Apulton, and William Giffard were unusual. As a group fullers do not appear to have gained any significant edge over weavers and, indeed, financially were on a par, if not marginally worse off (Figure 4.1 and Table 4.2).[226] Most of them were so poorly rewarded, and added such a relatively small proportion to the cost of cloth-making that, as discussed above, mechanising their work was rarely worthwhile.[227] Only a tiny proportion of them, indeed a smaller proportion than of spinsters, paid more than the standard 12d. poll tax in 1381. A smaller proportion of fullers, than weavers, did business with their immediate neighbours, perhaps reflecting the fact that they were spread over fewer parishes. Nevertheless, nearly 85 per cent of their debt disputes were of a purely local nature, with adversaries less than ten miles

221 TNA CP 40/636, m. 24v (AALT IMG 1085); TNA CP 40/716, m. 35v (AALT IMG 1105); TNA CP 40/834, m. atts (AALT IMG 861); TNA E/101/342/8; TNA E10/342/25; TNA E101/343/4; Grimwade and Serjeant (1984), p. 13.

222 TNA CP/736, m. 300v (AALT IMG 1632).

223 *CPR 1446–1452*, pp. 338, 343, 356, 359.

224 Oldland (2007), pp. 101–2; Oldland (2008), p. 172.

225 In the late 1460s no more than six recorded fullers from Hadleigh, four from Long Melford, and two from Lavenham presented cloth to the alnager. For further discussion about who became a clothier see chapter 6 below. In the London alnage accounts for the 1370s fullers were, by far, the most numerous of taxpayers, but by the 1430s they had almost disappeared: Quinton (2001), pp. 117–20.

226 With the use of tax records, Kowaleski found the same in Exeter: Kowaleski (1995), p. 128; as did Swanson, with the use of testamentary records, in York: Swanson (1989), p. 157.

227 Swanson (1989), p. 41.

from home (Table 4.3). Furthermore, nearly half those disputes, a higher proportion than was the case with weavers, involved the minimum sum of 40s. The Babergh Military Survey of 1522 showed some improvement in their fortunes, indicating that in general fullers were slightly more prosperous than weavers.[228] Nevertheless, for most of them, the old saying, that 'where there's muck, there's money', was just empty words.

Structures of Specialisation

If the geographical spread of fullers was similar to weavers, then the ratio between one and the other, and with dyers, in any given town was not (Table 4.4). From an early date, as Britnell put it, 'there were structures of specialisation [...] within the clothmaking region'.[229] Members of one craft gravitated towards certain towns, while those of another were attracted to others. John Robert of Glemsford was a successful weaver, but when his son of the same name decided to carry on business as a fuller he moved to Long Melford.[230] In some places, like Bury, weavers heavily outnumbered fullers, as they also did in Lavenham and Sudbury. In others, the reverse was true. In Hadleigh, throughout the fifteenth century, there were nearly three fullers for every weaver, and in Long Melford nearly two. In 1522 weavers continued to outnumber fullers and dyers in Lavenham by a ratio of five to one, while in Long Melford there were still more fullers than weavers.[231]

Several possible explanations for such specialisation present themselves. It may have depended on the extent to which cloth-making was concentrated in the hands of an elite. By the late 1460s half a dozen wealthy clothiers dominated cloth-making in Hadleigh in a way that may have eliminated, and/or pushed out of town those independent weavers most likely to be litigants in the court of common pleas. These same clothiers may have had their better-quality cloth heavily fulled, so requiring the services of more fullers. In Lavenham control of cloth-making at that time was more diffuse, with its top seven clothiers responsible for only a third of total production.[232] This argument, however, becomes less persuasive fifty years later when Thomas Spring III dominated cloth-making in the town.

Where weavers outnumbered fullers, it may be that many of those weavers were engaged in the production of worsteds that required little or no fulling.[233] Such was

228 Pound (1986), p. 133.
229 Britnell (2003), p. 88.
230 TNA CP 40/919, m. 164v (AALT IMG 1062).
231 Pound (1986), pp. 75–87.
232 Amor (2004), pp. 419–20.
233 Worsteds were sometimes fulled for the purpose of degreasing, rather than felting: Munro (1999), p. 115.

TABLE 4.4: NUMBER OF FULLERS AND DYERS FOR EVERY WEAVER 1350–1500

Town	*No. of fullers*	*No. of dyers*
Bury St Edmunds	0.46	0.71
Hadleigh	2.94	1.89
Ipswich	1.08	1.42
Lavenham	0.58	0.14
Long Melford	1.91	0.45
Nayland	2	1.5
Stoke-by-Nayland	0.78	0.56
Stowmarket	0.56	2.22
Sudbury	0.59	0.38

Source: TNA CP 40.

the case in Bury St Edmunds and may also have been so in Lavenham. We have already met Thomas Parker who made vesses, Henry Galyot who probably called himself a 'vesser' and William Rysby who used settcloth slays. All three came from Lavenham and all three may have produced worsteds as well as woollens.

The best explanation is the local availability of flowing water so essential to both fullers and dyers. In Lavenham the unusually high water table feeds springs and wells, but does not float away the detritus of cloth-making.[234] In Hadleigh the Brett runs conveniently west of, and close to the High Street, providing a ready source of flowing water, and is a far more mature river than it is in Lavenham (Plate 4.9).

Long Melford lives up to its name by being one of the longest towns, for its size, in England and abutting an unusually lengthy stretch of river along both the Stour and the Chad Brook. Following test pitting in 2011, Access Cambridge Archaeology reported that:

> The most notable pattern observable in the data from Long Melford, however, is that of the 15th and 16th centuries: the volume of pottery of this [period] recovered is extremely high, from all parts of the present village. All bar eight of the excavated pits produced material of this date, many more than in the 12th–14th century, and in nearly all of the pits, considerable quantities of pottery was recovered.This is in stark contrast to nearly all other University of Cambridge CORS in the eastern region where test pitting has been carried out, where

234 Personal communication with Jane Gosling of Lavenham Guildhall, 2 September 2015.

> the pattern is more commonly of a considerable drop in the volume of pottery recovered. In the later medieval period, Long Melford really does seem to have been as long as it is today. It is well known that Long Melford was somewhere which derived considerable wealth from the cloth-making industry in the later medieval period, but to see this prosperity reflected so very strongly in the excavated data, which reveals a vibrantly thriving and expanding settlement, is striking – especially as this trajectory appears to be so very different to most other rural communities in the eastern region of England.[235]

It is almost as if the town stretched to its present length in order to accommodate all its fullers who, by the end of the century, outnumbered those anywhere else in the country outside London (Appendix B). A rental of 1441/42 for the manor of Melford Hall explicitly describes as fullers five tenants in Hall Street, south of the Chad Brook. By comparing names in the rental with those for the relevant period in the plea rolls, another six or seven fullers can be identified with reasonable confidence, making a total of at least eleven in just one street.[236] Their holdings were tiny, certainly not big enough to support subsistence agriculture, with only three holding more than an acre and none more than two. However, John Barker had two shops on his land and John Waryn the Elder a small tenter-yard. Medieval Hall Street is reminiscent of a small industrial estate (Plate 4.7). Nearly all these fullers' recorded business was conducted with their immediate neighbours in Long Melford, or with residents of towns such as Great Waldingfield, Kersey and Lavenham less than ten miles away.

Suppliers, Customers and Investors

With whom did weavers and fullers do business? (Table 4.5) The plea rolls provide us with clues. As we saw in chapter 3, both crafts numbered among the customers of woolmen. In the 1420s we find two fullers buying substantial quantities of wool. Richard Ram of Ipswich bought 190 stone in his own town for £31 12s. 4d., while William Tayllour of Cavendish bought 114 stone at Sudbury for £19.[237] In each case this equated to about 3s. 4d. per stone. Later, John Hervy, fuller of Sudbury, referred in his will to 'all my wolle and cloth dyed & not dyed'.[238] Weavers, more often than fullers, dealt with London merchants. Although this business was not

235 http://www.access.arch.cam.ac.uk/reports/suffolk/long-melford/2011. See also Beresford and St Joseph (1979), p. 139.

236 SROB J/523.

237 TNA CP 40/656, m. 229v (AALT IMG 1423); CP 40/676, m. 241v (AALT IMG 1449).

238 TNA PROB 11/18/273.

TABLE 4.5: OCCUPATIONS OF PARTIES IN DEBT LITIGATION WITH SUFFOLK FULLERS AND WEAVERS 1350–1500

	Weavers			*Fullers*		
Occupation	*No.*	*Percentage*	*Mean value of claims*	*No.*	*Percentage*	*Mean value of claims*
Clergy (inc. Clerk)	10	7.4	£5 5s. 2d.	14	13.5	£7 13s. 11d.
Clothier	10	7.4	£8 7s. 6d.	5	4.8	£10 8s. 2d.
Cloth-worker	11	8.1	£7 7s. 2d.	10	9.6	£5 16s. 11d.
Husbandman/Yeoman	9	6.7	£7 15s. 4d.	10	9.6	£2 19s. 8d.
Member of London company	13	9.6	£4 17s. 5d.	2	1.9	£6 1d.
Lord/Gentry	28	20.7	£6 13s. 6d.	26	25.0	£4 11s.
Merchant/Mercer	20	14.8	£4 7s. 11d.	10	9.5	£8 8s. 10d.
Other	29	21.5	£6 10s. 8d.	23	22.1	£5 18s. 2d.
Woolman	5	3.7	£6 18s. 8d.	4	3.8	£6 15s.
Total	135			104		

Source: TNA CP 40.

particularly lucrative, there were a few high-value claims. The London grocer, Edmund Maupas, sued the Bury weaver Thomas Wode for £16 16s. 8d., probably for payment for the supply of dyes.[239]

Weavers and fullers commonly dealt with lords/gentry, many of whom lived in the same, or nearby town or village to the cloth-workers. This may reflect sales of their cloth, or alternatively investment by the higher orders in the industry. The Long Melford weaver, Robert Talbon, defended a claim for £6 by John Clopton Esquire, very probably the owner of Kentwell Hall and a major benefactor of the magnificent parish church.[240] Both John and William Clopton sued Long Melford fullers.[241] Trade with husbandmen and yeomen, who would also have figured prominently among end-users of the cloth, was much more limited. Perhaps they were more likely to buy from clothiers and drapers than direct from cloth-workers.

Within the textile industry itself, the highest value transactions of both weavers and fullers were with clothiers, although this business gave rise to only a small

239 TNA CP 40/756, m. 28r (AALT IMG 58).

240 TNA CP 40/853, m. 196r (AALT IMG 382); Dymond and Paine (2012), pp. 14–20, 193.

241 TNA CP 40/837, m. 85r (AALT IMG 140); CP 40/883, m. 290v (AALT IMG 539d); CP 40/931, m. 283r (AALT IMG 577f).

fraction of their total disputes. Of their dealings with fellow cloth-workers, more than half were with dyers. Only one dispute between a weaver and a fuller was recorded. This suggests either that direct trade between the two crafts was surprisingly limited, or that any trade was of low value and below the 40s. threshold. Quite possibly, when operating independently but on a modest scale, both weavers and fullers completed in-house all stages of cloth-making except perhaps dyeing.

In 1415 John Roundes of Boxford deserted his position as a weaver to work for the ploughman Nicholas Brounyng. Had he been incompetent, his former master, William Kyng, would not have taken the steps he did to force John back to the loom.[242] One explanation for John's flight is that he had found life as a weaver involved too much hard work and saw a return to the land, strolling in the fresh air behind ox or horse, as the easier option. With the exception of the horizontal loom, that had been widely adopted, technology had made only limited impact on the textile industry by 1500. Cloth-making was by any standards hard labour. The spinster walked many miles to turn her spinning wheel. The weaver endured long hours in the back-breaking tasks of throwing the shuttle and beating the weft. The fuller spent his day wading up to his waist in noxious liquids and inhaling vapours that must have slowly crippled his respiratory system. Theirs is the human story behind the success of the Suffolk cloth industry. Without their toil there would have been no woollens or worsteds to wear, and no export earnings to fill merchant coffers.

If they were lucky and stood the pace then, at least until the final third of the fifteenth century, weavers, fullers and even some spinsters could enjoy a degree of prosperity as independent operators. They turned out a variety of types and sizes of cloth. After 1470 things became more difficult. Bed weavers in Bury St Edmunds continued in business under the protection of their guild. Elsewhere, an increasing number of craftsmen and women became outworkers of rich clothiers. They might earn a little more each day than an agricultural labour, but few acquired riches. Furthermore, many fullers lost their jobs as mill fulling mechanized their function. Instead of an eclectic range of cloths, a majority of them produced higher quality, heavier fabrics for overseas markets. Although the law required otherwise, much of this cloth exited the country in the unfinished condition in which it had left their hands.[243] The remainder was finished by dyers and shearman in the manner explained in the next chapter.

242 TNA CP 40/616, m. 407r (AALT IMG 813).

243 A statute of 1487 that prohibited the export of unfinished woollens worth more than 40s was more often honoured in the breach, particularly by London mercers: *Statutes*, 2, p. 520; Sutton (2005), p. 336.

– 5 –

Cloth Finishing

Robert Gardener was a member of one of the most distinguished cloth-making families in medieval Bury St Edmunds, albeit probably only of a cadet branch. His forebears Thomas (died 1420) and Thomas's son Roger (died 1440) had been dyers before him and he too was in business as a dyer.[1] In September 1476 he took the road to Chelmsford in Essex where, on the twentieth day of that month, he met and traded with John Kyng a prominent member of the London company of grocers.[2] He bought a cask of wine called 'a botte of wyne sek', by which was meant the fortified white wine of Spain much loved by Shakespeare's Falstaff. That was probably for his personal consumption. More pertinently, he bought for his business '100 pounds of cere, 200 pounds of madder, a barrel of black soap [. . .] and three bales of woad'.[3] The Latin term 'cere' can be translated as silk, but in this context must mean arsenic – sometimes used by dyers insensible to its lethal side effects. This instance teaches us much about medieval dyers. At least in England, their principal raw materials were madder and woad, and their trade in woad was much the bigger of the two. They used soap as a cleansing agent. They were ready to employ other materials, even arsenic, that might give their coloured cloth an edge over that of competitors.[4] In this chapter, we first consider these dyes and dyestuffs in more detail, before looking at the art of the dyer, their distribution around the county, their trade with Londoners, and their relationship with other suppliers and customers. The chapter closes with an examination of shearing and shearers.

---

1 SROB 152, 250 Osbern.

2 TNA CP 40/883, m. 213v (AALT IMG 387 dorse).

3 Rigby stated that a bale was 'a quantity which varied considerably in weight but which was often around 4–5 [hundredweight]' and that, in the case of woad, a hundredweight equated to 112 pounds: Rigby (2005), pp. 259–60. In his will William Tyll of Sudbury referred to a bale weighing 3 hundredweight, see below.

4 Italian dyers used at least fourteen different dyes and mordants in the fourteenth century: Edmonds (2003), p. 30

Dyers and Dyeing

Dyes and dyestuffs

Dyes were nearly all made from plants, lichens, insects or molluscs until the invention of aniline dyes in the mid-nineteenth century.[5] The principal medieval dyes were woad and madder, and several others were used that together turned cloth a wide variety of hues. Indeed, in his London shop the draper Thomas Gylbert sold cloth in no less than eighteen different colours and shades.[6] In addition, dyers used a variety of mordants – metallic solutions that are easily absorbed by wool and act as a form of adhesive. Molecules in a mordant cling to the surface of the fibres while simultaneously attaching and holding fast to molecules in the dye. They were essential in fixing most dyes other than woad and those extracted from lichen.

Before examining woad and madder, let us look briefly at some less well documented dyestuffs. Weld, known colloquially as 'dyer's rocket', is a flowering plant that grows abundantly in England. Its chopped leaves and flower stems are rich in the chemical luteolin and produce a brilliant yellow dye. The mulch was probably dried out so that it could be used throughout the year.[7] Weld was one of the three most common dyes, along with woad and madder, mentioned by Geoffrey Chaucer in his minor poem *The Former Age*.[8] When cloth is dyed with woad and then weld it emerges in a Lincoln green colour. The use of weld in Suffolk may be illustrated by the case of John Talbot of Hadleigh who, in November 1411, sold to the Bury draper John Hoog two green pieces of cloth, as well as four red and two blue.[9] Kermes (otherwise known as grain) is a red dye derived from the dried bodies of female shield lice that live in the Mediterranean region on the sap of the Kermes oak tree. Grain was used to colour scarlet, the most prized of all woollen cloth that was taxed particularly heavily. The Suffolk alnage accounts for the 1390s make it clear that none of the cloth produced then was scarlet, something that would not change during the fifteenth century.[10] The name of Brazil, the world's fifth largest country, derives from brazilwood, another source of red dye, which the Portugese found there. Brazilwood had previously been sourced from the Far East, but, with the fall of Constantinople in 1453, the trade had been disrupted and it had ceased to be available to English dyers. A reference to 'brasyle' in the late-thirteenth century customs of Ipswich quay means that it was known to the town authorities, but,

---

5 Crowfoot et al (2001), p. 19.
6 C. Dyer (2005), p. 144.
7 Personal correspondence from Jean Rogers 12 October 2015.
8 http://www.bartleby.com/258/62.html
9 TNA CP 40/616, m. 181r (AALT IMG 361).
10 Carus-Wilson (1987), p. 618; Spufford (2002), p. 333; Munro (2003), pp. 212–6; TNA E 101/342/8.

owing to its high cost and scarcity even before 1453, it would have been used only sparingly in late medieval Suffolk's cloth industry.[11] Vermillion was another scarce red dye that was mentioned in a mid-thirteenth century Ipswich murage grant and derived from red crystalline mercuric sulphide found on the shores of the Red Sea.[12] Those who wanted the most expensively dyed cloth had to look to London producers and retailers. When Sir Roger Chamberleyn of Gedding paid the colossal sum of 53s. 4d. for four yards of violet cloth, he bought it from Ralph Holand, a tailor who carried on business in the parish of St Mary Aldermary, in the ward of Cordwainer Street, in the heart of the City.[13]

Woad is a flowering plant the leaves of which provide the raw material of indigo dye (Plate 5.1). These leaves were stripped from the plant, macerated, and the pulp thrown into small heaps and allowed to ferment and drain. The pulp was then hand-kneaded into balls 2 to 6 inches in diameter and left for up to four weeks to dry on airing racks. The next stage was the critical one of couching. The woad balls were crushed into powder, piled into a heap up to a metre deep, sprinkled with water and left in woad-houses to ferment for weeks if not months. During this period the heaps were regularly turned and raked. Once it had fermented into a dry, dark clay-like substance, the woad was ground up, sifted and packed into barrels. As a result of this lengthy process of fermentation, the clay contained twenty times as much indigo as the same weight of leaves.[14] References to woad-houses in medieval Suffolk are not unknown. They were probably used for storage of the barrels and also for making the dye. The Bury clothier Walter Copynger bequeathed one to each of his sons Robert and Thomas, while John Halle of Ipswich left one to his son William.[15] In 1454 Thomas Benalde of Hadleigh left a woad-house to his wife Joan for life which, on her death, reverted to their son Thomas. The son no doubt made good use of his inheritance in pursuing a career as a clothier. He presented twenty broadcloths to the alnager in 1467/68, and another four in 1468/69. He was still in business and travelling to London in the early 1480s.[16]

Woad was the main source of blue dye and could also be used as the base for other dyes to produce various colours. Dyeing with woad and then madder produced a black, or, in old vats, a green colour.[17] Skilful use of the two dyes could also produce various olives, reds, purples and violets. For these reasons it became

11 Spufford (2002), pp. 333–4; Twiss (1873), p. 189.
12 Carus-Wilson (1967), pp. 218–19.
13 TNA CP 40/776, m. 416r (AALT IMG 821).
14 Hurry (1930), pp. 24–7; Crowfoot et al (2001), p. 19; Edmonds (1998), pp. 13, 24–5.
15 SROB 182 Pye; NRO NCC, Reg. Popy, f. 275.
16 TNA PROB 11/4/23; TNA E101/343/4–5; TNA CP 40/891, m. 406r (AALT IMG 724).
17 Clark (1972), p. 118. 'Dyes would normally not be mixed, but the cloth would be immersed in a series of vats': Edmonds (2003), p. 30.

'"the universal dye of the Middle Ages" of which more was required than all the other dyes combined'.[18] The value of woad imported into Ipswich between 1459 and 1465 was more than four times the value of madder.[19] Woad was big business, partly because the fashion of the fifteenth century was for darker clothes.[20] In his will the prominent Bury draper, John Odeham, distributed numerous black gowns among family and friends. In Ipswich Margaret Fastolf owned a gown of murrey, and Joan Freborne a gown and tunic both of violet cloth, as well as a gown of musterdevillers (a shade of grey), while Alice Pipho left one black gown and another of musterdevillers.[21] One in ten Suffolk dyers called themselves woaders, including John Halle and Thomas Benalde. They used woad in their own dyeing businesses, but also probably sold it on to other dyers working on a more modest scale.

In the woad trade supply side factors were as important as demand. Although some woad had been grown in England since classical times, its domestic cultivation during the Middle Ages has been described as 'desultory' and most was imported.[22] Until 1430 some came from Picardy, but the main source was south-west France particularly Toulouse.[23] The Mediterranean climate allowed the woad to dry quickly and minimised the risk of the woad balls becoming mouldy or even fly blown.[24] John Howard sold the clothier Robert Sergeaunt and William Howet 4 sets of Toulouse woad for £8, while William Tyll of Sudbury instructed his executors to sell 2 bales of Toulouse woad, each weighing 3 hundredweight.[25] A long shelf life and a long-haul voyage to this country, meant that woad could be imported in bulk, and, indeed, needed to be in order to justify the cost and risk of transit. Very substantial loads of woad arrived regularly in the port of Ipswich and the customs of the quay set labour rates for loading, unloading and carrying it into town.[26] In April 1398 14 tons from south-west France worth £70 were unladen from the *Saint Maryship*. Later in the same year John Seman of Hadleigh shipped 8 coupe with a customs value of £5 4s. 1d. on the *Skonausyt*. Following England's loss of Gascony in 1453, an alternative source of woad was found in Hamburg from whence, in 1459, it was carried by Hanseatic merchants down the river Elbe

18 Hurry (1930), p. 49.
19 TNA E 122/52/42, 43, 45, 46, 47, 48.
20 Veale (2003), pp. 140–1.
21 SROB 129 Hawlee; SROI IC/AA1/135, IC/AA2/5, IC/AA2/77.
22 Hurry (1930), p. 60. A view confirmed by Kerridge (1972), p. 20.
23 Hurry (1930), p. 84; Childs (1978), p. 107. The export trade in woad from Picardy suffered a 'profound crisis' in 1430, probably for political reasons, from which it never recovered: Edmonds (1998), pp. 14–5.
24 Edmonds (1998), p. 12.
25 Crawford (1992), II, p. 302; TNA PROB 11/14/577.
26 Twiss (1873), p. 181.

and across the North Sea. That year Henry Vanderheth shipped into Ipswich 14 pipes valued at £42.[27]

Having arrived in this country in bulk, woad was stored in large quantities. Thomas Fulmarde held 7 tons and 4 quarters of woad worth £68 in a granary in Hadleigh. In Ipswich the woad man Andrew Coke had invested so heavily in the dye that he had over stretched himself and his creditors sought to seize his stock worth £128 10s. 1d.[28] The domestic trade was buoyant throughout the later Middle Ages. In 1405 John Cousyn sold 5 quarters in Boxford to John Tylle for £9, and 3 quarters in Hadleigh to John Levyng for £5 13s. 4d. Forty years later the Ipswich merchant John Deken sued the woader, Robert Derehawe of Blythburgh, for payment of £4 6s. 8d. In 1459 Robert Cake the Elder bought in Stowmarket from Nicholas Taverner 2,080 pounds of woad and madder for £16 16s.[29]

When, for political reasons, French woad was hard to come by, dyes were made from various species of lichen, particularly *roccella tinctoria*. Some dyes known as litmus and others as orchil, were used as cheaper substitutes for the production of blue or purple colorant respectively. The necessary chemicals were extracted to make orchil by macerating, soaking and stirring the lichen in a bath of ammonia and allowing the solution to ferment. Human urine would do the trick. They were extracted to make litmus in a similar way except that extra lime, chalk or gypsum was added to produce the blue dye. Unlike fermented woad, orchil and litmus did not have a long shelf life.[30] Peter Garne was one of several merchants involved in the trade in litmus from the Low Countries in the late 1450s and early 1460s. Almost certainly a Lowlander himself, he was resident in Ipswich for several years.[31] At the same time orchil was shipped by Genoese merchants from the Canary Islands.[32] Sixty years later, with the deterioration of relations and the outbreak of war with France in 1522, the orchil trade revived due to disruption of the sea route along the French coast. It provided the foundation for the fortunes of a young John Smith (alias Dyer) of Ipswich who described himself as 'orchil maker' in seeking payment in the court of common pleas from numerous customers in Suffolk and beyond. He went on to become one of the town's most prominent sixteenth-century

27 TNA E 122/193/33; E 122/52/42. A coupe was 'a measure of capacity of uncertain size used for woad', while a pipe was a weight that varied according to the commodity but often half a ton: Rigby (2005), pp. 260–2. Even for the initiated the different weights and measures used for woad are confusing.

28 *CCR 1389–92*, p. 374; SROI C/2/3/1/41.

29 TNA CP 40/576, m. 313r (AALT IMG 645); CP 40/576, m. 308v (AALT IMG 1668); CP 40/736, m. 365r (AALT IMG 716); CP 40/871, m. 217v (AALT IMG 1335). A quarter was a quarter of a hundredweight and, in the case of woad, presumably equalled 28 pounds.

30 Kok (1966), pp. 260, 264; Edmonds (2003), p. 40.

31 TNA E 122/52/42; Amor (2011), pp. 135, 250.

32 Childs (1978), p. 107.

burgesses, serving as bailiff and MP.[33] Although lichen dyes could be used without a mordant they tended to fade and were often frowned upon by the authorities. From 1483/84 the use of orchil was prohibited unless and until the cloth had been dyed with woad or madder.[34]

Madder was the main source of red and russet dye and, in the Middle Ages, was second in importance only to woad.[35] The crucial chemical, alizarin, was extracted from a flowering plant, dyer's madder, but in this case from the roots rather than the leaves. Following harvest in September, the plant was dried before a fire, its roots crushed, and the powder from the tough inner part of the root then used for dyeing.[36] Alizarin could produce both red and yellow dye, or a mixture of the two, depending upon the solution in which it was dissolved. In alkaline water it produced a bright brick-red colorant, in acidic water a bright orange one and in boiling water a brown one.[37] Madder was grown in small quantities in Suffolk, for instance at Great Waldingfield, Mildenhall and Clare. In 1336/37 the reve of Clare generated income of £1 3s. 6d. from its sale.[38] Most, however, came from France or the Low Countries, in particular from Flanders and Zeeland, and was variously described as cropped, coarse or powder.[39] Perhaps because it was not in such demand as woad, or perhaps because the distance it had to travel and the risks that had to be taken in transit were so much less, madder does not appear to have been bought and sold in bulk quantities. It generally arrived in Ipswich in smaller and less valuable consignments. Peter Garne imported 3 bales valued at £6.[40] Evidence of a domestic trade in the dye in late medieval Suffolk is sparse. Robert Cokerell sought payment from William Sparhauk of £1 for the purchase of 200 pounds of madder, the amount that a dyer might reasonably require for the purposes of his own business.[41] Without doubt, madder was saleable and, indeed, gave its name to one of the principal markets of Norwich.

33 TNA CP 40/1046, mm. 88v, 487r (AALT IMG 163 dorse, 989 front); http://www.historyofparliamentonline.org/volume/1509-1558/member/smith-john-149899-1571.

34 *Statutes*, 2, pp. 487–8. Kok suggested that using orchil with other dyes did not improve its fastness: Kok (1966), p. 267.

35 Crowfoot and her team subjected 351 samples of medieval textiles from London to dye analysis and found traces of madder in far more samples than traces of woad. She proposed the explanation that 'weak dyeings with woad fade and are difficult to detect after long-term burial': Crowfoot et al (2001), pp. 199–201.

36 Edmonds (2003), p. 38; Kerling (1954), p. 123.

37 Edmonds (2003), p. 44.

38 *Nonarum Inquisitiones* (1807), p. 103; Ward (2014), p. 106.

39 Kerling (1954), pp. 122–5.

40 TNA E 122/52/42.

41 SROI C2/3/1/48 (24.09.1415).

Nevertheless, at least in Suffolk, it was not a tradable commodity in quite the same way as woad.

Unlike woad, madder and weld lack the necessary chemical qualities to fix on wool without the use of a mordant. In medieval cloth-making, alum, a double sulphate of aluminium and potassium found in the mineral kalinite, was by far the most important mordant, although green copperas, an iron sulphate, was an alternative for use in production of darker shades.[42] Alum was mined in Asia Minor and shipped to western Europe by Italian merchants. With the fall of Constantinople the supply temporarily dried up and it became scarce until, in 1462, a fresh source was discovered in the Papal States north of Rome.[43] By circumventing a papal monopoly, Henry VII was to make a fortune from an illicit trade in alum in the early-sixteenth century.[44] Both alum and copperas were mentioned in the customs of the Ipswich quay, small quantities of both were imported into the town in the late 1390s, and alum was imported, again in small quantities, in 1459.[45] At the same time as Robert Cokerell sought payment from William Sparhauk for madder, he also claimed from him 10s. for 100 pounds of alum.

Potash was another substance central to the dyer's craft. Comprised of a potassium salt and produced by leaching the ashes of trees or other vegetable matter to remove the charcoal, it was used to degrease wool in readiness for dyeing and also, as an alkali, to dissolve the indigo in the woad vat.[46] In 1400 in the wood pasture region of High Suffolk timber was still relatively plentiful, so avoiding any need for imported potash. In the late 1390s none arrived by sea in Ipswich. Conversely, between 1459 and 1466 nine consignments, almost certainly from the forests of eastern Europe, were unloaded in the town.[47] John Howard sold the clothier Robert Sergeaunt eight 'barelles of wode ashes to receive' at the port of Harwich.[48] Quite possibly, by the mid-fifteenth century Suffolk's domestic supply of ashes had been depleted by the demands of shipbuilding and of the textile industry itself. Nevertheless, in 1480 Thomas Ille of Bildeston still earned a living as an ash burner and his occupation remained a viable one in the county until at least the seventeenth century.[49]

---

42 The nature of the iron in the mordant determines the shade because mordant dyes are polygenic: Clark (1972), p. 119.

43 Spufford (2002), p. 334.

44 Penn (2011), pp. 201–4.

45 Twiss (1873), pp. 187, 189; TNA E 122/193/33; E 122/52/42.

46 Kerridge (1985), p. 165; Edmonds (1998), p. 7; Munro (2003), p. 211.

47 TNA E 122/52/42, 43, 45, 46, 48.

48 Crawford (1992), I, p. 370.

49 TNA CP 40/871, m. 214r (AALT IMG 398); Kerridge (1985), p. 165.

The Art of the Dyer

William Tyntewell of Stowmarket was very attached to his *Maderyngled,* so much so that it was the only item that he specifically mentioned in his will.[50] Such attachment was understandable as this was the vessel in which he prepared madder for dyeing and so earned his living. He did not name a beneficiary for this precious belonging, but it may well have descended to his kinsman, perhaps son, Robert Tyntewell who, in the 1480s, was in business as a dyer in the town.[51] The art of the dyer was a skilled and demanding one. The selection, purchase and preparation of dyes, of good quality and at a fair price, were only a few of many tasks. Dyes were expensive, accounting for between 10 and 18 per cent of total cloth production costs.[52] So, the amounts of money involved were substantial, far more than was the case with weavers or fullers, and a wrong decision could be very costly. Nearly a third of all dyers' disputes were for sums of £10 or more (Table 5.1).

The dyer had to travel many miles, much further than weavers or fullers, to meet his suppliers and customers (Table 5.2). Only a small fraction, less than one in eight, were immediate neighbours, at least half lived more than twenty miles away. As discussed below, more than one in four were with London merchants. On return from his travels, the dyer had to fetch, store and dispose of many gallons of water which explains why so many based their operations close to rivers. In Bury St Edmunds they congregated, from as early as the thirteenth and into the fifteenth century, in the north-east of the town close to the Lark around Eastgate and Northgate, making use of the running stream known as 'The Spanne'.[53] In Hadleigh John Garleberd was granted an exclusive right to wash his dyed cloth along the bank of the millpond. William the Dyer rented land for his work on the Tyefield beside the river Brett, Vincent the Fuller rented a dye-house nearby, and William Geffrey built another dye-house near the river to the south of the town in Benton.[54] Thomas Benalde's woad house was probably on land known as Clobs Croft close to the mill-race that served Toppesfield Mill.[55]

Having collected water, the dyer had to gather many armfuls of wood, kindle and sustain a fire.[56] He then had to stand for long hours in a dyeing house over his vat, stirring the hot noxious liquid and judging just how much dye to add and when

---

50 SROB 290 Baldwyne.

51 TNA CP 40/888, m. 5r (AALT IMG 12).

52 Munro (2003), p. 216.

53 SROB 250 Osbern (Roger Gardener); 208 Hawlee (John Gerad); 328 Hawlee (William Lyenge); 19 Pye (William Honybourn); V. Redstone (1908), p. 198.

54 CCA-DCc-CR/17 (courtesy of Margaret Woods).

55 Personal correspondence from Sue Andrews 1 August 2015.

56 When dyeing with madder, 'good colour can also be achieved without the use of heat': personal correspondence from Jean Rogers 12 October 2015.

TABLE 5.1: VALUE OF DEBT LITIGATION OF SUFFOLK DYERS AND SHEARMEN 1350–1500

	Dyers		*Dyers' disputes with Londoners*		*Shearmen*	
Value of claim	*No. of claims*	*Percentage of claims*	*No. of claims*	*Percentage of claims*	*No. of claims*	*Percentage of claims*
£20 or more	34	9.6	6	12.8	1	2.6
£10 or more but less than £20	71	20.1	15	31.9	4	10.5
£5 or more but less than £10	74	21.0	13	27.7	5	13.2
Less than £5 but more than 40s.	68	19.3	8	17.0	7	18.4
40s. or less	98	27.8	5	10.6	21	55.3
Unknown	7	2.2	0	0	0	0
Total	352		47		38	

Source: TNA CP 40

best to stop (Plate 5.2). The more concentrated the dye in the vat, the darker the shade, so a skilled dyer could achieve a subtle blending of colour.[57] Although he would not have known it, the woad dyer in particular relied on a good working knowledge of organic chemistry. He 'worked with a small number of dyes using recipes of great complexity involving chemical reactions only understood in recent years'.[58] After filling the vat with *eau-sure* – a mixture of water, grated madder and bran – he had to maintain a constant heat of about 50° C. By judiciously adding potash, he kept the level of alkalinity in the solution steady for thirty hours over a period of two to three days. Years of experience taught him to test this by feel, smell, taste and sample dyeing. Only in this way would the indigo reduce to its soluble form, flow out from the sludge at the bottom of the vat as a green ink, fix to the cloth and finally turn blue when removed from the liquid and exposed to the oxygen in the air.[59] The ultimate shade of blue depended on the concentration of the dye. Blues were twice as dark as azures which, in turn, were twice as dark as plunkets.[60] With time and financial success, the dyer could afford to hire help to

57 Betterton and Dymond (1989), p. 33. An orchil bath for wool or cloth was prepared in a similar way: Kok (1966), p. 264.

58 Ponting (1971), p. 11.

59 Edmonds (1998), pp. 7, 23–6; Munro (2003), p. 211. For a fifteenth-century Italian recipe for the woad vat: Edmonds (2003), pp. 51–6.

60 Kerridge (1985), p. 17; Betterton and Dymond (1989), p. 33.

TABLE 5.2: DISTANCE BETWEEN PARTIES TO DEBT LITIGATION WITH SUFFOLK DYERS 1350–1500

Distance	*No. of claims*	*Percentage of claims*	*Mean value of claims*
Same town or village	22	12.4	£5 2s. 11d.
Less than 6 miles	23	13.0	£10 15s. 2d.
More than 6, but less than 10 miles	9	5.1	£10 5s. 2d.
More than 10, but less than 20 miles	31	17.5	£9 15s. 3d.
London	47	26.6	£12 8s. 2d.
Norwich	21	11.9	£9 11d.
Overseas	3	1.7	£9 14s. 5d.
Other more than 20 miles	21	11.9	£8 13s. 5d.
Total	177		

Source: TNA CP 40

perform the more menial tasks, but such help was not always reliable. William Middilton employed the woad worker Edmund Bele in Hadleigh until Bele absconded without completing his term of service.[61] The life of the dyer was never an easy one.

Dyeing could be done at any one or more of four stages of cloth production: in the wool; in the yarn; before fulling, in the say; or after fulling and first shearing, in the piece.[62] Cloth was often dyed at more than one stage, most frequently with a woad base in the wool, and another dye later in the piece to give it the requisite colour. If dyed in the yarn the hanks were suspended from a pole just above the surface of the vat and moved manually around the pole and through the solution so that the yarn was dyed uniformly.[63] If dyed in the piece, the cloth was then passed through ringers and hung on racks to dry.[64] In his will Stephen Gardiner of Bury St Edmunds referred to his woad vat and his cloths hanging to dry in his hall.[65]

In Suffolk there were good reasons for dyeing at an early stage. The indigo dye in processed woad could not easily penetrate woven cloth. The county's hard water made achieving a uniform shade problematic when dyeing in the piece.[66] If dyed in

61 TNA CP 40/717, m. 362r (AALT IMG 729).

62 Generally, woollens were dyed more often than worsteds. However, 'Blewe worsted' was sold in London in 1440: TNA CP 40/736, m. 326v (AALT IMG 1682).

63 Clark (1972), p. 122.

64 Munro (2003), p. 212.

65 TNA PROB 11/6/76.

66 Suffolk water is hard because chalk is the county's dominant underlying rock: Wymer (1999), pp. 16–7.

the wool any unevenness could be mitigated in the subsequent carding, spinning and weaving.[67] Finally, some woollens known as medleys and rays could not possibly be dyed after weaving because in medleys the warp and weft were of two different coloured wools, while in rays the cloth was striped in different colours.[68] In some cloths three colours of yarn were used, one for the warp, another for the main weft and a third for the pattern weft.[69] If dyed in the wool, a woad base was preferred because this did not require the addition of a mordant that would have made the wool less supple and more difficult to work.[70] Bequests of wool were uncommon and references to its colour even more so, but those references were usually to blue. John Place of Lavenham owned white and blue wool and yarn; and William Orderych of Hitcham 13 pounds of blue coloured wool. Every rule has an exception. As well as blue wool, William Eve of Little Waldingfield and James Spring of Lavenham both bequeathed red, in Eve's case 'madered'.[71]

For those who succeeded, particularly those who were able to combine their craft with a trade in dyes, the rewards were considerable. Indeed, the dyer might well be regarded as the prince of artisans.[72] William Honybourn of Bury St Edmunds was one such, transacting high-value business with the London grocer John Peyntour. When William died in 1493 he left a house full of treasured possessions that included silver, pewter, bedding, furniture, clothing, utensils for personal washing and hygiene, and a well-equipped kitchen. Among the choice items were feather beds, silks, beads of jet garnished with coral paternosters and embroidered cushions. He generously remembered his wife Katherine, his son Thomas and his daughter Agnes, but his other daughter Annis was the apple of his eye and received the largest share.[73]

Successful dyers were rewarded not only with riches, but with high status. In Ipswich, among fifteenth-century clothworkers, only dyers and shearmen are ever recorded as acquiring freedom of the borough.[74] We can trace the rise of John Halle who became a freeman in 1477, a chamberlain in 1479/80, a member of the

67 Kerridge (1985), pp. 17, 163; Edmonds (2003), pp. 35, 38.

68 Chorley (1987), p. 374. Bridbury (1982), p. 73; Crowfoot et al (2001), p. 18; C. Dyer (2005), pp. 144–5. The customs of Ipswich quay refer to rays: Twiss (1873), p. 197. William Eve bequeathed medley cloth to his servant John Fen: SROB 333 Baldwyne.

69 Crowfoot et al (2001), pp. 53–4.

70 Munro (2003), p. 211.

71 SROB 24, 333, 499v Baldwyne; McClenaghan (1924), p. 11.

72 Carus-Wilson described dyers as being 'among the ruling class of the rapidly growing towns' of the twelfth and thirteenth centuries: Carus-Wilson (1967), p. 227. In Exeter in 1377: Kowaleski (1995), p. 128; and in York in the period 1390–1520: Swanson (1989), p. 157 dyers were the wealthiest of cloth-workers.

73 TNA CP 40/907, m. 79v (AALT IMG 986); SROB 19 Pye.

74 BL, Ms Add. 30,158.

borough council by 1480, a portman in 1488 and a bailiff in 1489/90 and again in 1495/96.[75] In Bury St Edmunds Andrew Starbott was appointed by John (alias Jankin) Smith as one of the original Guildhall Feofees in 1470, and served as an alderman of the town.[76]

Even if only a few dyers ever scaled such heights, most remained in business on their own account due, perhaps, to their rare skill and experience. Spinsters, weavers and fullers were remembered in clothiers' wills because they had been drawn into dependency upon them as out-workers. Suffolk dyers were not, suggesting that most had retained their independence, if not always their prosperity. And, when Shakespeare came to write about the Suffolk uprising of 1525 against the king's subsidy, the duke of Norfolk reported the desperation of the 'spinsters, carders, fullers, weavers' who were dependent on the rich clothiers, but made no mention of dyers.[77]

The Distribution of Dyers

Two hundred and seventy five dyers were recorded in the plea rolls, stirring their vats in sixty-three Suffolk parishes (Map 5.1). The distribution of weavers and fullers within the county, and the distribution of dyers were significantly different (compare Map 5.1 with Maps 4.2 and 4.4). Nearly three quarters of weavers and nearly two thirds of fullers lived in Babergh and Cosford, while only just over a third of dyers did so. Lavenham is famous for its blue cloth and there is evidence of dyeing in the town. For instance, Roger Crytott owned two dye-houses and, later, John Fermer owned both a dye-house and, perhaps for the storage of potash, an ash house.[78] Nevertheless, while forty-three Lavenham weavers were cited in the plea rolls, only five dyers were named in total and just one in the rolls of twelve terms between 1480 and 1500. In Long Melford thirty-one fullers, contrasted with only ten dyers in total and just three between 1480 and 1500 (see Table 4.4). Weavers were recorded in twenty-three Babergh parishes, dyers in thirteen. Slightly later, the 1522 Military Survey of Babergh records 112 weavers and forty-one fullers, but only thirteen dyers, few of whom prospered.[79]

The number of dyers in Suffolk was in decline from the mid-fifteenth century, but particularly so in Babergh (Figure 5.1). Between 1450–70 and 1490–1510 the proportion resident in the hundred halved from 27 to 13 per cent. This may be a quirk of the records, but various more credible explanations present themselves.

75 BL, Ms Add. 30,158, ff. 34r, 37r; *Annals* (1884), pp. 142, 157, 159, 169.

76 Statham, pp. xx, 283–4; Lobel (1934), p. 28.

77 Henry VIII, Act 1, Scene 2.

78 SROB 130 Pye; TNA PROB 11/21/251.

79 Pound (1986), p. 133.

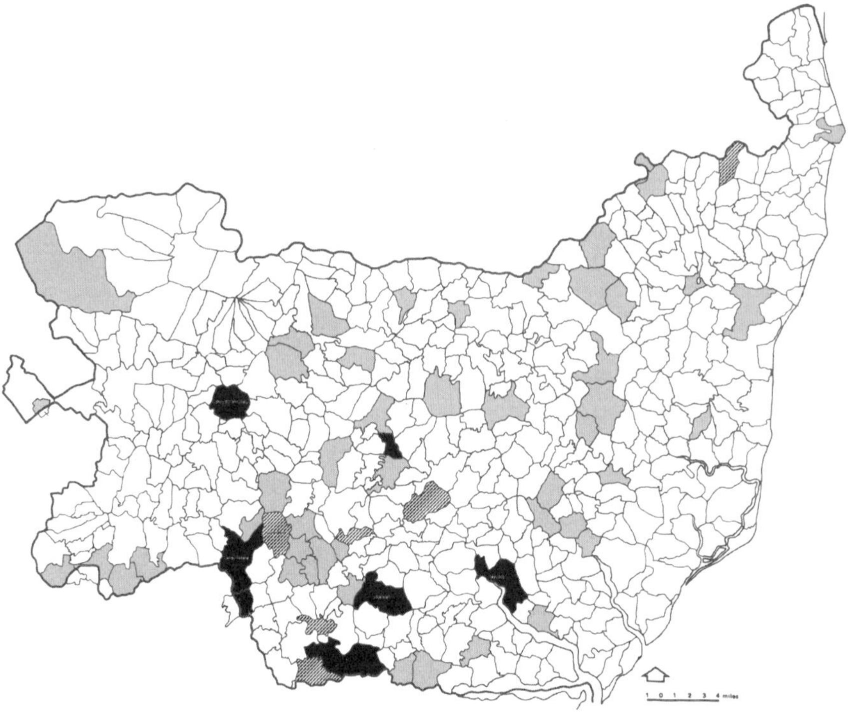

Map 5.1: *Suffolk Dyers 1350–1500*

Notes: Grey marks a parish in which between one and four dyers were recorded. Hatched dark grey marks a parish in which between four and nine dyers were recorded. Black marks a parish in which not less than ten dyers were recorded.

Source: TNA CP 40.

Map format attribution: David Dymond and Edward Martin.

Firstly, Babergh dyers may have lost their independence and so disappeared from the rolls of the court of common pleas. As well as his dye-house and ash-house, John Fermer owned a tenement in Lavenham in which lived 'great John the dyer', perhaps his servant but certainly someone who could not afford a place of his own. Among great John's fellow townsfolk in 1522 were Robert Russhebroke and Jamys Deynes both dyers who had 'no substaunce in goodes'.[80] However, this explanation sits uneasily with the evidence above that dyers maintained their independence. Secondly, many of the more successful dyers chose to call themselves clothiers, including John Fermer himself.[81] Thirdly, at least some Lavenham cloth was dyed

80 Pound (1986), p. 81.
81 TNA CP 40/911, m. 218r (AALT IMG 423).

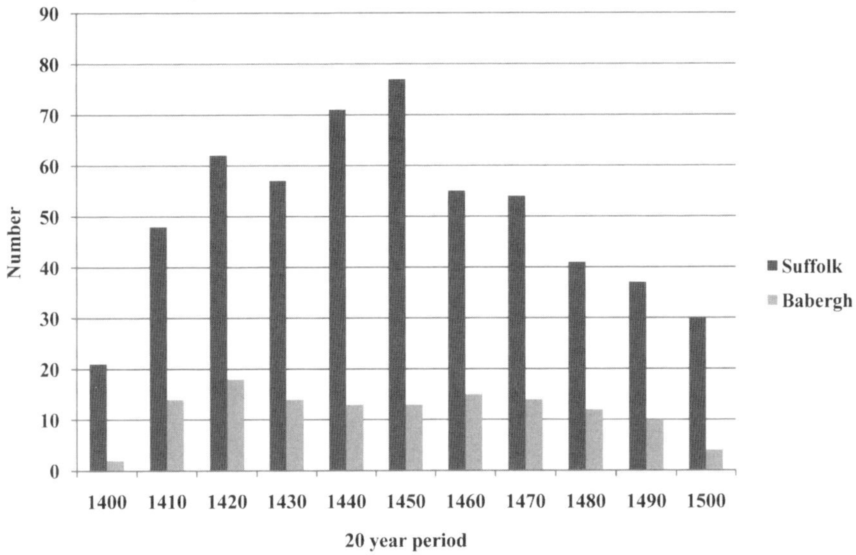

Figure 5.1: *Number of Suffolk and Babergh Dyers*

Notes: Each dark column records the number of Suffolk dyers, and each light column the number of Babergh dyers in the twenty years centred on the given date.

Source: TNA CP 40A

outside the town. In 1470 William Spring was commissioned in Colchester to dye blue 100 pounds of wool on behalf of John Spring of Lavenham.[82] Nevertheless, the most convincing explanation is that the major Babergh cloth-making towns increasingly produced unfinished cloth for export through London to Antwerp, a subject that will be considered further in the next chapter, or sent their cloth to the Capital for finishing. As Oldland stated 'the growing concentration of the cloth trade through London allowed small-town and rural clothmakers to bypass the provincial towns and have their cloth finished in London'.[83]

The Dyers of Ipswich and Stowmarket

Even as cloth production migrated to the countryside in the fifteenth century, provincial towns retained a hold on cloth finishing. Urban craftsmen had the necessary skills, including the remedial ability to disguise any shortcomings in the workmanship of rustic spinners, weavers and fullers.[84] This was true in Bury St Edmunds, Ipswich and Stowmarket.

82 Britnell (1986), p. 165.

83 Oldland (2007), pp. 100, 106–7.

84 Munro (2003), p. 275.

Ipswich was a major haven for the import of dyes and dyestuffs – the value of such trade rose to over £1,000 between 1459 and 1466 before falling away again at the end of the fifteenth-century – and was also home to a community of dyers.[85] In the mid-1400s Thomas Roberd and John Bunt were two of them. Thomas Roberd may have been the son of John Roberd, a successful merchant, who traded with Gascony and presented twenty-two narrow cloths to the alnager in 1396/97. While Thomas does not appear to have ventured like his father on the high seas, he faced high-value claims for £15 and £20 in the court of common pleas. Moreover, he generated sufficient custom to justify taking on John Bowyer as an apprentice. He was penalized for polluting the common well in Brook Street by washing his dyed cloth. As well as being a dyer, he was a dealer in woad. He became a freeman and held minor borough offices. Like most burgesses, in order to generate additional income he was a man of many parts. Between 1415 and 1424 he brewed and sold ale contrary to the assize. He was dead by 1440 when his widow was left tidying up his affairs.[86]

We first hear of John Bunt in 1434 when he was fined for non-attendance at the leet court of south ward. Although he may have been a dyer for a longer period, his career as such can be traced in the records from 1440 to 1460 during which time he faced a series of relatively low value claims in the court of common pleas.[87] His business dealings were not as substantial as those of Thomas Roberd. In 1451 he was in serious trouble with the town authorities and nearly lost his freedom of the borough on account of '*his manifest offences transgressions and derelictions*' – perhaps like Roberd he had polluted the water supply. John brewed ale contrary to the assize, as well as running a guesthouse. Another dyer bearing his surname, Thomas Bunt, may have been a son. Thomas was penalized by the same leet court in 1465 for selling ale contrary to the assize, and in 1475 for fly tipping dung that may have been the waste product of his trade. That same year he became a freeman and thereafter held minor borough office. He was probably in business partnership with John Whetyngstone, another freeman, because the two of them were sued together in the Ipswich petty court on four occasions between 1489 and 1492. The largest claim was for £6 and the final one was made by a London grocer John Smert. Thomas died in 1513 leaving a widow Margaret, but no apparent issue.[88]

---

85 Amor (2011), pp. 118, 179.

86 TNA E 122/93/33; TNA CP 40/696, m. 155v (AALT IMG 1260); TNA CP 40/716, m. 24r (AALT IMG 48); BL, Ms Add. 30,158, f. 5v; SROI C/2/8/1/2, 3, 5, 6, 7.

87 SROI C/2/8/1/8; BL, Ms Add. 30,158, f. 15r; TNA CP 40/716, m. 145r (AALT IMG 289); TNA CP 40/736, m. 82v (AALT IMG 1198); TNA CP 40/796, mm. 243r, 261v (AALT IMG 499, 1362).

88 SROI C/2/10/1/2, C/2/8/1/16; BL, Ms Add. 30,158, ff. 33r, 37v; SROI C/2/10/3/1 ff. 159, 299; SROI C/2/10/3/2 ff. 14, 24, 92; SROI IC/AA7/23.

At the end of the fifteenth century Stowmarket bucked the general trend towards production of unfinished cloth. Compared with three recorded weavers and just one fuller, at least ten dyers were in business there between 1480 and 1500. It is not unreasonable to assume that most of them were contemporaries in competition with each other. The business of dyeing in the town can be traced back to at least 1381 when the '*tinctore*' Stephen Sarle paid 3s. poll tax on behalf of himself and his wife, while his servants Robert and Margaret paid a further 12d. between them.[89] The success of his business can be measured not only by his high tax bill, but also by the high value (£16) of his dispute in 1390 with the prominent Ipswich merchant Gilbert Boulge who, no doubt, supplied his raw material.[90] A generation later Thomas Ingham, who described himself as a woader, was claimant in no fewer than eight disputes in 1415. In four of those cases he pursued payment from yeomen of Debenham, Kersey, Oakley and Wetheringsett for sums of between 40s. and £9. In another two the defendants were dyers of Bury St Edmunds and Harleston in Norfolk to whom he may have supplied woad. Ten years later he, in turn, was defending a claim for £4 by another prominent Ipswich merchant John Deken.[91]

In the third quarter of the century Robert Cake was pre-eminent. He may have been the draper of that name who was mentioned, as early as 1446, in the will of the wealthy widow Marion Fenkele of Gipping as her cousin and husband of her housekeeper Agnes.[92] He next entered the records in 1460 as a dyer, by 1470 was called a draper and ten years later a clothier. In the late 1460s in Stowmarket he presented fifty-six broadcloths and 174 narrow to the alnager, and in Bury St Edmunds another twelve narrow cloths, making him Stowmarket's leading clothmaker and fortieth in the county.[93] He had business with Norwich drapers, suing William Swayn in 1480.[94] He died in 1481 leaving a large family and extensive landholdings. The only clue to his earlier life as a dyer was the woad he bequeathed to his daughter Joan. One of his sons, Robert the Younger, became a clothier and another, Edward, a dyer.[95]

In the final twenty years of the century neither Edward, nor any of the other nine dyers who were his contemporaries, played in the same league as Robert. Only

89 Fenwick (2001), p. 523.

90 TNA CP 40/516, m. 59r (AALT IMG 159); Amor (2011), p. 240.

91 TNA CP 40/616, m. 407r (AALT IMG 813); TNA CP 40/656, m. 342v (AALT IMG 1651); Amor (2011), p. 245.

92 SROB 72 Baldwyne.

93 TNA E 101/342/25; E 101/343/2; E 101/343/4; E 101/343/5.

94 TNA CP 40/796, m. 62v (AALT IMG 967); CP 40/834, m. 214r (AALT IMG 381); CP 40/871, m. 207r (AALT IMG 403).

95 SROB 212 Hervye.

four of them faced a substantial claim in the court of common pleas, and none of the others contested more than 70s. Only one was in dispute with a Londoner and four were litigating with men from Needham Market or Stowmarket itself. Theirs was probably a relatively low value and local business, catering for the residents of the town's immediate hinterland who were sufficiently prosperous to afford their services. By this time the town's more ambitious families, such as the Gowles and the Baldrys, sent their sons to London to learn and practice their trades.[96] Even in Suffolk's two most important centres of dyeing, Ipswich and Stowmarket, the business was no longer a source of great riches.

London Company Men

In contrast to the parochialism of Stowmarket dyers, others were involved in long-distance business with wealthy Londoners (Tables 5.2 and 5.3). Among these more adventurous souls were, in addition to Robert Gardener, his kinsman Stephen Gardener, William Honybourn and Thomas Welham all of Bury St Edmunds; Geoffrey Spryng, John Talbot and John Wynlove of Hadleigh; and William Bolthood, William Danyell and John Juwell of Stoke-by-Nayland. Between the ten of them they transacted recorded business with Londoners worth in excess of £143, the price of, say, 15 tons of woad. Bolthood was the most active of the ten and quite probably the busiest dyer in Babergh hundred. A John Bolthood had been a dyer in Stoke-by-Nayland in the second quarter of the fifteenth century, but was dead by 1455.[97] William may have been his son. We know of eight of his disputes in the court of common pleas between 1465 and 1489, four of them with Londoners and one with Arnold le Stall a Hanseatic merchant who sued him for £11 13s. 4d., a debt that had been incurred in London.[98] Perhaps Bolthood had tried to cut out the City middlemen by buying dyes from a shipper. He may have leased his business premises from Sir John Howard who, in 1463/64, was drawing a very substantial rent of 66s. 8d. from a tenement called 'Bolthoodis'.[99]

In all forty-seven London cases involving Suffolk dyers, the Londoner was the claimant, so presumably the supplier or, at least, the creditor. In nearly 80 per cent of cases the contract had been made in the City. In accordance with the anti-fair policy that prevailed for most of the fifteenth century, members of London's merchant companies did not dirty their boots with provincial soil, but expected their customers

96 Sutton (2010), pp. 168–70.

97 TNA CP 40/776, m. 541v (AALT IMG 2183).

98 TNA CP 40/895, m. 59v (AALT IMG 658). Bolthood was not the only dyer to deal direct with Hanseatic merchants, Thomas Welham of Bury St Edmunds being another: *CPR (1467–77)*, p. 202.

99 Arundel Castle MS G1/3 (see Crawford (1992), p. xlv).

to come to them.[100] Suffolk dyers duly complied, becoming increasingly dependent on them for raw material as the flow of imported dyes through Ipswich abated. Such dependency was symptomatic of London's stranglehold on that trade, as on all overseas commerce, which became ever tighter. The drapers were one source of supply. For example, John Brickles sold thirty bales of woad to John Person of Lavenham.[101] However, at least until the end of the fifteenth century, the drapers' principal interest lay not in dyes, but in buying provincial cloth at Blackwell Hall.[102] The mercers were another company who dealt with Suffolk dyers. Many of them came from East Anglia, so they could build on pre-existing relationships. John Stile was born in Ipswich, in 1478 completed an apprenticeship in London with the mercer John Shelly and was soon dealing with Thomas Reynold, dyer of East Bergholt.[103]

The grocers and, indeed, many of their leading lights, were the merchants most engaged in Suffolk's dye trade. Of the thirty-nine cases involving members of the greater London companies, more than half involved grocers and the mean value of their transactions (£10 11s. 7d.) was significantly higher than those of drapers (Table 5.3). William Venour was lord mayor of London. In the last quarter of the fourteenth century his business empire extended to many parts of England.[104] Among his Suffolk customers were some of the county's leading merchants, such as Geoffrey Starling and Robert Waleys of Ipswich who together owed him the enormous sum of £535 12s. At least two were dyers, namely John Whepstede, probably of Monks Eleigh, who owed him the lesser, but still substantial sum of £120, and John Tylle of Boxford who owed £27.[105] Venour was an exceptionally early venturer into the Suffolk market. Nearly two-thirds of disputes between grocers and Suffolk dyers were in or after 1445 when 'individual advances by the richer members of the Company to provincial dyers increased in size [and they] recaptured in the 1450s much of the market for dyes and alum'.[106] In 1455, for instance, we find Richard Lee, twice lord mayor, and his son-in-law and business partner John Stokes, later an alderman of London, in dispute with John Fuller dyer of Long Melford.[107]

---

100 Nightingale (1995), pp. 396, 472; Barron (2004), pp. 82–3; Sutton (2005), pp. 118, 201. Parliament annulled the anti-fair policy in 1487 so that citizens of London might 'carry all manner of wares to foreign markets': *Rot. Parl.*, vi, p. 402; *Statutes*, 2, p. 518. This annulment made no appreciable difference to the London centred pattern of trade of either dyers or clothiers as recorded in CP 40, although there is some evidence that more London merchants were visiting the port of Ipswich: Amor (2011), pp. 199–200.

101 TNA C 1/70/102.

102 Quinton (2001), pp. 137–8, 144–5.

103 TNA CP 40/891, m. 388r (AALT IMG 692); Sutton (2010), pp. 172–3.

104 Nightingale (1995), passim.

105 TNA CP 40/477, m. 123r (AALT IMG 252); CP 40/516, m. 327v (AALT IMG 817); CP 40/541a, m. 369v (AALT IMG 793 dorse).

106 Nightingale (1995), pp. 477, 485.

107 TNA CP 40/796, m. 24v (AALT IMG 891); Nightingale (1995), p. 498.

Yeomen and Drapers

Suffolk dyers dealt with a wide variety of suppliers and customers (Table 5.3). Their least-valuable business was with fellow cloth-workers. John Tylle pursued two weavers from Hartest and Long Melford, who each owed him 40s, and also three Long Melford fullers.[108] One can imagine that the value of most transactions between dyers and fellow cloth-workers fell below the 40s. threshold for common pleas actions. The small number of disputes between dyers and clothiers is less surprising when one learns that they populated different parts of the county and operated in increasingly divorced commercial arenas. Relatively few richer clothiers sold finished cloth. The mean value of disputes between dyers and yeomen, in which the dyer was almost always claimant, was over 70 per cent higher than those with lords/gentry and, indeed, higher than those with any other occupational group. Here one most clearly finds evidence of a newly affluent class buying medium quality finished cloth for themselves and their families. Edmund Dykeman of Great Wratting was one, dealing with Andrew Starbot and owing him 8 marks at the time of Starbot's death.[109]

Among others in dispute with dyers were five drapers, four of whom were claimants. This litigation probably represented quarrels over the sale and purchase of unfinished cloth. The Bury St Edmunds draper John Odeham sued John Hillary of Hadleigh for 24 marks (£16).[110] Both men were prominent in their respective towns. Odeham was in business for at least forty years up to the date of his death in 1469, with customers as far afield as Barking in Essex, Harwich, Ipswich, King's Lynn, London, Norwich, Sudbury and Thetford. In his final years he presented eighty-four broadcloths to the alnager, but failed to seal and so forfeited stocks of red kersey, as well as blue, green and russet straits.[111] The Hillarys were a dynasty of dyers in Hadleigh for even longer. Robert was first cited in the court of common pleas in 1415, and had died by 1445. Two of his executors, William and John, were family members, probably sons. William was a merchant, and John another dyer. Among John's nine recorded disputes between 1440 and 1460 was one with William Babington, abbot of Bury St Edmunds (1446–1453) who sued him for 20 marks (£13 6s. 8d.).[112] The mean value of these claims was very nearly £16, so, while John's commercial interests were not as far flung as Odeham's, he was clearly a man of substance. John Hillary the Younger was the third generation of family dyers to be cited in 1465.[113]

---

108 TNA CP 40/776, m. 198v (AALT IMG 1505).

109 TNA CP 40/853, m. 196r (AALT IMG 382).

110 TNA CP 40/716, m. 426v (AALT IMG 1891).

111 TNA E 101/342/23; E 101/343/4; E 101/343/5.

112 TNA CP 40/756, m. 247v (AALT IMG 1426); http://www.odnb.com/view/article/54429?docPos=2.

113 TNA CP 40/814, m. 234r (AALT IMG 452).

TABLE 5.3: OCCUPATIONS OF PARTIES IN DEBT LITIGATION WITH SUFFOLK DYERS 1350–1500

Occupation	*No.*	*Percentage*	*Mean value of claims*
Clergy (inc. Clerk)	15	10.1	£5 12d.
Clothier	3	2.0	£7 11s. 1d.
Cloth-worker	18	12.2	£5 2s. 8d.
Yeoman	8	5.4	£14 3s. 8d.
Member of London merchant company	39	26.4	£10 13s. 8d.
Draper	10		£7 11s. 8d.
Fishmonger	1		
Grocer	23		£10 11s. 7d.
Mercer	5		£12 10s. 7d.
Lord/Gentry	13	8.8	£8 6s. 8d.
Merchant/non-London Mercer	28	18.9	£10 10s. 8d.
Other	23	15.5	£7 3s. 4d.
Woolman	1	0.7	£13 6s. 8d.
Total	148		

Notes: The great London merchant companies are as described by Thrupp (1962), p. 43.

Source: TNA CP 40

Shearing

Cloth was ready for shearing once it had dried from its tentering and wet napping, or, if dyed in the piece, from its final dyeing.[114] The figure responsible for this critical task was the shearman. Having secured the cloth with hooks to a curved, padded and downward sloping bench, known as a horse, his job was to raise with teazels and closely clip the nap with giant size, razor sharp shears. He did this by sliding the weighted lower blade slowly down the bench and over the cloth, while using a lever to open and shut the upper blade. In working on the best quality cloth, each side might be sheared three or four times with four to seven nappings per shearing. This labour of love could take five days to complete one broadcloth. Furthermore, worn cloth could be returned for re-napping, so inspiring Shakespeare's lines that 'Jack Cade the clothier means to dress the commonwealth and turn it, and set a new nap on it'.[115] Provided the shearmen did his job well, the

114 The shearing of cloth before it had been 'fully wette' was prohibited: *Statutes*, 2, pp. 486–7

115 Henry VI, Part 2, Act 4, Scene 2, lines 4–6.

surface was left uniform and smooth; and the finished product soft and flexible, indeed almost silky, with a draping quality that caught the buyer's eye.[116] One slip, however, could ruin weeks of work. Leonardo da Vinci may have dreamt of mechanizing this process, but then he dreamt of flying machines too, and the world had to wait several more centuries for either.[117] Once sheared, the cloth was pressed in a wooden vice and was ready for sale.[118]

A pair of blunt-ended shearman's shears, generally more than a meter long and with blades nearly half that length, is illustrated in the Christ of the Trades painting in Hessett church (Plate 5.3). They were prized possessions. Thomas Deye of Ipswich owned a pair worth 10s.[119] Edmund Dameron of the same town bequeathed at least two pairs to his widow Cassandra who, in turn, left one to their son John and another to their servant Harry. The dyer William Wilson of Bury St Edmunds left a pair to his shearman friend Thomas Broun. His contemporary and fellow townsman, the clothier John Gardener, left two 'next the best' pairs to his son John, but reserved for his widow Agnes a right to use them during her lifetime. Outside Ipswich and Bury such references were rare. Of those testators whose wills were copied in the Register Baldwyne, only John Cryketot of Buxhall mentioned shears, leaving two pairs of 'scheremanysscheris' to his son Ed.[120] In east Suffolk the fuller Geoffrey Coone of Mendham left his best pair of shears to one son and the second best to another, while the shearman Richard Bayman of Woodbridge left instructions for the sale of his best two pairs to pay for his funeral and his month's mind – a direction that suggests that they were of significant value.[121] Keeping these shears razor sharp was essential and enabled Robert Fuller to earn a sufficiently good living as a shear grinder to be admitted as a freeman of Ipswich in 1483.[122]

Shearers

Successful shearing produced a smooth finished cloth, and so was a highly skilled art that could make or break fine broadcloth. In Norwich anyone wanting to practice as

116 Carus-Wilson (1987), p. 639; Munro (2003), pp. 210–1; Oldland (2007), pp. 98–9.

117 Leonardo's notebooks illustrated a shearing machine not greatly different from the one eventually invented in the early-nineteenth century: Ponting (1972), p. 172.

118 Munro (2003), p. 211. There is no evidence of calendaring (ie pressing with mechanically powered rollers) in medieval Suffolk, although the technique was known in London from the second half of the fifteenth century: Oldland (2011), p. 197.

119 SROI C/2/10/3/3 f. 146.

120 SROI IC/AA3/197; NRO NCC 148–9 Multon; SROB 284 Baldwyne; 155 Pye; 196 Pye.

121 SROI IC/AA3/197; NRO NCC 148–9 Multon; SROB 284 Baldwyne; 155, 196 Pye; SROI IC/AA7/120; SROI IC/AA8/14.

122 BL. Ms Add. 30,158, f. 41r.

a shearman had first to complete a seven-year apprenticeship.[123] Were the rewards of the craft commensurate with the dedication required? In London they certainly were. In 1400 shearmen earned about 12d. per standard broadcloth, in 1408 the rate increased to 16d. and in 1452 rose again to 24d. From 1482 they enjoyed a monopoly on shearing cloth made in the Capital. In that year one Suffolk shearman did even better, earning 5d. from John Howard for shearing 5 yards of narrow kersey.[124]

Ipswich shearmen were also doing well in the closing years of the fifteenth century. As well as pairs of shears, Edmund and Cassandra Dameron owned a well-appointed home full of bedding, clothes, brass pots and basins, silver spoons and mazers.[125] Another Edmund Dameron, perhaps a grandson, became a burgess in 1500 following an apprenticeship with the wealthy mercer Thomas Alvard.[126] Six shearmen appeared before the borough petty court in the final twelve years of the century, during which time three became freemen. Richard Faryngdon was among them. Admitted in 1490, he subsequently transacted business with some of the town's leading burgesses, defending a claim for £40 by the merchants William Baker and Thomas Drayle.[127] He served as one of the four wardens of the Corpus Christi guild. A little further north, in Woodbridge, Richard Bayman bequeathed a number of precious possessions to various members of his family, and also, to his wife, the contents of his shop that included cloth, madder, cork (alias orchil) and alum. Across in west Suffolk, in Nayland, Thomas Warde was busy at work for at least ten years and, between 1455 and 1465, embroiled in not less than five disputes in the common bench; one for £4, a second for 4 marks and the remaining three for 40s.[128] Fifty years later Richard Davy of the same town was a rare success story whose will was proven in the Prerogative Court of Canterbury. He bequeathed 10s. to his local church for tithes forgotten, as well as his shearing shears and handles to his cousin Thomas.[129] Nevertheless, most shearmen did not fare so well. The demand for high-quality shearing was insufficient to make many fortunes. The value of their disputes in the court of common pleas was low relative not only to dyers, but also to weavers and fullers (Tables 4.2 and 5.1). More than half their litigation was for the minimum sum of 40s.[130] The wills of three

123 Lipson (1921), p. 111.
124 Crawford (1992), II, p. 319.
125 NRO NCC 148–9 Multon.
126 SROI C/2/10/3/5, f. 226.
127 SROI C/2/10/3/1, ff. 220, 236, 289; C/2/10/3/5, f. 74.
128 TNA CP 40/776, m. 472r (AALT IMG 936); CP 40/796, mm. 260r, 261r, 196v (AALT IMG 532, 534, 1233); CP 40/814, m. 238 (AALT IMG 459);
129 TNA PROB 11/18/50.
130 In Exeter in 1377: Kowaleski (1995), p. 128; in York in the period 1390–1520: Swanson (1989), p. 157; and in Babergh in 1522: Pound (1986), p. 133, shearmen were the poorest of cloth-workers. In York in 1517 the shearmen were unrepresented on the common council: Heaton (1920), p. 59.

shearmen resident in Bury survive from the second half of the fifteenth century. None of them left more than 2s. to their local church, nor any other precious items to their family.[131]

Despite their relative poverty, most shearmen probably remained independent operators rather than becoming outworkers.[132] As with dyers they were rarely named as a group among the beneficiaries of clothiers' wills, although Thomas Paycocke of Coggeshall remembered them in his.[133] Some shearmen sought to improve their lot by diversifying into other cloth-making work. In the 1460s Thomas Wysman of Ipswich was also a card-maker. Twenty years later John Wryte of Lavenham and William Chaumberleyn of Bury St Edmunds were both described as 'alias fuller'.[134] Their skills as shearmen would equip them well to carry out wet napping on the tenter frame.

In comparison to other cloth-workers, shearmen were rare beasts – forty-one recorded in the plea rolls, set against hundreds of weavers, fullers and dyers. Eighteen shearmen, nearly half of them, were found in the older established cloth-making towns of Bury St Edmunds, Ipswich and Sudbury where a tradition of cloth finishing lived on. The newer centres of Lavenham and Long Melford were home to just three shearmen, strengthening the argument that those towns concentrated on producing unfinished cloth for the London and export markets. For them by the late-fifteenth century the demands of these markets were paramount and their buyers were not generally interested in cloth that had been finished in the provinces. Intriguingly, a small community of shearmen lived in and around Beccles, as far south as Covehithe and as far north as Blundeston. So distant from the main clothmakers of Suffolk, quite probably they serviced those of eastern Norfolk.[135]

Even more so than weavers and fullers, many dyers and a few shearmen enjoyed a long period of prosperity in the late Middle Ages. So long as finished cloth continued to be shipped by Hanseatic merchants from Colchester and Ipswich, they benefited from this export trade. With his additional interest in the sale of dyes, particularly woad, the dyer was truly the prince of artisans. After 1470 times were harder. The production of unfinished cloth in Babergh hundred became the dominant sector of the Suffolk textile industry. So, dyers and shearmen were rarely needed or engaged as outworkers. Instead, they focused their activities in towns such as Ipswich and Stowmarket, concentrated their efforts on finishing cloth for the domestic market and adjusted their financial expectations accordingly.

---

131 SROB 120 Hawlee; 5, 37 Pye.

132 Zell (1994), p. 186.

133 Power (1924), p. 156.

134 TNA CP 40/885A, m. 5r (AALT IMG 10); TNA CP 40/895, m. 76r (AALT IMG 149); SROI C/2/10/1/2, m. 4v.

135 Luxury double worsted was sheared: Oldland (2011), pp. 186–8.

MAP 5.2: *Suffolk Shearmen 1350–1500*

Notes: Grey marks a parish in which one shearman recorded. Hatched dark grey marks a parish in which between two and four shearmen were recorded. Black marks a parish in which not less than five shearmen were recorded.

Source: TNA CP 40.

Map format attribution: David Dymond and Edward Martin.

By the time the shearman made his last cut something of great beauty had been created. As the product of medieval craft, fine broadcloth stands comparison with the best metal and stone work of the Middle Ages. Since the moment that the wool had been taken from the sheep's back, hard labour, rare skill and a good grasp of practical chemistry had been essential to turn it into a product that was in demand throughout Europe. Cloth was durable enough to pass down the generations, but not sufficiently so to travel through the ages. Unlike metal or stone, as an organic material it has only survived in small sodden scraps and in the most exceptional circumstances. So, sadly, while we can still gaze with wonder on the jewellery and the churches of our medieval forebears, we can only admire the quality of their cloth from afar.

In the next chapter we look at a new breed of men, by the name of clothiers, who emerged at the very end of the Middle Ages to organise the production, marketing and sale of cloth.

– 6 –

The Clothier

The merchant-clothier looks down on the nave of Clare church from on high (Plate 6.1). With his keen eyes, plump cheeks and slight smile he is the very picture of affluence and confidence. He wears his hat with style. His face is one of a shrewd businessman with just a hint of kindness in his soul. When he first took his place more than five hundred years ago he was the master of his world. And many of those in the congregation below owed allegiance to him rather to any feudal lord. Sitting among the congregation of Clare church were two real-life clothiers, Henry Borham and John Wright. They were contemporaries. They must have known each other, and, when their paths crossed after the service or in the market place, may even have exchanged pleasantries. Both were described as 'clothier' in their wills, but if these documents bear true witness to their business activities, they lived very different lives. Borham bequeathed looms to his wife and son, a gown to his son, a brass pot to his daughter and 12d. to the church – nothing more. He was evidently a hands-on, working cloth-maker who had made limited progress along the career route mapped out in an earlier chapter for John Kent. In contrast, Wright mentioned no cloth-making equipment or stock, but his estate included property at Nether Hall in Clare and at Braintree; as well as bedding, brass, various items of clothing, fine furniture, kitchenware, a mazer, pewter and two-dozen silver spoons. He was a successful entrepreneur who, if he had ever sweated at the loom, had long ceased to do so.[1] The term 'clothier' encompassed a broad church.[2]

---

1 SROB 59 Newton, 41 Johnson.

2 The terms 'clothier', 'clothmaker' and 'clothman' were used interchangeably and without any evident distinction. Indeed, in one plea roll John Stokton of Frome in Somerset was described in separate entries as each of the three: TNA CP 40/814, mm. 24r, 211v, 351v (AALT IMG 50, 1392, 1663). Although the Suffolk entries most commonly used the term 'clothmaker', this volume uses 'clothier' to distinguish him from others who contributed to the making of cloth.

A Variety of Clothiers

The archetypal clothier grew rich on profits of the putting out system. He was, so it has been said, in the vanguard of a new capitalist class at the dawn of modern society.[3] This archetype was personified in the lives of Thomas Springs II and III of Lavenham.[4] Borham and Wright illustrate that such an archetype is over simplistic. The difference between these two men, who were both labelled clothier, is repeated across the country. Heaton found that 'whilst the big man apparently predominated in Somerset and Wiltshire, "meaner clothiers" formed the greater part of the Yorkshire industrial army'.[5] Ponting considered that Suffolk clothiers held 'an intermediate position', some making cloth and others organising its manufacture.[6] Certainly, within the county both bigger and meaner men called themselves clothiers, as an analysis of their debt disputes shows (Table 6.1). These debts were generally of higher value than those of cloth-workers, particularly weavers and fullers, and nearly a third were for £10 plus. Reading the wills of the forty-five rich clothiers in Suffolk (those proven in the Prerogative Court of Canterbury before 1520 and listed in Appendix C) reveals a picture of prosperity, even affluence, with long lists of property and goods.[7] Fifteen of them mentioned silverware – bowls, mazers, plate, salters, spoons – compared with only four Bury weavers who did so (Table 4.1).[8] William Forthe bequeathed to his wife Margaret, among many other items, 86 ounces of silver plate for the stated purpose of ensuring 'more concord and amity' between her and Robert who was his son and probably her step-son. Yet, these richer clothiers were only a significant minority. Compared with the number of debt disputes for £10 plus, even more were for less than £5. Furthermore, no silver decorated Henry Borham's table.

We know a little more about the background of the clothiers Robert and John Reynham and Stephen Raynham who were resident in Nayland and Stoke-by-Nayland. They lived better than Borham without ever aspiring to the status of a Spryng. Each made wills that were proven in the Canterbury court and John was among a tiny minority of medieval testators who also left a probate inventory (Appendix D).[9] Robert and John may have been brothers, since the first appointed the second as his executor, and they both died in the 1490s. Stephen was John's son and executor,

3 For a discussion of the clothier as capitalist, see below.

4 http://www.oxforddnb.com/view/article/55402/61760?docPos=2

5 Heaton (1920), p. 92.

6 Ponting (1972), p. 19.

7 The wills of the forty-five are referenced in Appendix C rather than in footnotes to this chapter.

8 John Hyll of Long Melford did not mention silver in his will, but he did so in his probate inventory: TNA PROB 2/120. Quite possibly other rich clothiers owned silver that they similarly omitted to mention in their wills.

9 TNA PROB 2/87.

TABLE 6.1: VALUE OF DEBT LITIGATION OF SUFFOLK CLOTHIERS 1440–1500

Value of claim	Clothiers: No. of claims	Clothiers: Percentage of claims	Clothiers' disputes with Londoners: No. of claims	Clothiers' disputes with Londoners: Percentage of claims
£20 or more	49	11.7	15	11.5
£10 or more but less than £20	81	19.3	19	14.5
£5 or more but less than £10	97	23.1	30	22.9
Less than £5 but more than 40s.	96	22.9	36	27.5
40s. or less	80	19.0	28	21.4
Unknown	17	4.0	3	2.3
Total	420		131	

Source: CP 40

living until 1514. Robert and John had significant property interests, although Robert's were more widely spread in Colchester, Great Horkesley, Lawford and Manningtree, as well as in Nayland and Stoke-by-Nayland. John and Stephen both made substantial cash bequests to family members, with the latter leaving the enormous sum of 500 marks to his widow Christine. All three remembered in their wills the socio-religious guilds of St Thomas and Holy Trinity in Nayland.

John had a large nuclear family with a wife Joan, who survived him, and in addition to Stephen, two sons and two daughters. One of each was under age when he died. He and his family lived in Nayland in a six-room home that boasted a hall, parlour, two bedrooms, buttery and kitchen. The hall and parlour were furnished sparingly, with a bench covering and twelve cushions being the only items of luxury. No ostentation there. As was usual at the time, more had been spent on the bedrooms and bedding with two featherbeds valued at 8s. each and two coverlets at 20s. each. In the buttery two pewter pots and thirty other pewter pieces were given relatively modest values that suggest they were well worn.[10] John's collection of silver was more impressive with two images of St Christopher, a plate, twelve spoons and two mazers being among the most valuable items in the house. As one might expect, his business stock included cloth, woad, alum and wool. The wool bore a high-value equivalent to over £7 a sack, suggesting a better quality than that

10 For comparative values, see Amor (2011), p. 221.

grown locally. The stock of wool was, however, quite modest – less than a sack. John's book debts, which accounted for the bulk of his personal estate, have already been remarked upon.

Indisputably, clothiers were men, and a few women, generally widows, from many different backgrounds. According to Carus-Wilson their ranks 'were recruited from all sorts and conditions of men'.[11] Some had been woolmen. Others had learned cloth-making skills, and been unusually successful the hard way, in one of the various stages of cloth-working – Henry Galyot as a weaver, William Alysaundre as a fuller and Robert Cake as a dyer.[12] Laurence Dowe of Long Melford was cited as a weaver in 1475 and as a clothier five years later.[13] John Horrold of Clare began his working life as a weaver too.[14] Five of the dyers who made Stowmarket and Needham Market centres of their industry presented cloth to the alnager in the late 1460s. Nevertheless, in each of their crafts, and particularly in weaving and fulling, they were among a small elite. Few of their colleagues marketed or sold cloth. Among the rich clothiers, three mentioned tenter-yards in their wills, and others retained an interest in dyeing. Thomas Fulsnape of Hadleigh, Roger Ponder of Great Waldingfield, Robert Sergeaunt of Stoke-by-Nayland and Alexander Sexteyn of Lavenham owned dye-houses. Nicholas Hach the Elder, John Reynham and Stephen Raynham all of Nayland held stocks of woad.[15] They may, however, have relied on skilled dyers to do their work, men such as Great John the dyer who leased a cottage from John Fermer of Lavenham.[16]

Some drapers were involved in cloth making before clothiers appeared on the scene. Indeed, according to Latham, the Latin word '*pannarius*' can be translated as both 'clothier' and 'draper'.[17] Nevertheless, the draper performed a different role. He was primarily concerned with buying and selling cloth, rather than organising its manufacture. Hadleigh was home to over a third of Suffolk drapers cited in plea rolls between 1350 and 1460. At the end of the fourteenth century Richard Loveman pursued the London grocer, William Brykles, for payment of £10 13s. 4d. and defended a claim by Robert Scarlet of Lavenham for 46s.[18] It is, however, Loveman's contemporary, John Kempston, who stands out.[19] We can trace his career over more

11 Carus-Wilson (1987), p. 685.

12 'Clothiers lived among the weavers and spinners and may well have evolved from their ranks': Clarkson (1985), p. 24. 'Enterprising local artisans and craftsmen transformed themselves from small masters to putting-out employers': Zell (1991), p. 4.

13 TNA CP 40/853, m. 455v (AALT IMG 1815); CP 40/871, m. 25r (AALT IMG 53).

14 TNA CP 40/716, m. 402v (AALT IMG 1843).

15 TNA PROB 2/87.

16 TNA PROB 11/21/251; Zell (1991), p. 166; C. Dyer (2005), p. 167.

17 Latham (1965), p. 330.

18 Britnell (2003), p. 88; TNA CP 40/536, m. 31r (AALT IMG 75).

19 In the plea rolls Kempston was described in 1390 as a 'pannarius', and in 1396 as a 'chapman': TNA CP 40/516, m. 223r (AALT IMG 567); TNA CP 40/541a, m. 85v (AALT IMG 194).

than thirty years from 1379 until 1410, by which time his widow Maria was seeking recovery of chattels from John Lyons of Hadleigh.[20] During that period Kempston entered into a number of high-value transactions, at least two of which were worth in excess of £100, with suppliers and customers not only from Hadleigh itself, but from Colchester, Ipswich, London and Pulham Market in Norfolk.[21] He produced thirty dozens to the alnager in 1394/95.[22] Sadly, he eventually appears to have run into financial difficulties. In 1395 he was pardoned for failing to answer the Londoner Gilbert Maghfeld and others for debts amounting to £80, and from that date was defendant in all his recorded disputes.[23] Kempston may have been the first known victim of the Hadleigh textile boom.

While some drapers became clothiers, it would be an error to believe that the two were the same under different names. Only eight known drapers presented cloth to the alnager in the late 1460s. Most of them relied on cloth-makers for their supplies. The clothiers Robert Flegge and William Forthe each numbered at least three drapers among their customers.[24] Furthermore, the majority of clothiers came from a different part of the county to drapers. Whereas over half of all clothiers operated from the hundred of Babergh, little more than one in eight of known drapers did so. The difference between the two was a substantive one, not simply a matter of semantics.

Other clothiers began their commercial lives, and indeed continued to operate as general merchants. They had acquired experience, expertise and contacts that could be put to good use in the marketing and sale of cloth. John Stanesby and John Motte provide good examples. Both could trace their Bildeston roots back into the early-fifteenth century.[25] Both were admitted to the London company of fishmongers and traded as stockfishmongers. Stanesby had Italian connections and Motte Spanish ones.[26] In the late 1460s they were first and fifth respectively among Suffolk's most prolific cloth producers.[27] Even though he was resident in Candlewick

---

20 TNA CP 40/596, m. 381r (AALT IMG 779).

21 TNA CP 40/516, m. 352v (AALT IMG 871); Britnell (2003), p. 96.

22 TNA E 101/342/8.

23 *CPR 1391–6*, p. 675. For details of Kempston's trading with Maghfeld, see James (1971), pp. 204–5.

24 TNA CP 40/883, mm. 189r, 247v (AALT IMG 331, 455 dorse); TNA CP 40/888, m. 53r (AALT IMG 108); TNA CP 40/871, m. 24v (AALT IMG 985); TNA CP 40/931, mm. 5r, 172v (AALT IMG 11, 1098).

25 A Robert Stanesby of Bildeston forfeited cloth to the alnager in 1434/35: TNA E 101/342/21. Motte's father, another John, was an agriculturalist and merchant in Bildeston who died in 1444: TNA CP 40/736, m. 69v (AALT IMG 1171); TNA PROB 11/3/496.

26 *CCR 1461–68*, p. 183; *CCR 1468–76*, pp. 222, 361; *CCR 1485–1500*, p. 77; Childs (1978), pp. 187, 212; Amor (2012), pp. 9–11.

27 Amor (2004), p. 426.

Ward in the City of London by the time he died, Motte still bequeathed 6s. 8d. to each spinner, weaver and fuller who had worked for him in the county. Stanesby employed eleven alien weavers in his Bildeston workshop.[28] John Kyng was another example. He was a member of the London company of grocers and, indeed, its warden in 1459/60.[29] He had extensive commercial and property interests both in Suffolk and further afield. In his will he described himself as a citizen of London, yet he chose to be buried in the church of his home parish of Hadleigh.

Stanesby, Motte and Kyng all trod a successful career path that took them to London. As discussed below, trade between Suffolk clothiers and London merchants was substantial (Figure 1.3). As we saw in chapter 2, London merchants 'fed commercial capital into the cloth industry'.[30] Nevertheless, during the Middle Ages there is little evidence that London born merchants ever became clothiers in Suffolk. One possible exception was Richard Cok who, in 1470, described himself as 'of London alias Hadleigh', and as a 'grocer alias clothmaker' in seeking payment of various debts. However, several later references to a Richard Cok in the 1480s describe him as 'of Hadleigh'.[31] He, like Kyng, may have been a Suffolk boy who went to London in his early career, only to return home in later years.

Charles, duke of Orleans, was captured at the battle of Agincourt and spent the next twenty-five years in captivity in England, latterly in Suffolk. He knew the eastern counties well. As the purported author of the *Heralds Debate,* he had his French herald point out that in 'In England your clothiers dwell in great farms abroad in the country, where as well they make cloth and keep husbandry, as also graze and feed sheep and cattle'.[32] In Suffolk Thomas Warburton of Lavenham and Stoke-by-Nayland was described as 'yeoman alias clothier'.[33] This reference, and others like it, provides evidence that Bailey's 'active, dirty boot farmers' were another group who assumed the role of clothier.[34] They dedicated to commerce the same drive and energy as they had devoted to the acquisition of property, and converted the profits back into more land. For them it was a virtuous cycle. They could look to their tenants and servants in quiet times of the agricultural year to help make cloth. In times of financial stringency, they could put their property up as security for loans.

28 TNA E 179/180/111.

29 Nightingale (1995), p. 510.

30 Thrupp (1962), p. 311.

31 TNA CP 40/834, m. 4v (AALT IMG 872); CP 40/871, m. 250r (AALT IMG 463).

32 Carus-Wilson (1987), p. 686; http://www.nature.com/nature/journal/v1/n14/abs/001352a0.html

33 TNA CP 40/951, m. 222r (AALT IMG 454).

34 Swanson identified a similar development in the West Riding of Yorkshire: Swanson (1989), p. 143; as did Zell in the Kentish Weald: Zell (1991), p. 191.

At a more exalted level, as we saw in chapter 2, John Howard had a long-standing interest in cloth making. By the time of his death, at the battle of Bosworth Field in 1485, he had acquired very extensive estates.[35] The enormous landed wealth of Thomas Spring III included at least sixteen manors in Suffolk, seven in Norfolk, two in Essex and one in Cambridgeshire.[36] If no other clothier quite matched him, several owned significant acreages. John Kyng acquired manors in Shelley, Raydon and Polstead; while John Hyll of Long Melford held that of Bower Hall in the village of Pentlow, on the southern bank of the river Stour opposite Cavendish. Clothiers used their land for sheep pastures, fulling mills, dyeing establishments, and warehouses; but also for more extensive agricultural operations.[37] Kyng bequeathed to his wife and son his grain, carts, ploughs, harrows and beasts.

Having considered the type of men who became clothiers, let us now look at how they rose to prominence.

The Rise of the Clothier

The clothier is first recorded in the plea rolls in the early fifteenth century, but then only in the West Country.[38] Not until 1440 do we hear of Suffolk clothiers, such as Robert Grineer of Lavenham and John Strut of Bildeston.[39] Over the course of the next fifty years their numbers increased sharply (Figure 6.1). In twelve terms of the court of common pleas between 1480 and 1500 very nearly two hundred engaged in litigation. Their rise was most apparent in the hundred of Babergh (Map 6.1). One hundred and thirty seven (54.3 per cent) of the 250 recorded fifteenth-century Suffolk clothiers lived there, as did thirty-four (75.6 per cent) of the forty-five rich clothiers (Appendix C). Five of the nine parishes with more than ten clothiers, and fourteen of the thirty-three with any clothiers at all lay in Babergh. Thirty came from Lavenham, twenty-nine from Stoke-by-Nayland, eighteen from each of Nayland and Sudbury

35 http://www.oxforddnb.com/view/article/13921?docPos=3

36 McClenaghan (1924), pp. 86–8.

37 Zell (1991), p. 192; C. Dyer (2002), p. 326.

38 As early as 1418 several West-Country litigants were described in the plea rolls as clothier or clothmaker which might mean that the putting out system evolved there at least twenty years before it did so in Suffolk. This might explain why the West Country was England's leading textile region in 1400. Hare commented that cloth-making in Somerset 'reached full maturity by the end of the fourteenth century': Hare (1999), p. 13.

39 TNA CP 40/716, mm. 162r, 24v (AALT IMG 322, 1084). The earliest Suffolk wills in which the testator described himself as a clothier, clothmaker or clothman was that of Stephen Gardiner of Bury St Edmunds in 1469 (SROB 125 Hawlee). Britnell shared with the author his view that the Suffolk clothier did not emerge until the mid-fifteenth century: personal communication with Richard Britnell on 7 July 2013.

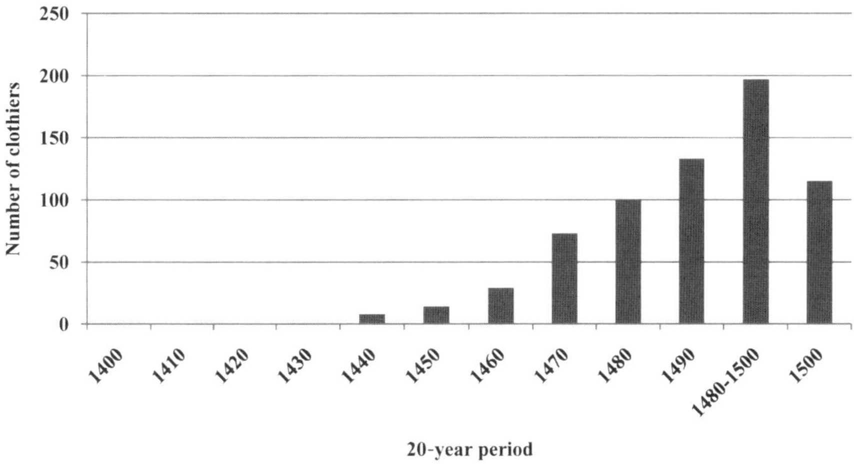

Figure 6.1: *Numbers of Suffolk Clothiers*

Notes: Each column records the number of Suffolk clothiers in the twenty years centred on the given date, with the exception of the penultimate column which records the number of Suffolk clothiers recorded in twelve TNA CP 40 plea rolls between 1480 and 1500.

Source: TNA CP 40A, CP 40B

and seventeen from Long Melford.[40] This is unsurprising. Babergh had long been the centre of gravity for cloth-making. In the period 1466–69 284 (49.2%) of the 577 named persons who presented cloth to the alnager did so in Babergh hundred, and together were responsible for 43.9% of sealed cloth.[41] Thereafter, as we shall see, textile production was increasingly concentrated in fewer hands and directed at the London market ready for export. This drew more work to the main cloth-making towns of the hundred and to the well established outliers of Bildeston, East Bergholt and Hadleigh.

Of course, production continued elsewhere in the county, but of a different nature. Fourteen clothiers were based in Bury St Edmunds, but only four of them had business in London. Just over a third of common bench disputes involving a Bury clothier were with Londoners, compared with two-thirds of those involving a Sudbury clothier and over three-quarters of those involving a clothier from Bildeston or East Bergholt. As we have seen in chapter 4, in the final quarter of the fifteenth century Bury enjoyed its own mini boom in the weaving of cloth.

40 While every effort has been made to ensure that no clothier is counted twice in the data presented in this chapter, the risk of duplication cannot be entirely eliminated and is greatest in the case of neighbouring parishes. Wealth gave clothiers more mobility than other cloth-workers. So, while two clothiers with similar names, one from each parish, have been treated as separate, they may have been the same person.

41 TNA E 101/342/25; E 101/342/2, 4, 5.

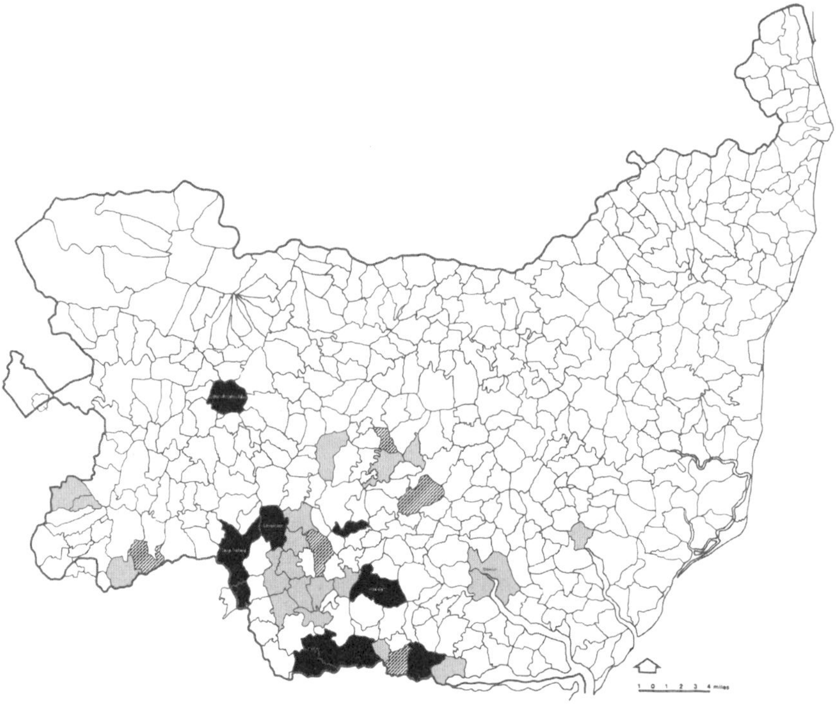

Map 6.1: *Suffolk Clothiers 1440–1500*

Notes: Grey marks a parish in which between one and four clothiers were recorded. Hatched dark grey marks a parish in which between four and nine clothiers were recorded. Black marks a parish in which not less than ten clothiers were recorded.

Source: TNA CP 40.

Map format attribution: David Dymond and Edward Martin.

Some of this went overseas. Robert Gardener traded in Norwich with John Bakeham of Cologne and the will of his kinsman John recorded book debts due from, among others, three merchants of King's Lynn.[42] However, most of the work was done by master craftsmen and aimed primarily at the domestic market. None of them remembered outworkers, such as spinners, weavers or fullers, in their wills. John Gardener's two stalls in the town's great market probably generated more profit than his overseas sales.

The well-funded clothier enjoyed many advantages over his more humble colleagues. He had the liquidity to buy better-quality wool, and to pay slightly higher weaving rates that enabled him to produce premium quality cloth for export.

42 TNA CP 40/883, m. 17r (AALT IMG 34); SROB 196 Pye.

In dealing with London merchants he could be trusted with loans and/or afford to carry debts, and so entered into a world of international finance that depended on credit. By operating on a larger scale, he acquired more negotiating muscle and could achieve better prices for his cloth. He could plan ahead and spread his production across the year, helping him to recruit and retain sufficient skilled outworkers.[43] He could make contacts and build networks with his fellow clothiers in Suffolk and in the wider world outside the county – not only to do business, but also to build trust, access credit and share market information. As evidence of this trust, clothiers remembered each other with bequests in their wills and appointed each other as executors and supervisors. John Motte bequeathed £5 to each of his fellow townsmen John Cooke and John Keggill. Robert Fen appointed Nicholas Hach the Elder as his executor, who in turn appointed John Reynham. Both Robert Fen and John Kyng chose Robert Sergeaunt, and Thomas Horrold looked to his son-in-law Thomas Paycock of Coggeshall. Stephen Raynham was supervisor of the estate of Thomas Tewe of Nayland. Robert Broke of Sudbury chose a London draper as his executor, as did John Motte. Having friends in the right places is always a blessing.

Inputs and Outputs

The rise of the clothier was closely linked to the growth of the industry and concentration of cloth-making in fewer hands. In neither of the two sets of alnage accounts from the late 1390s were the top five cloth producers responsible for more than 10 per cent of all sealed cloth. In the late 1460s the top five clothiers were responsible for over 20 per cent.[44] If the sources are to be trusted, cloth production in Lavenham in the 1470s was concentrated in fewer hands than it had been in the 1460s, a trend that continued into the sixteenth century.[45] The great wealth amassed by Thomas Spring III amounted to nearly 40 per cent of the combined assets of Babergh clothiers, and his family contributed 37 per cent of Lavenham's total tax bill in 1524.[46] The 1522 Military Survey recorded 119 clothiers in Babergh, of whom just twenty-six owned wealth of £100 or more.[47]

Assuming that all the entries in the 1394/95 alnage accounts in the name of William Fuller related to the same man, he was the foremost cloth producer in the

43 Zell (1991), pp. 207–8, 218.

44 TNA E 101/342/8, 10, 25; E 101/343/2, 4, 5.

45 TNA E 101/343/7, 9–11, 13; Britnell (2003), p. 99. The alnage accounts after 1473 are not as reliable as those for the late 1460s: Thornton (1928), pp. 148–9.

46 Cornwall (1988), p. 72; C. Dyer (2002), pp. 308–9. For a discussion of the degree of concentration of the cloth industry in various Essex and Suffolk towns, see Britnell (2003), pp. 88–90, 92; and Amor (2004), pp. 420–1.

47 Pound (1986), p. 133.

late-fourteenth century with 268 narrow, and twenty-four broad cloths. Even had he managed to maintain the same level of production over four years, and the evidence suggests otherwise, he would not have produced as much as any of the top forty clothiers in the late 1460s.[48] John Stanesby presented 2,400 narrow cloths in 1467/68, his most productive year, in excess of six fold more cloth than Fuller.[49] In chapter 4 we estimated that, each year, one spinster could spin enough yarn for five broad, or twenty narrow cloths; a master weaver and assistant could weave forty broadcloths, or 160 narrow cloths; and a master fuller with two assistants could full sixty to seventy broadcloths, or 260 narrow cloths. So, with hard work, Fuller could have achieved his output with a team of eighteen spinners, five weavers and four fullers. Stanesby would have needed 120 spinners, thirty weavers and twenty-eight fullers. The labour cost alone might have been £380. To do their work they would have needed as many as 133 sacks of wool that might have cost Stanesby another £570. If he achieved a price of say 12s. per narrow cloth then his total sales proceeds would have been £1,440. After allowing for finishing, storage, transportation and other transaction costs of say £190, he might have turned a profit of £300.[50] His profit margin would have been a modest 21 per cent.[51]

---

48 William Fuller's name appeared five times in the alnage accounts for 1394/95, but not at all in 1396/97.

49 TNA E 101/343/4.

50 These calculations are based on the following speculative assumptions:-

1. Stanesby paid each spinner 1½d. per day, each team of weavers (master and assistant) 10d. per day and each team of fullers (master and two assistants) 14d. per day;
2. Each of them, or someone in their place worked 200 days in the year;
3. A stone of wool cost 3s. 4d. and a sack £4 6s. 8d;
4. A sack of wool produced eighteen good-quality narrow cloths (Carus-Wilson and Coleman (1963), pp. 15–6);
5. Stanesby achieved a higher price for his cloth than the 6s. 8d. valuation recorded both in the 1414/15 customs accounts (TNA E 122/51/39) and in the 1438/39 Royal Wardrobe accounts (Munro (2003), p. 319), because:
   a. customs valuations were prone to be artificially deflated, while the Crown was in a strong position to secure preferential prices; and
   b. by the late 1460s, the quality of Suffolk narrow cloth was better than it had been earlier in the century owing to the demands of overseas markets and because, in 1468, a minimum prescribed weight of 9½ lbs was first established by statute (*Statutes*, 2, pp. 424–5);

   – if this assumption is incorrect and the correct price was, in fact, 6s. 8d. then the assumed cost of labour and materials is also too high;
6. The finishing, storage, transportation and other transaction costs were 50 per cent of production costs; and
7. The cloth was produced by workers in their own homes using their own equipment, so no allowance need be made for premises or depreciation of plant and equipment.

51 A profit margin of 21% is one a rich clothier might expect to achieve and is comparable with those cited by C. Dyer (1989), p. 193.

Buying the wool, organising and paying this small army, and disposing of the cloth they produced, must have been a major logistical exercise well beyond the most capable of master craftsmen. It was also fraught with risk – sub-standard wool, shoddy workmanship, damage by fire or water, loss in transit, market disruption or a fall in prices could quickly wipe out the profit. Indeed, such an ambitious enterprise may eventually have become too much even for Stanesby. In 1474 he gifted to Robert Dene, a gentleman of Nowton, and John Smert, a grocer of London, all his goods, chattels and book debts. As a symbolic affirmation of this gift, he delivered to them a silver bowl weighing 6 ounces. Was this gift security for a much-needed loan, or a device to put his assets beyond the reach of his creditors? Whichever, it was a sign of financial distress.[52] Happily, we know from the record of aliens working in his weaving shop that he was still in business in 1483, so he appears to have survived the crisis.

The London Connection

Between 1460 and 1470 the number of recorded Suffolk clothiers more than doubled from 29 to 73, just as the Hanseatic merchants downsized their export operations from the ports of Colchester and Ipswich (Figures 2.3 and 6.1).[53] When the Hanseatic trade was at its height, a master craftsman could deal direct with a merchant or his local agent.[54] For instance, in 1458 Thomas Peverell of East Bergholt agreed to full four vesses and two new greys on behalf of Eberhard Kryt of Cologne, while two years later another two fullers, William Alysaunder and Richard Neulond, faced claims for £20 and £16 respectively by Roger Ryng also of Cologne.[55] When he died in 1463 Thomas's kinsman William Peverell in his will not only instructed his son and daughter-in-law to collect and distribute £20 due from a Maynard Hamburghman, but also referred to other book debts due in England and overseas.[56] Even after 1470, once the Germans had focused their trade on London, such dealings were still possible. In 1486 two weavers from Stoke-by-Nayland, Thomas Brigge and John Debenham, each faced a debt claim by Conrad van Elsign.[57] Nevertheless, this case was a rare exception. Suffolk's export-grade cloth was increasingly channeled through the City by local clothiers selling to London merchants, rather than direct to overseas buyers. Indeed, this may have been their original *raison d'être*.

52 *CCR (1468–76)*, p. 361; Nightingale (1995), pp. 522–3, 542; Bolton (2012), p. 282.
53 Carus-Wilson and Colman (1963), pp. 102–3.
54 Sutton (2005), pp. 283–4.
55 Britnell (1986), p. 172; TNA CP 40/796, m. 9v (AALT IMG 861).
56 SROI IC/AA2/85.
57 TNA CP 40/895, m. 38v (AALT IMG 618).

More than two in every five Suffolk clothiers' disputes, as recorded in the plea rolls, were with Londoners. Of these, over 85 per cent arose from bonds or contracts concluded in the Capital (Table 6.2). The lure of the City was irresistible. The direct distance from most of Suffolk's principal cloth-making centres to London was less than sixty miles. Although this was a longer journey in the Middle Ages than it is now, it could be completed with a heavy laden cart in three to five days.[58] The same journey by boat might be preferred on the grounds of cost, but was further and would take longer.[59] Cloth-makers contributed towards the cost of maintaining these links. John Motte bequeathed £240 for the repair of roads including that between Colchester and London. Anyone travelling from Suffolk to London almost inevitably had to cross the river Stour. John Caves visited Hadleigh to collect alms for the upkeep of the bridge and causeway over the Stour at Cattawade, for which he had been made responsible by the Crown.[60] Travel by road might entail the

TABLE 6.2: DISTANCE BETWEEN PARTIES TO DEBT LITIGATION WITH SUFFOLK CLOTHIERS 1440–1500

Distance	*No. of claims*	*Percentage of claims*	*Mean value of claims*
Same town or village	16	5.3	£10 10s. 10d.
Less than 6 miles	29	9.7	£6 13s. 1d.
More than 6, but less than 10 miles	27	9.0	£8 5s. 4d.
More than 10, but less than 20 miles	35	11.7	£9 18s. 3d.
London	131	43.7	£9 6s. 10d.
Norwich	17	5.7	£7 4s. 5d.
Overseas	2	0.7	£11 4s. 2d.
Other more than 20 miles	43	14.3	£13 4s.
Total	300		

Notes: Almost certainly, the claimant's residence was more likely to be stated in disputes with Londoners than in disputes with those from elsewhere. So, the proportion of disputes with Londoners may thereby be inflated.

Source: TNA CP 40

58 Childs (2006), p. 261.

59 Ogilby's road maps of 1675 show the various routes running through, in and out of Suffolk: http://www.campop.geog.cam.ac.uk/research/projects/occupations/transport1379-1729/

60 TNA CP 40/402, m. 76v (AALT IMG 1768). As Stenton noted, Cattawade Bridge was of sufficient importance to be illustrated on the Gough Map: Millea (2007), pp. 70–1.

payment of tolls to the authorities of towns *en route*. Happily, the carriers of Hadleigh, such as John Page and John Preston, enjoyed the benefit of the town's market charter, re-confirmed in 1414/15, which exempted them from tolls on goods taken to London for sale or export. The tenants of the earl of Oxford in Lavenham enjoyed a similar right of exemption from tolls on all goods throughout the realm.[61]

The mean value of clothiers' disputes with Londoners was higher (£9 6s. 10d.), but not much higher, than that of local disputes with protagonists less than twenty miles from home (£8 14s. 9d.). These mean figures do, however, hide variations that are revealed by a closer analysis of each Londoner's occupation (Table 6.3). Over half the Londoners were members of the three great City merchant companies of mercers, grocers and drapers. In the late-fifteenth and early-sixteenth centuries mercers, by now trading as Merchant Adventurers, eclipsed the grocers to enjoy the lion's share of the booming overseas trade in unfinished cloth – 'their trade and wealth was reaching a zenith'.[62] Suffolk clothiers dealt with them more often than with members of the other two companies and those dealings were easily the most valuable, nearly double the value of ones with drapers. The twenty-eight disputes involved nineteen different mercers, and clothiers from no less than sixteen towns and villages, suggesting that the production of unfinished cloth in the county was widespread. John Cooke of Bildeston dealt with at least three mercers over a period of ten years between the mid-1480s and 1490s, with the value of his transactions ranging between 40s. and £26.[63] Both John Horrold and his son Thomas the Younger traded with mercers too. When he died in 1478 John was indebted to John Salford in the sum of £22, while Thomas pursued John Baron for payments of £16 17s. 3d. and £13 9s. 6d. and Thomas Bryce for £15.[64] Although, no mercer could hope to achieve a stranglehold on a trade as large as that in Suffolk cloth, Thomas Lok, Thomas Nyche and Robert Southwood were each involved in three disputes spaced over several years, indicating a continuing interest.[65] Both Nyche and Southwood were prominent within their company. They were apprenticed in 1430 and 1446 and rose to become wardens in 1461 and 1484 respectively. During his career Nyche trained at least five apprentices. In 1463/64 Southwood was the largest

---

61 TNA CP 40/636, m. 469v (AALT IMG 1976); CP 40/676, m. 202r (AALT IMG 412); McIntosh (2013), p. 38; *CChR 1326–41*, p. 122.

62 Sutton (2005), p. 318.

63 TNA CP 40/891, m. 319r (AALT IMG 558); CP 40/919, m. 300v (AALT IMG 1339); CP 40/931, m. 209r (AALT IMG 431).

64 TNA CP 40/895, m. 142r (AALT IMG 261); *CPR 1485–94*, pp. 2, 187. John Horrold had two sons named Thomas.

65 TNA CP 40/895, m. 88v (AALT IMG 713); CP 40/919, m. 319v (AALT IMG 1371); CP 40/834, m. 84v (AALT IMG 1031); CP 40/837, m. 97r (AALT IMG 155); CP 40/871, m. 250r (AALT IMG 462); CP 40/883, m. 216v (AALT IMG 394 dorse); CP 40/907, m. 201r (AALT IMG 378).

TABLE 6.3: OCCUPATIONS OF PARTIES IN DEBT LITIGATION WITH SUFFOLK CLOTHIERS 1440–1500

Occupation	*No.*		*Percentage*	*Mean value of claims*
Clergy (inc. Clerk)	10		3.5	£17 14s. 3d.
Clothier	8		2.8	£7 2s. 4d.
Cloth-worker	29		10.1	£6 3s. 7d.
Husbandman/Yeoman	17		5.9	£10 11s. 7d.
Lord/Gentry	40		14.0	£9 7s. 3d.
Member of greater London company	85		29.7	£8 19s. 3d.
Draper		15		£6 6s. 10d.
Fishmonger		5		£8 1s. 2d.
Goldsmith		1		
Grocer		25		£7 10s. 3d.
Ironmonger		1		
Mercer		28		£12 4s.
Salter		2		
Tailor		8		£7 3s. 7d.
Merchant/Mercer	25		8.7	£14 6s. 10d.
Other	64		22.4	£7 14s. 6d.
Woolman	8		2.8	£8 8s. 1d.
Total	286			

Source: TNA CP 40

shipper of all his colleagues, exporting 349 cloths from London – one may wonder how many of them had been made in Suffolk.[66] They were both men to be reckoned with.

The small proportion of trade with drapers and its relatively low value is surprising. They controlled the great market at Blackwell Hall, which once stood adjacent to the Guildhall as the main collection point for provincial cloth.[67] One must assume either that other merchants were poaching on the drapers' territory, or that a lot of Suffolk cloth was being sold outside Blackwell Hall.[68] According to Quinton:

66 http://www.londonroll.org; Sutton (2005), pp. 293 (n. 108), 557.

67 McClenaghan (1924), p. 24; Quinton (2001), pp. 134–40.

68 All provincial cloth had to be sold at Blackwell Hall 'between noon on Thursday and Saturday': Carus-Wilson (1987), p. 682. Thornton gave the opening and closing times as 11.00, but opined that in the fifteenth century, just as they did in the sixteenth, Suffolk clothiers almost certainly disregarded this regulation and sold at private markets: Thornton (1928), p. 163.

> As drapers became more heavily involved in the export of cloth and the import of raw materials, taking over some of the distributive roles previously enjoyed primarily by London grocers, not only did their commercial arrangements with cloth suppliers in the countryside become more diverse, but their overall levels of business increased [...] It can be no coincidence that by this decade [1480s], the proportion of Common Pleas debt cases involving London drapers and cloth-related tradesmen of the countryside had risen significantly.[69]

Ten of the fifteen disputes between Suffolk clothiers and London drapers were recorded between 1495 and 1500, so only in these final five years did the proportion of their cases rise significantly. By that time drapers had won a significant share in the export trade of finished cloth to Gascony and Iberia.[70] Much of this would have been sheared and dyed in the City, but some may have been both produced and finished in the provinces. One draper, Piers Joye, shipped to Bordeaux, Portugal and Spain cloth that he had bought from John Hyll.[71] Another, Laurence Tester, in 1500 owed William Jacob of Lavenham £20, a sum that was still in default three years later when he sought a pardon for not answering to the debt.[72] A third, Thomas Boterell, was doing rather better. In 1493 he took on one apprentice; in the following year, another successfully completed his training to became a freeman of the City; and in 1503 a third was remembered with a bequest in Boterell's will. Boterell enjoyed a close association with the Spanish merchant Pedro Valladolid. At the same time he traded with William Hill, John Peverell and Thomas Raynold, all clothiers of East Bergholt.[73] Nevertheless, despite the endeavours of Botterell and his colleagues, both domestically and overseas, in 1500 London drapers had some way to go to catch up with mercers and grocers in penetrating the Suffolk cloth trade.

Of course, not all trade between Suffolk clothiers and London merchants was one way. Wealthy Suffolk residents almost certainly bought fine linen and other mercery from the mercers; and from the grocers they could obtain dyes for their business and spices for their table. Some of those grocers who dealt with Suffolk dyers also traded with clothiers. Men like John Stokes had fingers in many pies for many years. As we have seen, earlier in his career, in 1460, he had been in dispute with John Fuller, a dyer of Long Melford. In his final years, in the early 1480s, he

69 Quinton (2001), pp. 140, 145.

70 Quinton (2001), pp. 276–8.

71 TNA C 1/209/42.

72 TNA CP 40/951, m. 78v (AALT IMG 154 dorse); *CPR 1494–1509*, p. 330.

73 http://www.londonroll.org; Childs (1978), p. 231; TNA CP 40/931, m. 33r (AALT IMG 67); TNA PROB 11/3/467.

traded with John Flegge of Needham Market who sometimes styled himself 'dyer' and at others 'clothier'.[74]

William Forthe of Hadleigh

However important the London market may have been for Suffolk clothiers, it was not the only place in which they did business. Indeed, the greatest margins could probably be achieved elsewhere. In the Capital, Suffolk men faced seasoned merchants expert in striking a hard bargain. In the provinces they enjoyed much greater negotiating power. The mean value of their trade with those living more than twenty miles away (excepting London and England's second city of Norwich) was highest of all. In the round, their dealings with husbandmen and yeomen, who were keen to sell them wool and buy their cloth, were of higher value than with members of the great London companies (Table 6.3).

One man who seems to have understood this well was William Forthe, perhaps England's greatest clothier in the late-fifteenth century.[75] Assuming (as seems likely) that all references to Forthe were to the same man, he enjoyed a remarkably long and successful career. He lived and died in Hadleigh and his mansion may have been the fine house then known as *The Cross* and now as Edge Hill (Plate 6.2).[76] At the time of his death he was married to Margaret, although she may not have been his first wife. He had a son, Robert, and a daughter, Elizabeth. Elizabeth married well, no doubt because of her father's connections. She wed Thomas Baldry who came from Thorney, near Stowmarket, moved to London for apprenticeship, became a successful mercer, and was made Lord Mayor of London and knighted in 1523. William may possibly have been apprenticed himself in London, in 1439/40, to the extremely successful mercer and adventurer Thomas Dunton.[77] Initially William described himself as a merchant, but from 1480 as a clothier. In 1461 he presented twenty-eight broadcloths to the alnager in Colchester and later in the decade he was the second biggest cloth producer in Suffolk, in just four years presenting 426 broadcloths, and 814 narrow ones.[78] He was one of six clothiers in Hadleigh who produced nearly three-quarters of the town's sealed cloth.

---

74 TNA CP 40/871, m. 251v (AALT IMG 1401); CP 40/891, m. 289v (AALT IMG 1241).

75 As is so often the case when studying people with relatively common names, one can never be certain that all entries relate to the same person. It may be that there were two William Forthes of Hadleigh in the fifteenth century, perhaps father and son. This would not detract from the points made in this section.

76 Personal conversation with, and supporting documents from, Susan Andrews, 2 May 2014 and 1 August 2015.

77 Sutton (2010), pp. 166, 169–70.

78 TNA E 101/342/21, 25; TNA E 101/343/2, 4, 5.

William is first recorded as appearing before the common bench in 1440 and, by the time of his death more than sixty years later in 1504, he had engaged in at least ninety disputes, in nearly all of which he had been the claimant. Most of William's trade was of high value. In 15 per cent of disputes the amount in question was £20 plus, and in over half it was £10 plus. Although he had family and owned property in the City, none of these disputes was with a Londoner.[79] With the exception of Colchester and Ipswich, William did very little business in the larger regional settlements such as Bury St Edmunds, King's Lynn and Norwich. He traded throughout the small towns and villages of Essex, Norfolk and Suffolk. More than a third of his disputes were with people living further than twenty miles from Hadleigh and these had a mean value of more than £13, nearly double that of his litigation with those living less than ten miles from home. Carriage and storage costs, tolls and transit risks such as robbery and storm meant that, in longer distance trade, only high value transactions would generate sufficient profit. As we saw in chapter 2, many of William's long-haul destinations could be reached by boat, down the river Stour, up the coast to Blythburgh and Kessingland, and further on up the Yare and Wensum to Pirleston, Eaton and Norwich.[80] He relied on mariners such as William Blak of Ipswich and Thomas Tyte of St Osyth.[81] For him retail trade was more important than wholesale. More than 40 per cent of his disputes were with husbandmen and yeomen, with a mean value of nearly £11. He dealt with an extraordinary variety of people, including an ashburner, a baker, a barber, a beer brewer, a carpenter, a draper, a glover, an ironmonger, a mariner, a miller, a pewterer, a sawyer, a smith and a tailor. His name must have been known the length and breadth of East Anglia.

Early Capitalists?

In his classic study, *Religion and the Rise of Capitalism*, Tawney surely referred to clothiers when he wrote of those 'who fled, precocious individualists, from the fellowship of borough and craft, that they might grow to what stature they pleased in rural isolation' and, in doing so, were responsible for 'the outburst of capitalist enterprise in [...] textiles'.[82] His great contemporary and collaborator Power described the cloth manufacturing areas as 'pockets of capitalism in a pre-capitalistic world'.[83] Continuing the debate, Hilton wrote that 'the continuing

---

79 CCR *1485–1500*, p. 77.

80 TNA CP 40/834, m. 214r (AALT IMG 381); CP 40/871, m. 24v (AALT IMG 985); CP 40/883, m. 173v (AALT IMG 311 dorse); CP 40/911, m. 4v (AALT IMG 10 dorse); CP 40/931, m. 5r (AALT IMG 11).

81 TNA CP 40/834, m. 226r (AALT IMG 403); CP 40/919, m. 263v (AALT IMG 1265).

82 Tawney (1938), pp. 79–80.

83 Power (1941), p. 4.

organization of production on the basis of the artisan family unit prevented the development of capitalist relations of production', but believed that 'modern capitalism derived its initial impetus from the English textile industry'.[84] Were Tawney and Power correct in heralding clothiers as early capitalists, or was Hilton right to be more circumspect?

The Oxford English Dictionary defines a capitalist as 'a person who possesses capital assets; *esp.* one who invests these esp. for profit in financial and business enterprises'.[85] In answering his own question 'Were there any Capitalists in Fifteenth Century England?', Dyer set out three possible definitions of capitalist, but preferred the following Marxist interpretation as being most 'precise':

> A system of relationships in production, in which the ownership of the means of production is concentrated in the hands of entrepreneurs. They are able to employ a free labour force, who have themselves become separated from other means of production. Capitalists buy the labour of the workers, and sell the goods at a profit.[86]

Let us apply the OED and Dyer's definitions to the rich Suffolk clothiers considered above. Certainly, they were entrepreneurs, they took risks and, if these paid off, made handsome profits. They helped transform the basis of industry from small, isolated family workshops to scattered yet connected ones, and, in doing so, took a major step towards the evolution of the factory system.[87] In Suffolk, and some other regions of the country, large numbers of people became dependent upon them for the whole or part of their living.[88] They brought about a paradigm shift in the organization of production. Clothiers were certainly new beasts in the economic jungle. What is far less clear is the extent to which they were capitalists in the sense of owning capital assets/means of production and employing labour.

Their wills show that they invested in stock including wool, cloth, dyes and dyestuffs and often held significant book debts. When he died in 1507 John Gardener listed moneys due to him in a total sum of £64 12s. In 1522 Thomas Spryng III owned 'dettes sperat and desperate [ie doubtful]' worth a massive £2,200.[89] As Dyer himself stated, 'the bulk of the capital of the clothiers was tied up in trading costs'.[90] No testamentary evidence survives of any of them owning spinning shops or even

84 Hilton (1978), p. 156.
85 http://www.oed.com/view/Entry/27455?redirectedFrom=capitalist#eid
86 C. Dyer (1994), p. 306. See also C. Dyer (2005), p. 41.
87 Braudel (1982), pp. 298–9.
88 Britnell (1996), p. 234; Lee (2005), p. 12. Britnell was generally sceptical of the theory that the late Middle Ages was a period of transition from feudalism to capitalism.
89 SROB 196 Pye: Pound (1986), p. 75.
90 C. Dyer (2005), p. 167. Zell found the same in Wealden Kent: Zell (1991), pp. 199–200, 237.

spinning wheels. Many weavers owned their own looms and only three rich clothiers mentioned these in their wills. As mentioned above, another three referred to tenters and four owned dye-houses. A small but growing number invested in fulling mills, particularly towards the end of the fifteenth century. Nevertheless, we are left with the strong impression that they aped the nobility and gentry in preferring the acquisition of estates and fine silver to plant and equipment.[91] As Clarkson stated:

> Cottage industry also made minimal demands on fixed capital, for no special industrial premises were required and the machinery used – spinning wheels, looms and anvils for example – was usually small and inexpensive: small enough to be housed in domestic cottages and cheap enough to be owned by poor farmers.[92]

Even though Suffolk's cloth production was concentrated in fewer hands, no clothier could build a business empire of the type that would become common in later centuries. The aggregate demand for cloth in the late medieval economy was insufficient. Furthermore, as we saw in chapter 2, overseas markets were prone to disruption by war. It may be significant that the most prolific producer of cloth in the 1460s, John Stanesby, was in financial trouble in the 1470s. The second most prolific, William Forthe, steered clear of the great markets in London and Continental Europe, concentrating on retail trade in the provinces. A third clothier who tried to break into the big time, John Hyll, ended up defending a Chancery law suit when he failed to deliver on all the £1,000 worth of woollen cloths that he had promised to supply to the London draper Piers Joye.[93]

The relationship between clothiers and other cloth-workers, such as weavers and fullers, took various forms. The number of disputes between them was not insignificant (Table 6.3). Although thirteen were with dyers and shearmen of London, all but one of the remaining sixteen disputes were with Suffolk residents – one card-maker, four fullers, six weavers, three dyers and one shearman. Of low mean value relative to other clothier disputes, they were of high value relative to the generality of cloth-worker disputes, particularly those of weavers and fullers (Table 4.2). For instance, two claims against the weavers Thomas Hobert of Fen Street in Monks Eleigh and John Robert the Elder of Glemsford were for £11 and £10 respectively.[94] We may safely assume that these protagonists were master craftsmen rather than outworkers.

---

91 Such acquisitiveness often provoked jealousy among the gentry: Ramsey (1982), p. 50.

92 Clarkson (1985), pp. 15–6. See also Coleman (1983), p. 445; Britnell (1996), p. 234; Munro (2003), p. 219.

93 TNA C 1/209/42.

94 TNA CP 40/837, m. 83r (AALT IMG 139); TNA CP 40/919, m. 165v (AALT IMG 1062).

The essential features of a modern employment relationship are control by the employer of the employee in a way that generally involves integration within the employer's organization; and also mutual obligations on the employer to pay a wage and on the employee to do the work personally.[95] In studying clothiers' wills, one rarely comes across evidence of a relationship that has all these features other than with household servants whose cloth-making duties, if any, were probably subsidiary to their main jobs. Both John Kyng and Stephen Raynham remembered apprentices with bequests of £1 and 33s. 4d. respectively, but it is most unlikely that either apprentice had been waged during their master's life. More commonly one encounters outworkers who operated at home and used their own spindles, looms, vats, tenters or other equipment to make cloth. They had an opportunity to do such work as the clothiers gave them, but no right to a regular wage. When the cloth industry was depressed, as it was in the mid-1520s, many of them were redundant and went hungry. In such troubled times our old friend, Jack of all Trades, fared best. He had not become separated from other means of production and had retained an interest in husbandry.

How close was the relationship between the clothiers and their outworkers? Although historians have commented on their bequests to spinsters, weavers and fullers, in fact such provision was unusual. Thomas Spryng II left a very generous sum of 100 marks, but, out of his almost boundless wealth, his son Thomas Spryng III left nothing in his will and his executors' accounts record no such payment.[96] Indeed, of the other rich clothiers, only five – John Golding, John Motte, Stephen Raynham, William Spryng and Thomas Sturmyn – remembered their outworkers.[97] Even these men enjoyed a different and much closer relationship with their household servants. Raynham bequeathed 6s. 8d. to each servant, but only 20d. to each weaver, and 8d. to each spinster. Sturmyn's spinsters, who lived in Glemsford and Stoke-by-Clare, received an even more miserable 2d. each. We must conclude, as did Ramsey, that it would be 'misleading' to equate clothier and outworker with employer and employee.[98]

If, as Dyer suggested, in the fifteenth century East Anglian clothiers 'took the first tentative steps towards an early form of industrial capitalism', they still had a long way to go. And Bailey reminded us that 'it is possible for commercial activity and entrepreneurial attitudes to exist without full-blown capitalism'.[99]

95 Slade et al (2014), p. 359.
96 Betterton and Dymond (1989), pp. 12, 14–5, 100–2.
97 TNA PROB 11/11/213, 11/6/171, 11/18/55, 11/16/289, 11/10/42.
98 Ramsey (1982), p. 24.
99 C. Dyer (2000), p. 325; Bailey (2007), p. 294.

The Clothier and Welfare

In modern times capitalism has proven to be the most resilient of economic systems partly because, in many societies, its unacceptable face has been mitigated by the provision of social welfare for vulnerable groups – the young, the old, the sick and disabled, and the poor. In the late Middle Ages there were several good reasons for providing such welfare. The Catholic doctrine of purgatory taught that, after death, most souls had to undergo a period of purification before ascending to heaven, and that the term in purgatory could be shortened by the intercessory prayers of the living. The risk of permanent incarceration in hell could be reduced in the same way. The prayers of the needy were particularly efficacious. Accordingly, a substantial minority of testators made provision for the distribution of alms to the poor at funerals and obits – the commemorative services often held one week (the 'week's mind'), one month (the 'month's mind') and/or one year (the 'year's mind') after death. Generally, such distributions were indiscriminate in order 'to spread the good work of prayers for their souls as widely as possible'.[100] They were a form of after-life insurance.[101]

The Catholic church also encouraged congregations to divide their moveable goods, on death, into three parts, leaving one third as religious or charitable alms to be distributed 'for the good of their soul'. As Langland recorded in *Pier's Ploughman*:

> Truth sent the merchants a letter in which he urged them to buy and sell freely, save the profits, and put them to good use. They should repair hospitals and help the sick; be active in repairing rutted roads or bridges that had collapsed; bring about suitable marriages for orphaned girls [...] provide food for the destitute and imprisoned, enable boys to be educated at school or apprenticed to a trade.[102]

However, even those testators who complied with this injunction did not necessarily bestow the one third on their vulnerable neighbours. The funds could, and often were, used for the beautification of the parish church or the improvement of local roads. Betterton and Dymond listed the numerous bequests that Lavenham people made towards these ends, and the merchant mark of Thomas Spryng II on the outside wall of the tower of Lavenham church still stands as a reminder of his bequest (Plate 6.3).[103]

---

100 McIntosh (2012), p. 10.

101 Duffy offered an alternative explanation for such bequests, 'intercession for the dead was less like fire insurance than the extension of the obligations of friendship and family and neighbourhood into the dark world of the dead': Duffy (2015), p. 58.

102 Langland (1992), p. 75.

103 Betterton and Dymond (1989), pp. 103–9, 113–5.

Religion was not the only motivation for good works. Self-promotion was another. Generous bequests to the church re-enforced the image of the donor amongst the local community as 'a person of status and wealth'.[104] Civic leaders were, like Thomas Spryng II, keen to earn earthly immortality by attaching their name or mark to a building, institution or trust. Nor should true altruism be ruled out. Although the image of medieval squalor lingers long in folk memory, civic leaders took pride in their towns and villages and 'espoused a growing commitment to matters of communal welfare'.[105] This included looking after their own poor, if not the poor of other communities. The century after the Black Death was one of relative affluence, but from the mid-fifteenth century life for some became much tougher. The Great Slump, poor harvests in the early 1480s and rising mortality levels brought the problem of poverty back into focus. A family could be devastated by the loss of employment, injury or sudden death of the main breadwinner, leaving them without obvious means of support. Much better to care for them at home than to allow them to swell the growing ranks of mighty beggars who wandered the country in search of alms.[106] Thus more long-term social welfare projects were launched, and by 1500 between 1.4 and 1.8 per cent of GDP was devoted to formalised poor relief – a higher rate than at any time over the next two hundred years.[107] Almshouses were among the most obvious physical manifestations of such relief. These provided private living quarters, usually one room but sometimes two, very occasionally coupled with other benefits such as food or small incomes. They were intended primarily for the honest elderly poor, particularly those who had fallen on hard times through no fault of their own. Blocks of almshouses were relatively small, accommodating on average eight or nine residents. A few had a vegetable garden. A disproportionate number were to be found in the towns and market centres of East Anglia, the very places where rich clothiers lived.[108]

If fifteenth-century Suffolk clothiers were taking the first steps towards capitalism, were they also active in providing welfare? Dyer found, in the wills copied in the Register Baldwyne (1439/74), evidence of what he called 'the Lavenham project' – 'a co-ordinated campaign [lifting people out of poverty for a whole year] which invites comparisons with experimental schemes in social welfare in the sixteenth century.'[109] Of the fourteen testators whose longer-term bequests funded this

104 Sear (2015), p. 246.
105 Rawcliffe (2013), p. 4.
106 Smith (2012), pp. 80–2; McIntosh (2011), p. 43. In 1482 the prior of Ixworth fined some residents for harbouring ignominious and suspicious men and 'ryotous mytybeggars' in the village: Amor (2002), p. 140.
107 Van Bavel and Rijpma (2016), pp. 171–4, 176.
108 McIntosh (2011), pp. 59–65, 70, 71, 75.
109 C. Dyer (2012 (B)), pp. 59, 68.

project, at least five were engaged in cloth-making. For instance, Thomas Spryng I left 12d. per week to be distributed among twelve poor persons every Wednesday for four years, which John Place capped by leaving 24d. per week for the same period.[110] Let us consider what provision the rich clothiers made for the poor in the late-fifteenth and early-sixteenth centuries, and the extent to which they continued and broadened this Lavenham project outside that town. Nearly all of them left money for alms at funerals and obits. John Horrold gave 1d. each to the honest paupers who interceded for his soul at his funeral, and a smock to four of them who, in addition, held torches around his corpse. John Kyng and John Hache both made a very similar provision of cash and gowns for the poor men who, at their funerals and month's minds, prayed and held torches. John Smyth of Long Melford left a kirtle to each of five poor women who attended his month's mind.

Twelve of the forty-five (26.7 per cent) rich clothiers made provision for social welfare that went beyond the bare minimum of after-life insurance for a period of at least a year after their deaths (Table 6.4). Among the most notable of these bequests were those of Roger Ponder, Robert Reynham and Robert Sergeaunt. Ponder directed that the rent from two messuages, called Colmans and Rawlyns, should be used to pay taxes and other charges imposed on the poor of Great Waldingfield. Reynham similarly directed that the rent from two properties in Newlands Lane, Nayland should be used to indemnify the poor against rents and other customary payments. Sergeaunt not only provided five almshouses, but set aside the rent from a cottage to generate funds for his daughter and son-in-law to keep the accommodation wind and watertight. He also bequeathed 20s. to each of six poor maidens to provide dowries and 'help them to their marriage'. Thomas Spryng II bequeathed £280 7s. 6d. for the poor of Lavenham and other towns and, with another gift of 100 marks (£66 13s. 4d.), endowed exhibitions for scholars at the universities of Oxford and Cambridge that might have opened the doors of higher education to boys from modest backgrounds.[111]

In addition to the twelve, Robert Cowper the Elder of Nayland remembered the poor of numerous neighbouring parishes with significant cash gifts, as, indeed, did Roger Ponder. Perhaps they were rewarding the towns and villages where they bought their wool, sold their cloth or hired outworkers. If so, then Ponder had a long reach because he included in his list Luton in Bedfordshire and Hoddesdon in Hertfordshire. Others left bequests conditional upon the failure of another gift or upon the discretion of their executors to favour the poor over churches and highways.[112] Walter Copynger of Bury St Edmunds was outside the magic circle

110 SROB 19, 24 Baldwyne.
111 TNA PROB 2/14.
112 See the wills of John Rote the Elder of Nayland and Thomas Horrold of Clare.

TABLE 6.4: LONG-TERM SOCIAL WELFARE PROVISION BY RICH CLOTHIERS

Name	*Town/Village*	*Provision for the poor*
John Archer	Sudbury	12d. every Friday for six years
William Forthe	Hadleigh	8 marks a week for a year
Thomas Fulsnape	Hadleigh	12d. every Sunday for ten years
John Golding	Glemsford	6d. a week until residuary bequest exhausted
John Horrold	Clare	5d. every Friday for five years
Roger Ponder	Great Waldingfield	Permanent endowment
Robert Reynham	Stoke-by-Nayland	Permanent endowment
Robert Sergeaunt	Stoke-by-Nayland	Five almshouses
Alexander Sexteyn	Lavenham	12d. every Wednesday for two-hundred weeks
John Smyth	Long Melford	10s. each year at Christmas
Spring Thomas II	Lavenham	100 marks to fund exhibitions for scholars at the universities of Oxford and Cambridge
Thomas Sturmyn	Lavenham	6d. every week for two years

because his will was not proven at Canterbury. He bequeathed £5 for the purchase of firewood for the poor between Hallowmas (1 November) and Candlemas (2 February) for each of the five years following his death, an early form of winter fuel allowance.[113]

Notwithstanding the generosity of these twelve individuals, some may find the benevolence of the rich clothiers generally disappointing. It is perhaps more impressive when placed in context. By comparison to the twelve, Dyer found only two-dozen testators (1.0 per cent) who made provision out of 2,324 whose wills were copied in the Register Baldwyne. Very few Ipswich burgesses left anything at all to the poor. Of course, more might have been expected from the rich clothiers than from their humbler neighbours. If they were taking only their first tentative steps towards becoming capitalists, they were not much further down the road to becoming philanthropists.

Suffolk clothiers followed many different paths in life, but it is the rich and successful whom we remember best. In the putting out system they found a novel model of business organisation that they imposed on their neighbours in the small towns and villages of south Suffolk, particularly those of Babergh hundred, and

113 SROB Pye 182.

whom they recruited as spinsters, weavers and fullers. Rather than lamenting the departure of the Hanseatic merchants from Ipswich and Colchester as a disaster, they seized it as an opportunity. With notable exceptions they looked to London as the focus of their trade and found their most important customers among mercers who bought their heavily fulled, but unfinished broadcloth for export to Antwerp. The most enterprising clothiers made themselves rich, acquired lands and furnished their homes with fine silver. In some ways they looked like capitalists, and a few behaved like philanthropists. They were, nevertheless, men of their time, with horizons and worldview quite different to Victorian captains of industry. One must beware applying to them a false veneer of modernity.

– 7 –

Conclusion

By the close of the Middle Ages Suffolk was by far the leading textile region in England. In terms of production, the county's clothiers had triumphed over those elsewhere. This success story may be divided into three distinct periods – the first up to 1400, the next from 1400 until 1470 and the last up to 1500.

In the earliest period cloth-making was initially concentrated in the towns of Bury St Edmunds and Sudbury. In Ipswich it was of little consequence. From the late thirteenth-century the industry gathered momentum in Hadleigh and, thanks to the archives of the priory of Christchurch Canterbury, its progress is well documented. The years between 1350 and 1400 witnessed significant expansion in Suffolk cloth production, although not by as much as the alnage accounts suggest. By the end of that period Hadleigh was almost certainly the most important textile centre in the county, although in many other small towns and villages along the banks of the rivers Stour and Brett cloth-makers were already hard at work. They produced an eclectic range of fabrics, both woollens and worsteds, but more narrow cloths were manufactured than broadcloths. These were of a relatively poor quality and probably most were made from the wool of Suffolk sheep. If they were dyed at all, then woad was preferred to madder. The major technological innovation during this period was the horizontal loom that later morphed into the broadloom.

In the second period between 1400 and 1470 cloth-making spread across the county and was well established in each of its river valleys. South Suffolk towns did best and by 1470 Lavenham had out-stripped all others. Many weavers and fullers enjoyed independence and some became prosperous. In the late 1460s more than five-hundred cloth-makers presented their wares to the alnager for his seal of approval, and aggregate production was twice what it had been in the 1390s. Some, such as the weaver Roger Crytott of Lavenham and the fuller William Alysaunder originally of Monks Eleigh, enjoyed extraordinary business success. They increasingly targeted their production at the export market and sold to the Hanseatic merchants who traded through Colchester and Ipswich. Better quality broadcloth was more in demand now and woolmen were busy in procuring premium wools from beyond the county boundary. The Hanse wanted finished cloth to carry in their ships to northern Europe, enabling dyers and shearmen to earn a good living too. The Hillary family of dyers became prominent in Hadleigh. Although they and their kind relied heavily on City grocers for the supply of dyestuffs and dyes, at least some of this was shipped into local ports. The Great Slump of the mid-fifteenth century hit Suffolk cloth production less severely than elsewhere in England. Its effect was most evident in Bury St Edmunds where it impoverished many weavers. Technological change during this period of growth was surprisingly limited.

In the third and final period, after 1470, the focus of production turned to the London market. The Hanseatic merchants moved their operations to the Capital, giving rise to a sharp reduction in the export of cloth, and the import of dyestuffs and dyes through Colchester and Ipswich. The impact on the county's cloth-makers was dramatic. Production did not decline, but was organized quite differently. The clothier was the key figure, relying on his contacts and good credit in London. While not, as sometimes claimed, the first capitalist, he was a new beast in the commercial jungle. He used his networks and resources to buy wool direct from the grower without the aid of local woolmen; and introduced the putting out system by engaging spinsters, weavers and fullers to work for him. He increasingly fulled his cloth by mill rather than by hand or foot. The number of recorded clothiers increased dramatically, while that of woolmen, weavers and fullers went into sharp decline. In the hundred of Babergh, where most clothiers operated, the emphasis was on producing heavily fulled, but unfinished broadcloth for sale to London merchants, particularly Merchant Adventurers, and onward shipment to Antwerp. Not all clothiers were rich or surrounded by large teams of out workers. The most successful, such as Thomas Spryng II of Lavenham, did, however, make enormous fortunes. When they died they left some of their wealth to fund church building, road repairs and the relief of poverty.

Even in the late fifteenth-century not all Suffolk cloth was made for export. William Forthe of Hadleigh, in his day perhaps England's foremost clothier, appears to have avoided London altogether and concentrated on the domestic market. The weavers of Bury St Edmunds, such as John Roket, found new prosperity by successfully exploiting a growing demand for bedding. Cloth-finishers in Stowmarket and Ipswich, such as members of the Cake family, earned confortable livings by servicing that part of the industry still targeted at domestic consumption.

Textile industries elsewhere in England followed similar paths, in the West-Country a little earlier, and in the Kentish Weald a little later than in Suffolk. Common factors were at play – cold winters which increased demand for heavily felted, dark clothes; the availability of flowing water which was essential for fulling and dyeing; a shift from arable to pastoral husbandry which was less labour intensive and so gave smallholders and labourers more time to make cloth; engrossment and enclosure which created a pool of landless peasants in need of an income from cloth-working; demographic changes brought about by plague which left their more fortunate neighbours with spare cash to spend on clothes; overseas demand for English textiles; markets and fairs which facilitated the marketing and sale of cloth and the purchase of necessary raw materials; a transport network of rivers and roads to distribute those goods; technological and organisational innovation which facilitated higher levels of production; and the spread of skills and knowledge, not least thanks to émigré Flemish weavers.

This volume began with a reference to jigsaw puzzles. It is for the reader to judge what pieces have been added to our picture of the medieval woollen cloth industry. The author hopes, above all, that he has brought into sharper focus the lives of those who made cloth in the Middle Ages – their hard work, skills and expertise, aspirations, successes and failures, material possessions and family ties. This is not a book about kings, queens, battles, or executions. It is 'history-from-below' which tries to explain the rise of a major industry from the view-point of those who worked in it.

– Appendix A –

Cloths Subject to Alnage in the 1350s and 1390s

The increase in the number of cloths sealed in Suffolk between the 1350s and the 1390s was nearly 500 percent. Large increases were recorded for some other counties. The extent to which this reflected the actual increase in cloth production depends on the nature of the cloths that were included and excluded from the alnage during those two periods.

Statutes, Parliamentary Rolls and Letters Patent

Edward III's statute of 1353 included within the alnage cloths and half cloths of assize. A coloured cloth of assize should measure not less than 26 yards by 6½ quarters of a yard, and a ray not less than 28 yards by 6 quarters of a yard. The statute made special provision for cloths that exceeded the length of the cloth or half cloth of assize by more than 3 yards. It expressly exempted 'cloth which containeth not Half a Cloth'.[1]

Edward III's further statute of 1373 amended the 1353 statute by stipulating a slightly narrower width for coloured cloth that should be not less than 6 quarters of a yard. It went on, unfortunately, to provide that 'If any cloth be put to sale after [1 August] that is not of the said assize it shall be forfeit to the king'.[2]

As a result, in 1376, a petition was presented to Edward III's Parliament:

> To our lord the king and his good council; the commonalities of the counties of Essex and Suffolk declare: that whereas, in the statute made in the forty-seventh year of the reign of our present lord the king, it was ordained that all coloured cloths made for sale in England would be 26 ells in length measured by the back, and at least 5 quarters (sic) in width; and half cloth of the length and width according to the rate, on forfeiture of the same cloths.
>
> May it please our lord the king and his noblest council to grant to your said communities your gracious letters patent and to declare by the same in this present parliament that the cloths called cogware and kersey made in the said counties, and other such narrow cloths made there and in other regions also, should not be included in the said statute, in aid and relief of the said commonalty.

The king's answer was that:

> The king wills that they should have such letters in which it should be declared that the narrow cloths called cogware and kersey, usually made in

1 *Statutes*, 1, pp. 330–1.

2 *Statutes*, 1, p. 395.

> the said counties, are not intended to be included in the said statute or under the penalty of the same.[3]

Edward III then issued letters patent stating that 'at the request of the commonalties of the counties of Essex and Suffolk [...] the cloths called "cogware" and "kereseyes" made in the said counties, and other like strait cloths made there and in other counties, be not comprised in the statute'. This was confirmed by Richard II's letters patent of 1381.[4]

In 1394 another petition was presented to Richard II's Parliament:

> 'The commons of Essex and elsewhere. Also the commons of Essex, Suffolk, and Norfolk request: that it may please you to ordain that they shall have licence to make their cloths as they used to in the past; so that each piece is of a length of twelve yards and a width of one yard of four quarters, and that the same cloths pay custom to the king according to their size, as well all other cloths in the country of England pay, notwithstanding any statute or ordinance made to the contrary. And may it please you to understand that the said cloths are made of wool of a far worse quality than that of other cloths, and most of the wool from which the same cloths are made could not be exported overseas for the making of other cloths. Nevertheless the said supplicants pray of you that they be not in a worse degree than all the other people of the land, considering that the bulk and majority of people from the said lands of Essex, Suffolk, and part of Norfolk do not know how to do anything else except work in the said craft; and they are lesser people, so that it would be a great pity to destroy their livelihood.'

The king's answer was that:

> 'The king wills that every man of the kingdom may make cloths, both kerseys and others, of such a length and width as he chooses, paying ulnage, subsidies, and other dues, namely, on each piece of cloth proportionately, notwithstanding any statute, ordinance, proclamation, restraint, or prohibition made to the contrary. And no one shall sell or put on sale any cloths before they have been measured by the king's ulnager and sealed with the seal so ordained, on pain of the penalties contained in the statutes thereon. And that no one shall add or mix lime or commit any other fraud in the cloths of kersey, on pain of forfeiting the same.'[5]

Richard II's statute of 1393/94 brought within the alnage 'as well Kerseys as others, of such length and breadth as [every man of the realm] please'.[6] The particular alnage accounts of the 1390s confirmed this by including straits.[7]

---

3 *Rot. Parl.*, ii, p. 347.

4 *CPR 1374–77*, p. 396; *CPR 1377–81*, p. 615.

5 *Rot. Parl.*, iii, p. 320.

6 *Statutes*, 2, p. 88.

7 TNA E 101/342/8, 10.

Commentary

Gray states of the 1350s and 1390s that 'both totals would include kerseys, since cloths of more than 3 yds. long were not exempt from ulnage dues'. His was clearly a misinterpretation of the 1353 statute.[8] Heaton states that 'The great majority of cloths made in the country districts of Yorkshire were narrow cloths, "streit" cloths, kerseys &c, which rarely exceeded 12 yards, and therefore escaped the payment of subsidy and the supervision of the ulnager. In 1393–4, however, a change in the policy of cloth regulation broke down this evasion'.[9] What was true of Yorkshire cloths was also the case with Suffolk cloths. Heaton's interpretation was almost certainly correct. Short narrow cloths such as kerseys and straits were not included in the 1350s alnage accounts. Although kerseys came, in the course of time, to measure 18 yards, if not longer, by 4 quarters (of a yard), and so possibly be half cloths, in the fourteenth century, as the 1394 petition implies, they were almost certainly shorter.[10]

One suspects that the cloth-makers of Essex and Suffolk petitioned Parliament in the 1370s and that the king issued letters patent granting relief because over-enthusiastic alnagers had taken the view that the 1373 statute prohibited the making of narrow cloths when, in fact, the intention of that statute was probably simply to confirm that cloths and half cloths of assize had to conform to certain dimensions. It was an early case of poorly drafted legislation.

Conclusion

Kerseys, straits and other narrow cloths were not included in the 1350s alnage accounts, but were included in those of the 1390s. Any comparison between cloth production in the 1350s and that in the 1390s, both in Suffolk and elsewhere, must take this into account and recognize that that the increase in sealed cloths between the two periods exaggerates the increase in actual cloth production.

8 Gray (1924), pp. 28–9.

9 Heaton (1920), p. 69.

10 *Statutes*, 2, p. 404. A 1433 statute required that straits be 14 yards 'unwatered' and 12 yards 'watered' (ie fulled): *Statutes*, 2, p. 285. Oldland (2014), p. 37.

– Appendix B –

Cloth-workers and Clothiers in England and Wales 1480–1500

(This data is extracted from ten plea rolls of the court of common pleas, namely TNA CP 40/883, 885A, 888, 895, 907, 909, 911, 919, 931, 943 and 951 that have been indexed by a volunteer team under the auspices of the University of Houston, Texas and spot checked by the author. The database is defined in this volume as TNA CP 40B)

	Card-makers	*Spinsters*	*Weavers*	*Fullers*	*Dyers*	*Shearmen*	*Clothiers*	*Total*
Bedfordshire								
Ampthill	0	0	1	0	0	0	0	1
Bedford	0	0	0	0	4	0	0	4
Biggleswade	0	1	0	0	0	0	0	1
Bromham	0	0	0	1	0	0	0	1
Dunstable	0	0	0	1	0	1	0	2
Eaton Bray (?)	0	0	0	1	0	0	0	1
Leighton Buzzard	0	0	0	0	2	0	0	2
Potton	0	0	1	1	0	0	0	2
Sandy	0	0	0	1	0	0	0	1
Westoning	0	0	1	0	0	0	0	1
Wyboston	0	1	0	0	0	0	0	1
Total	0	2	3	5	6	1	0	17
Berkshire								
Abingdon	0	0	3	2	1	1	4	11
East Hendred	0	0	0	0	0	0	1	1
Faringdon	0	0	1	1	0	0	0	2
Hampstead Marshall	0	0	0	0	0	0	2	2
Hungerford	0	0	0	0	0	0	1	1
Long Wittenham	0	0	1	0	0	0	0	1
Newbury	0	0	6	4	3	0	12	25
Reading	0	0	4	9	7	2	3	25
Wantage	0	0	0	0	0	0	3	3
Wokingham	0	0	0	3	1	1	4	9
Total	0	0	15	19	12	4	30	80
Buckinghamshire								
Amersham	0	0	0	0	1	0	0	1
Aylesbury	0	0	1	0	0	0	1	2

Buckingham	0	0	0	0	5	0	0	5
Burnham	0	0	0	1	0	0	0	1
Chepping Wycombe	0	0	2	1	0	0	0	3
High Wycombe	0	0	0	0	0	0	1	1
Norton Blossomville	0	0	0	1	0	0	0	1
Stony Stratford	0	0	0	0	0	1	0	1
Thornton	0	0	0	1	0	0	0	1
Wendover	0	0	0	0	0	1	0	1
Wooburn	0	0	0	1	0	0	0	1
Total	0	0	3	5	6	2	2	18
Cambridgeshire								
Abington Piggots	0	0	0	0	0	0	1	1
Cambridge	0	0	0	0	1	1	1	3
Ickleton	0	0	0	1	0	0	0	1
Leverington	0	0	1	0	0	0	0	1
Linton	0	0	1	0	0	0	0	1
Newmarket	0	0	1	0	0	0	0	1
Sawston	0	0	0	1	0	0	0	1
Wisbech	0	0	0	1	3	0	0	4
Total	0	0	3	3	4	1	2	13
Cheshire								
Congleton	0	0	0	0	0	0	1	1
Cornwall								
Bodmin	0	0	3	0	0	0	0	3
'Dunhenetburgh' (?)	0	0	1	0	0	0	0	1
Helston	0	0	1	0	0	1	0	2
Lawhitton	0	0	1	0	0	0	0	1
Liskeard	0	0	0	0	1	0	0	1
Padstow	0	0	0	1	0	0	0	1
St Clement	0	0	1	0	0	0	0	1
St Keverne	0	0	1	0	0	0	0	1
St Minver	0	0	1	0	0	0	0	1
Stratton	0	0	0	0	1	0	0	1
Truro	0	0	0	2	0	0	0	2
Warbstow Cross	0	0	1	0	0	0	0	1
Total	0	0	10	3	2	1	0	16
Cumbria								
Carlisle	0	0	1	0	0	0	0	1
Denbighshire								
Denbigh	0	0	0	0	0	0	1	1
Derbyshire								
Ashbourne	0	0	0	0	1	0	0	1
Belper	0	0	2	0	0	0	0	2

Chesterfield	1	0	0	0	2	0	0	3
Derby	0	0	2	1	1	0	0	4
Dronfield	0	0	0	0	1	0	0	1
Hopton	0	0	1	0	0	0	0	1
Kirk Ireton	0	1	0	0	0	0	0	1
Old Whittington	0	0	1	0	0	0	0	1
Wirksworth	0	0	0	2	0	0	0	2
Total	1	1	6	3	5	0	0	16
Devon								
Ashburton	0	0	0	2	0	0	0	2
Barnstable	0	0	2	0	1	0	1	4
Bideford	0	0	0	2	0	0	0	2
Bishop's Tawton	0	0	0	1	0	0	0	1
Boringdon	0	0	0	1	0	0	0	1
Bradninch	0	0	0	1	0	0	0	1
Braunton	0	0	0	1	0	0	0	1
Brixton	0	0	3	0	0	0	0	3
Buckland Monachorum	0	0	0	1	0	0	0	1
'Bykcombe' (?)	0	0	1	0	0	0	0	1
Chulmleigh	0	0	1	0	0	0	0	1
Colyford	0	0	0	0	1	0	0	1
Craddock	0	0	2	0	0	0	0	2
Crediton	0	0	2	1	0	0	0	3
Cullompton	0	1	0	2	3	0	0	6
Culmstock	0	0	0	1	0	0	0	1
Dolton	0	0	0	0	0	0	1	1
Dowland	0	0	0	0	0	0	1	1
Exeter	0	1	2	3	13	2	0	21
Fremington	0	2	0	0	0	0	0	2
Great Torrington	0	0	2	0	1	0	0	3
Hartland (?)	0	0	0	1	0	0	0	1
Hittisleigh	0	0	1	0	0	0	0	1
Honiton	0	0	1	0	2	0	1	4
Kingsbridge	0	0	1	0	0	0	0	1
Langtree (?)	0	0	1	0	0	0	0	1
Milton Damerel	0	0	0	1	0	0	0	1
Modbury	0	0	0	1	0	0	0	1
Newton Abbot	0	0	1	0	0	0	0	1
Newton Poppleford	0	0	1	0	0	0	0	1
North Molton	0	0	0	1	0	0	0	1
North Tawton	0	0	0	1	0	0	0	1
Oakford	0	0	0	1	0	0	0	1
Ottery St Mary	0	0	0	3	7	0	0	10
Pilton	0	0	1	1	0	0	1	3
Plympton	0	0	4	0	0	0	0	4
Shebbear	0	0	0	1	0	0	0	1

Sidmouth	0	0	1	0	0	0	0	1
Silverton	0	0	1	0	0	0	0	1
'Skyrydon'	0	0	0	1	0	0	0	1
'Smyllyngcote' (?)	0	1	0	0	0	0	0	1
South Moulton	0	0	0	3	0	0	0	3
Spreyton (?)	0	0	1	0	0	0	0	1
Staverton	0	0	0	1	0	0	0	1
Stockleigh English	0	0	1	0	0	0	0	1
Tavistock	0	0	1	0	1	0	0	2
Thornston (?)	0	0	0	1	0	0	0	1
Tiverton	0	0	1	1	0	0	0	2
Totnes	0	0	1	0	0	1	0	2
Uffculme	0	0	0	1	1	1	0	3
Umberleigh	0	0	1	0	0	0	0	1
Whitchurch	0	0	1	0	0	0	0	1
Willand	0	0	0	1	0	0	0	1
Witheridge	0	0	1	0	0	0	0	1
Unknown	0	0	0	0	1	0	0	1
Total	0	5	36	36	31	4	5	117
Dorset								
Bradfield Abbas	0	0	0	0	0	0	1	1
Bridport	0	0	0	0	1	0	0	1
Corscombe	0	0	0	0	0	0	1	1
Frampton	0	0	0	1	0	0	0	1
Lyme Regis	0	0	0	3	0	0	0	3
Shaftsbury	0	1	0	0	0	0	0	1
Sherborne	0	0	0	2	3	0	2	7
Sturminster Marshall	0	0	0	0	1	0	0	1
Thorncombe	0	0	0	1	0	0	0	1
Throop	0	0	0	1	0	0	0	1
Wimborne Minster	0	0	2	0	1	0	0	3
Total	0	1	2	8	6	0	4	21
Durham								
Durham	0	0	0	0	2	0	0	2
Essex								
Ardleigh	0	0	3	0	0	1	2	6
Billericay	0	0	0	1	0	0	0	1
Bocking	0	0	1	2	0	0	7	10
Braintree	0	0	0	0	0	1	5	6
Bulmer Tye	0	0	0	1	0	0	0	1
Burnham	0	0	0	0	0	1	0	1
Chelmsford	0	0	0	1	2	0	1	4
Coggeshall	0	0	3	8	0	1	7	19
Coggeshall Hamlet	0	0	0	1	0	0	0	1

Colchester	0	0	3	9	4	3	22	41
Dedham	0	0	1	2	0	0	9	12
Earls Colne	0	0	0	0	1	0	1	2
Feering	0	0	3	0	0	0	0	3
Felsted	0	0	0	0	0	0	1	1
Great Baddow	0	0	0	1	0	0	0	1
Great Burstead	0	0	1	0	1	0	0	2
Great Horkesley	0	0	0	0	1	0	0	1
Great Warley (?)	0	0	1	0	0	0	0	1
Halstead	0	0	2	1	1	0	0	4
Harlow	0	0	1	0	0	0	0	1
Heybridge	0	0	1	0	1	0	0	2
Hinxton (?)	0	0	0	1	0	0	0	1
Lambourne	0	0	0	0	1	0	0	1
Maldon	0	0	2	0	1	0	0	3
Manningtree	0	0	0	0	0	2	1	3
Moulsham	0	0	0	1	0	0	0	1
Nazeing	0	0	1	0	0	0	0	1
Rochford	0	0	1	0	0	0	0	1
Romford	0	0	0	0	1	0	0	1
Roxwell	0	0	0	1	0	0	0	1
Saffron Walden	0	0	1	1	1	1	0	4
St Osyth	0	0	0	2	0	0	0	2
Waltham Abbey	0	0	3	0	1	0	0	4
Walthamstow	0	0	0	1	0	0	0	1
West Ham	0	0	1	0	0	0	0	1
Witham	0	0	0	0	0	0	2	2
Writtle	0	0	0	0	2	0	0	2
Total	0	0	29	34	18	10	58	149
Gloucestershire								
Alderley	0	0	0	0	0	0	2	2
Bristol	0	0	9	2	10	0	2	23
Burford	0	0	0	0	0	0	1	1
Chipping Camden	0	0	3	0	0	0	2	5
Cirencester	0	0	2	0	1	0	4	7
Coln St Aldwyns	0	0	0	1	0	0	0	1
Dudbridge	0	0	0	1	0	0	0	1
Dursley	0	0	1	3	1	0	14	19
Eastington	0	0	0	1	0	0	0	1
Frampton on Severn	0	0	1	1	0	0	0	2
Gloucester	1	0	3	0	0	0	0	4
Horsley	0	0	0	1	0	0	0	1
Leonard Stanley	0	0	0	1	0	0	0	1
Lypiatt	0	0	1	0	0	0	0	1
Marshfield	0	0	0	0	0	0	1	1
Minchinhampton	0	0	0	1	0	0	1	2

Newent	0	0	0	0	1	0	0	1
Northleach	0	0	1	0	0	0	0	1
Rodborough	0	0	2	4	0	0	2	8
Stroud	0	0	0	9	1	1	13	24
Tetbury	0	0	2	0	0	0	0	2
Thornbury	0	0	1	0	0	0	0	1
Winchcombe	0	0	0	0	0	0	1	1
Woodmancote	0	0	1	0	0	0	0	1
Wotton under Edge	0	0	0	1	0	0	2	3
Total	1	0	27	26	14	1	45	114
Hampshire								
Alresford	0	0	1	0	0	0	0	1
Alton	0	0	2	0	2	0	1	5
Andover	0	0	0	0	1	0	0	1
Basingstoke	0	0	0	2	1	0	0	3
Bramshott	0	0	1	0	0	0	0	1
Charlton (?)	0	0	1	0	0	0	0	1
East Meon	0	0	0	0	2	0	0	2
Fordingbridge	0	0	0	0	1	0	0	1
Greywell	0	0	0	1	0	0	1	2
Hartley Wespall	0	0	0	1	0	0	0	1
Havant	0	0	1	0	1	0	0	2
Heckfield	0	0	1	0	0	0	0	1
Hillyfields	0	0	0	0	1	0	0	1
Kingsclere	0	0	2	0	0	0	0	2
Laverstoke	0	0	1	0	0	0	0	1
Lymington	0	0	2	0	0	0	0	2
Mattingley	0	0	1	0	0	0	0	1
Monk Sherborne	0	0	1	0	0	0	0	1
New Alresford	0	0	2	0	0	0	0	2
North Warnborough	0	0	0	1	1	0	0	2
Odiham	0	0	0	1	0	0	0	1
Overton	0	0	2	0	1	0	0	3
Petersfield	0	0	1	1	0	0	0	2
Ringwood	0	0	0	0	1	0	1	2
Romsey	0	0	0	3	1	0	1	5
Sherborne St John	0	0	0	1	2	0	0	3
Sherfield on Loddon	0	0	0	1	0	0	0	1
South Warnborough	0	0	1	0	0	0	0	1
Southampton	0	0	0	0	0	1	0	1
Southwick	0	0	1	0	0	0	0	1
Titchfield	0	0	0	0	0	0	1	1
Winchester	0	0	3	4	3	0	6	16
Wyck (?)	0	0	1	0	0	0	0	1
Total	0	0	25	16	18	1	11	71

Herefordshire								
Bodenham	0	1	0	0	0	0	0	1
Hereford	0	2	4	0	1	1	1	9
Leominster	0	0	0	1	1	0	0	2
Mordiford	0	1	0	0	0	0	0	1
Pembridge	0	0	1	0	0	0	0	1
Ross on Wye	0	0	1	1	1	0	0	3
Weobley	0	0	1	0	1	0	0	2
Unknown	0	0	0	1	0	0	0	1
Total	0	4	7	3	4	1	1	20
Hertfordshire								
Ashwell	0	0	0	1	0	1	0	2
Bishops Stortford	0	0	1	0	0	0	0	1
Cheshunt	0	0	0	1	0	0	0	1
Hatfield	0	0	0	0	1	0	0	1
Hitchin	0	0	1	0	0	0	0	1
Hoddesdon	0	0	1	0	0	0	0	1
St Albans	0	0	1	4	3	0	0	8
Sawbridgeworth	0	0	0	1	0	0	0	1
Wadesmill	0	0	0	0	1	0	0	1
Waltham Cross	0	0	1	0	1	0	0	2
Watford	0	0	0	1	0	0	0	1
Total	*0*	*0*	*5*	*8*	*6*	*1*	*0*	*20*
Huntingdonshire								
Bluntisham	0	0	0	0	0	0	1	1
St Ives	0	0	0	0	1	0	0	1
St Neots	0	0	0	1	0	0	0	1
Total	0	0	0	1	1	0	1	3
Isle of Wight								
Carisbrooke	0	0	0	0	1	0	0	1
Kent								
Benenden	0	0	1	0	0	0	8	9
Biddenden	0	0	1	0	0	0	10	11
Canterbury	0	0	3	0	6	0	0	9
Cobham	0	0	0	0	0	1	0	1
Cranbrook	0	0	6	3	1	1	20	31
Dartford	0	0	0	3	4	0	1	8
East Farleigh	0	0	0	1	0	0	0	1
East Peckham	0	0	0	1	0	0	0	1
Eastwell	0	0	1	0	0	0	0	1
Faversham	0	0	0	0	0	1	0	1
Frittenden	0	0	0	0	0	0	1	1
Garlinge (?)	0	0	1	0	0	0	0	1
Godmersham	0	0	0	1	0	0	0	1

Goudhurst	0	0	2	3	0	0	6	11
Great Chart	0	0	0	0	1	1	2	4
Hadlow	0	0	0	0	0	0	1	1
Hawkedon	0	0	0	0	0	0	1	1
Hawkhurst	0	0	3	0	0	0	1	4
Headcorn	0	0	4	0	0	0	0	4
High Halden	0	0	0	1	0	0	0	1
Hollingbourne	0	0	0	1	0	0	1	2
Horsmonden	0	0	0	1	0	1	2	4
Hothfield	0	0	0	1	0	2	0	3
Ickham	0	1	0	0	0	0	0	1
Kingsdown	0	0	0	0	0	0	1	1
Leeds	0	0	0	0	0	0	1	1
Lenham	0	0	1	0	0	0	0	1
Linton	0	0	0	0	0	0	1	1
Loose	0	0	0	1	0	0	0	1
Maidstone	0	0	4	2	2	0	0	8
Marden	0	0	2	3	0	0	2	7
New Romney	0	0	0	1	0	0	0	1
Rochester	0	0	0	1	0	0	1	2
Rolvenden	0	0	0	0	0	0	1	1
Sandhurst	0	0	0	1	0	0	3	4
Sandwich	0	0	1	0	0	0	0	1
Sevenoaks	0	0	1	0	0	0	0	1
Shoreham	0	0	0	1	0	0	0	1
Sittingbourne	0	0	0	1	0	0	0	1
Smarden	0	0	4	0	0	0	2	6
Staplehurst	0	0	2	1	0	0	5	8
Strood	0	0	1	1	0	0	0	2
Tenterden	0	0	1	0	1	0	7	9
Tonbridge	0	0	0	0	2	0	0	2
Tudley	0	0	0	0	0	0	1	1
Tunbridge Wells	0	0	0	0	5	0	0	5
Ulcombe	0	0	1	0	0	0	0	1
Warehorne	0	0	0	1	0	0	0	1
West Malling	0	0	0	0	0	1	0	1
Wye	0	0	0	1	0	0	0	1
Yalding (?)	0	0	1	0	0	0	0	1
Total	0	1	41	31	22	8	79	182
Lancashire								
Alkar (?)	0	1	0	0	0	0	0	1
Leyland	0	1	0	0	0	0	0	1
Total	0	2	0	0	0	0	0	2
Leicestershire								
Gaddesby	0	0	1	0	0	0	0	1
Leicester	1	1	0	0	1	2	0	5

Lutterworth	0	0	1	0	0	1	0	2
Pickwell	0	1	0	0	0	0	0	1
Sharnford	0	0	2	0	0	0	0	2
Total	0	1	3	0	1	1	0	6
Lincolnshire								
Aswarby	0	0	1	0	0	0	0	1
Barton on Humber	0	0	0	0	2	0	0	2
Baston	0	0	0	2	0	0	0	2
Baumber (?)	0	0	1	0	0	0	0	1
Beesby	0	0	1	0	0	0	0	1
Boston	0	1	4	1	4	1	0	11
Burgh le Marsh	0	0	0	0	1	0	0	1
Cadney	0	0	1	0	0	0	0	1
Caistor	0	0	1	0	0	0	0	1
Fulstow	0	0	1	0	0	0	0	1
Glanford Bridge	0	0	0	0	1	0	0	1
Grantham	0	0	0	1	0	0	0	1
Great/Little Steeping	0	0	1	0	0	0	0	1
Grimsby	0	0	1	0	0	0	0	1
Hagworthingham	0	0	1	0	0	0	0	1
Holbeach	0	1	1	0	0	0	0	2
Kirton	0	1	1	0	0	0	0	2
Lincoln	0	0	0	1	3	1	0	5
Louth	0	0	0	0	2	0	0	2
Marshchapel	0	1	0	0	0	0	0	1
Morton & Hanthorpe	0	0	1	0	0	0	0	1
Moulton	0	0	1	0	0	0	0	1
Oasby (?)	0	0	1	0	0	0	0	1
Partney	0	0	1	0	0	0	0	1
Saltfleetby	0	0	1	0	0	0	0	1
Sleaford	0	0	0	1	1	1	0	3
Spalding	0	0	0	0	0	1	0	1
Stamford	0	0	1	1	2	0	0	4
Tetney	0	1	0	0	0	0	0	1
Torksey	0	0	0	0	0	1	0	1
Tumby	0	0	0	1	0	0	0	1
Waynfleet	0	0	1	0	0	0	0	1
Total	0	5	20	6	14	5	0	50
London								
Total	0	12	11	19	22	45	2	111
Middlesex								
Brentford	0	0	0	0	1	0	0	1
Holborn	0	0	0	0	0	1	0	1
Islington	0	0	1	0	0	0	0	1
Tottenham	0	0	0	1	0	0	0	1
Uxbridge	0	0	1	0	1	0	0	2

Westminster	0	2	0	0	0	0	0	2
Total	0	2	2	1	2	1	0	8
Norfolk								
Aylesham	0	0	0	3	0	1	0	4
Barton Turf	0	0	1	0	0	0	0	1
Binham	0	0	1	0	0	0	0	1
Bradfield	0	0	0	1	0	0	0	1
Brancaster	0	0	1	0	0	0	0	1
Cawston	0	0	1	0	0	0	0	1
Cley next the Sea	0	0	0	0	0	1	0	1
Coltishall	0	0	1	0	0	0	0	1
Costessey	0	1	0	1	0	0	0	2
Cringleford	0	1	0	0	0	0	0	1
Cromer	0	0	0	0	0	1	0	1
Diss	0	0	1	0	0	0	0	1
East Dereham	0	0	0	1	1	0	0	2
East Harling	0	0	1	0	0	0	0	1
Emneth	0	0	1	0	0	0	0	1
Fakenham	0	0	1	0	0	0	0	1
Great Yarmouth	0	0	1	0	1	0	0	2
Harleston	0	1	0	0	0	1	0	2
Hilgay	0	0	1	0	1	0	0	2
Hingham	0	0	2	0	0	1	0	3
Honing	0	0	2	0	0	0	0	2
Horning	0	0	0	0	1	0	0	1
Horsham St Faith	0	0	1	0	0	0	0	1
Houghton	0	0	1	0	0	0	0	1
Kenninghall	0	0	0	1	0	0	0	1
King's Lynn	0	0	4	0	5	0	0	9
Kirstead	0	0	0	0	1	0	0	1
Little Melton	0	0	1	0	0	0	0	1
Little Walsingham	0	0	0	0	0	1	0	1
Marham	0	1	0	0	0	0	0	1
Mattishall	0	0	1	2	0	0	0	3
New Buckenham	0	0	1	0	0	0	0	1
Norwich	0	1	42	7	14	7	0	71
Pockthorpe	0	0	0	0	1	0	0	1
Pulham St Mary	0	0	1	0	0	0	0	1
Reepham	0	0	0	1	0	1	0	2
Sculthorpe	0	1	0	0	0	0	0	1
Sherringham	0	1	0	0	0	0	0	1
Shipdham	0	0	1	0	0	0	0	1
Sloley	0	0	0	0	1	0	0	1
Smallburgh	0	0	1	0	0	0	0	1
Snettisham	0	0	0	0	0	1	0	1
South Lynn	0	0	1	0	0	0	0	1
Starston/Sturston (?)	0	0	0	1	0	0	0	1

Stratton Strawless	0	0	1	0	0	0	0	1
Suffield	0	0	2	0	0	0	0	2
Swaffham	0	0	1	0	0	0	0	1
Swanton Abbott	0	0	1	0	0	0	0	1
Taverham	0	0	0	0	1	0	0	1
Thetford	0	0	1	0	4	0	0	5
Toftrees	0	1	1	0	0	0	0	2
Tottington	0	0	1	0	0	0	0	1
Tunstead	0	0	4	0	0	0	1	5
Upwell	0	0	1	0	0	0	0	1
West Winch	0	0	2	0	0	0	0	2
Westwick	0	0	4	0	0	0	0	4
Wood Dalling	0	0	1	1	0	0	0	2
Worstead	0	0	7	0	0	0	0	7
Worthing	0	0	0	1	0	0	0	1
Wymondham	0	0	1	0	1	0	0	2
Total	0	8	98	20	32	15	1	174
Northamptonshire								
Harringworth	0	0	1	0	0	0	0	1
Higham Ferrers	0	0	1	0	0	0	0	1
Kettering	0	0	0	0	2	0	0	2
Northampton	0	0	2	0	10	0	2	14
Oundle	0	0	0	1	0	0	0	1
Peterborough	0	0	0	0	1	1	0	2
Stanford on Avon	0	0	0	0	1	0	0	1
Syresham	0	0	1	0	0	0	0	1
Total	0	0	5	1	14	1	2	23
Nottinghamshire								
Blyth	0	0	0	0	1	0	0	1
Kingscliff	0	0	0	0	1	0	0	1
Loudham	0	0	0	1	0	0	0	1
Mansfield Woodhouse	0	0	0	1	0	0	0	1
Newark	0	0	4	0	2	0	0	6
Nottingham	0	0	0	2	4	0	0	6
Retford	0	0	0	0	3	0	0	3
Southwell	0	0	1	0	0	0	0	1
Warsop	0	0	0	0	1	0	0	1
Total	0	0	5	4	12	0	0	21
Oxfordshire								
Bampton	0	0	1	0	0	0	0	1
Banbury	0	0	0	2	2	0	0	4
Burford	0	0	2	0	1	0	0	3
Culham	0	0	0	1	0	0	0	1
Henley on Thames	0	0	4	1	6	1	3	15

Oxford	0	0	1	3	0	1	0	5
Thame	0	0	0	0	0	0	1	1
Watlington	0	0	2	0	0	0	0	2
Witney	0	0	2	2	5	0	0	9
Total	0	0	12	9	14	2	4	41
Pembrokeshire								
Tenby	0	0	0	2	0	0	0	2
Rutland								
Great Casterton	0	0	1	0	0	0	0	1
Shropshire								
Bridgnorth	0	0	1	1	0	0	0	2
Claverley	0	0	1	0	0	0	0	1
Cleobury North	0	0	0	1	0	0	0	1
Leebotwood	0	0	0	1	0	0	0	1
Leigh by Worthen	0	0	0	0	1	0	0	1
Ludlow	0	0	3	0	1	0	6	10
Newport	0	0	1	0	1	0	0	2
Shrewsbury	0	0	0	0	0	3	0	3
Total	0	0	6	3	3	3	6	21
Somerset								
Axbridge	0	0	5	1	1	0	3	10
Baltonsborough	0	0	0	0	1	0	0	1
Banwell	0	0	0	0	1	0	1	2
Barwick	0	0	0	2	0	0	0	2
Bath	0	0	1	1	1	0	5	8
Bathford (?)	0	0	1	0	0	0	0	1
Bridgwater	0	0	5	1	3	0	3	12
Bruton	0	0	0	1	1	0	0	2
Cannington	0	0	1	5	0	0	0	6
Chilcompton	0	0	0	0	0	0	1	1
Chipstable	0	0	0	1	1	0	0	2
Cole	0	0	0	0	0	0	1	1
Coleford	0	0	0	2	0	0	5	7
Compton Dando	0	0	0	1	0	0	0	1
Congresbury	0	0	1	0	0	0	0	1
Croscombe	0	0	0	0	2	0	0	2
Crowcombe	0	0	0	0	1	0	0	1
Dunster	0	0	1	0	1	0	0	2
East Harptree	0	0	0	0	0	0	5	5
Freshford	0	0	0	1	0	0	0	1
Frome	0	0	1	0	0	0	3	4
Glastonbury	0	0	0	0	2	0	2	4
Huish (Episcopi ?)	0	0	0	0	0	0	1	1
Ilminster	0	0	1	2	1	0	0	4

Kingstone	0	0	1	0	0	0	0	1
Langport	0	0	1	0	3	0	2	6
Leigh on Mendip	0	0	0	0	1	0	5	6
Marston	0	0	0	1	0	0	0	1
Mells	0	0	0	0	0	0	11	11
Milverton	0	0	1	0	1	0	0	2
Monksilver	0	0	0	0	1	0	0	1
Nether Stowey	0	0	1	0	1	0	0	2
Norton St Philip	0	0	3	0	0	0	0	3
Nunney	0	0	0	0	0	0	1	1
Pensford	0	0	0	1	0	0	6	7
Porlock	0	0	1	0	0	0	0	1
Publow	0	0	0	2	0	0	0	2
Shepton Mallet	0	0	4	1	0	0	5	10
Stoke St Michael	0	0	0	0	0	0	1	1
Stoke under Mendip	0	0	0	0	0	0	1	1
Stringston	0	0	1	0	0	0	0	1
Taunton	0	0	4	6	4	0	1	15
Timberscombe	0	0	0	0	0	1	0	1
Wellington	0	0	0	0	1	0	1	2
Wells	0	0	0	2	2	0	4	8
Westbury sub Mendip	0	0	0	0	0	0	1	1
Williton	0	0	1	0	0	0	0	1
Woollard	0	0	1	0	0	0	1	2
Worth	0	0	0	0	0	0	1	1
Wrington	0	0	0	1	0	0	2	3
Yeovil	0	0	2	0	0	0	2	4
Total	0	0	38	32	30	1	75	176
Staffordshire								
Burton on Trent	0	0	0	0	1	0	1	2
Gnosall	0	0	1	0	0	0	0	1
Huntington	0	0	1	0	0	0	0	1
Lichfield	0	0	2	3	5	0	0	10
Longdon	0	0	1	0	0	0	0	1
Newcastle under Lyme	0	0	1	0	0	0	0	1
Stafford	0	0	0	0	1	0	0	1
Tamworth	0	0	1	0	0	0	0	1
Tutbury	0	0	0	0	0	1	0	1
Uttoxeter	0	0	0	0	1	1	0	2
Walsall	0	0	1	0	0	0	0	1
Wolverhampton	0	0	1	0	0	0	0	1
Total	0	0	9	3	8	2	1	23
Suffolk								
Beccles	0	0	2	1	1	2	0	6
Bildeston	0	0	4	3	1	0	13	21

Boxford	0	0	2	1	0	0	1	4
Bradley	0	0	0	0	0	0	1	1
Brantham	0	0	0	0	0	0	1	1
Brent Eleigh	0	0	0	1	0	0	1	2
Bungay	0	0	0	0	2	1	0	3
Bures St Mary	0	0	1	1	0	0	0	2
Bury St Edmunds	0	0	27	7	10	4	10	58
Cavendish	0	0	0	2	0	0	0	2
Chelsworth	0	0	3	1	2	0	0	6
Clare	0	0	0	0	0	0	4	4
Creeting St Peter	0	0	0	0	0	0	1	1
East Bergholt	0	0	2	1	0	0	12	15
Edwardstone	0	0	0	0	0	0	1	1
Elmswell	0	0	1	0	0	0	0	1
Eriswell	0	0	1	0	0	0	0	1
Eye	0	0	1	0	0	0	0	1
Fornham All Saints	0	0	1	0	0	0	0	1
Glemsford	0	0	1	1	0	0	0	2
Great Waldingfield	0	0	0	1	0	0	3	4
Groton	0	0	0	0	0	0	1	1
Grundisburgh	0	0	0	0	2	0	0	2
Hacheston	0	0	0	1	0	0	0	1
Hadleigh	1	0	4	4	2	2	15	28
Harkstead	0	0	0	1	0	0	0	1
Haughley	0	0	0	0	1	0	0	1
Higham	0	0	0	1	0	0	2	3
Hitcham	0	0	1	0	0	0	0	1
Horringer	0	0	1	0	0	0	0	1
Icklingham	0	0	1	1	0	0	0	2
Ipswich	0	0	2	1	2	2	1	8
Kersey	0	0	3	4	1	0	1	9
Lavenham	0	0	5	9	1	2	17	34
Little Waldingfield	0	0	0	0	0	0	2	2
Long Melford	0	0	6	13	3	1	8	31
Metfield	0	1	0	0	0	0	0	1
Mildenhall	0	0	0	0	1	0	0	1
Monks Eleigh	0	0	1	1	0	0	2	4
Nayland	0	0	3	2	0	0	17	22
Needham Market	0	0	1	0	1	0	4	6
Newbourne	0	0	0	1	0	0	0	1
Orford	0	0	1	0	0	0	0	1
Preston St Mary	0	0	1	1	0	0	0	2
Snape	0	0	1	0	0	0	0	1
Stoke	0	0	0	0	0	0	1	1
Stoke by Clare	0	0	0	0	0	0	2	2
Stoke by Nayland	0	0	5	7	3	0	14	29
Stowmarket	0	0	2	1	7	2	8	20

Stratford St Mary	0	0	0	3	0	0	4	7
Sudbury	0	0	8	3	3	1	15	30
Waldingfield (?)	0	0	0	0	0	0	1	1
Walyngford (?)	0	0	1	0	0	0	0	1
Walsham le Willows	0	1	0	0	0	0	0	1
West Stow	0	0	0	3	0	0	0	3
Woodbridge	0	0	1	1	3	0	0	5
Woolpit	0	0	0	1	0	0	0	1
Total	1	2	58	65	32	10	136	304
Surrey								
Cove	0	0	1	0	0	0	0	1
Croydon	0	0	2	0	0	0	0	2
Egham	0	0	0	0	1	0	0	1
Farnham	0	0	2	1	0	0	2	5
Godalming	0	0	1	0	1	0	0	2
Guildford	0	0	0	2	2	0	0	4
Kingston upon Thames	0	0	0	0	1	0	0	1
Southwark	1	1	0	1	8	3	0	14
Wandsworth	0	0	1	0	0	0	0	1
Wimbledon	0	0	0	1	0	0	0	1
Woking	0	0	0	0	0	0	1	1
Total	1	1	7	5	13	3	3	33
Sussex								
Bramber	0	0	1	0	0	0	0	1
Burwash	0	0	0	1	0	0	0	1
Chichester	0	0	0	1	2	1	0	4
Ditchling	0	0	1	0	0	0	0	1
East Grinstead	0	0	0	0	0	1	0	1
Fittleworth	0	0	0	0	2	0	0	2
Framfield	0	0	0	0	1	0	0	1
Hastings	0	0	0	1	0	0	0	1
Henfield	0	1	1	0	0	0	0	2
Horsham	0	0	3	0	0	1	0	4
Lamberhurst	0	0	0	0	0	0	2	2
Lindfield	0	0	2	0	0	1	0	3
Midhurst	0	0	1	0	1	0	1	3
Petworth	0	0	0	0	1	0	0	1
Pulborough	0	0	0	1	0	0	0	1
Salehurst	0	0	0	0	1	0	0	1
Shoreham by Sea	0	0	1	0	0	0	0	1
Steyning	0	0	1	0	0	0	0	1
Trotton with Chithurst	0	0	0	0	0	0	1	1
Uckfield	0	0	1	0	0	0	0	1
Warbleton	0	0	0	0	0	0	1	1
Woodmancote	0	2	0	0	0	0	0	2
Total	0	3	12	4	8	4	5	36

Warwickshire								
Alderton	0	0	0	1	0	0	0	1
Ashow (?)	0	0	0	1	0	0	0	1
Atherstone	0	0	1	1	0	0	0	2
Birmingham	0	0	0	0	1	0	0	1
Coleshill	0	0	0	1	0	0	0	1
Coventry	4	0	11	2	15	10	2	44
Harborough Magna	0	0	0	0	0	1	0	1
Rugby	0	0	0	2	0	0	0	2
Solihull	0	0	1	0	0	0	0	1
Stoneleigh	0	0	0	1	0	0	0	1
Stratford upon Avon	0	0	1	5	0	0	0	6
Studley	0	0	1	0	0	0	0	1
Tanworth in Arden (?)	0	0	1	0	0	0	0	1
Warwick	0	0	2	0	0	0	0	2
Wolston	0	0	0	1	0	0	0	1
Total	4	0	18	15	16	11	2	66
Westmoreland								
Kendal	0	0	0	0	1	1	1	3
Wiltshire								
Ashton	0	0	3	0	0	0	0	3
Bradford upon Avon	0	0	1	0	1	0	2	4
Calne	0	0	1	0	0	0	0	1
Castle Combe	0	0	0	0	0	0	2	2
Chippenham	1	0	1	0	0	0	2	4
Christian Malford	0	0	1	0	0	0	0	1
Compton Bassett	0	0	1	0	0	0	0	1
Devizes	0	0	3	0	0	0	12	15
Donhead St Andrew	0	0	0	0	1	0	0	1
East Lavington	0	0	1	0	0	0	0	1
Horningsham	0	0	0	2	0	0	0	2
Keevil	0	0	1	0	0	0	0	1
Kingswood	0	0	0	1	1	0	0	2
Market Lavington	0	0	1	0	0	0	0	1
Marlborough	0	0	0	0	1	0	0	1
Norton Bavent	0	0	0	0	0	0	1	1
Potterne	0	0	0	0	0	0	1	1
Salisbury	0	0	9	0	2	0	2	13
Seend	0	0	0	1	0	0	0	1
Steeple Ashton	0	0	1	0	0	0	1	2
Sutton	0	0	1	0	0	0	0	1
Swallowfield	0	0	0	0	0	0	1	1
Trowbridge	0	0	0	1	2	0	6	9
Warminster	0	0	1	0	0	0	0	1
West Lavington	0	0	3	5	0	0	2	10
Westbury	0	0	2	0	0	0	0	2

Wilton	0	0	1	3	0	0	1	5
Unknown	0	0	0	0	0	0	1	1
Total	1	0	32	13	8	0	34	88
Worcestershire								
Bewdley	0	0	0	0	1	0	0	1
Bromsgrove	0	0	0	1	0	0	0	1
Evesham	0	0	0	0	1	0	0	1
Kidderminster	0	0	0	0	0	0	1	1
Pershore	0	0	0	0	0	0	1	1
Powick	0	0	1	0	0	0	0	1
Worcester	0	0	7	3	1	1	3	15
Wyre Pidell	0	0	1	0	0	0	0	1
Total	0	0	9	4	3	1	5	22
Yorkshire (East)								
Beverley	0	0	3	3	4	0	0	10
Blacktoft	0	0	1	0	0	0	0	1
Eastrington	0	0	1	0	0	0	0	1
Howden	0	0	0	0	1	0	0	1
Kingston upon Hull	0	0	0	2	0	1	0	3
North Cave	0	0	0	1	0	0	0	1
Pollington	0	0	1	0	0	0	0	1
Snaith	0	0	0	0	1	0	0	1
South Cave (?)	0	0	1	0	0	0	0	1
Wilsthorpe	0	0	0	1	0	0	0	1
Woodmansey	0	0	0	1	0	0	0	1
Total	0	0	7	8	6	1	0	22
Yorkshire (North)								
Aldwark	0	0	1	0	0	0	0	1
Boroughbridge	0	0	1	1	0	0	0	2
Carlton juxta Snaith	0	0	1	0	0	0	0	1
Farnley	0	0	1	0	0	0	0	1
Gilling East	0	0	1	0	0	0	0	1
'Hakney'	0	1	0	0	0	0	0	1
Helmsley	0	0	0	0	1	0	0	1
Hensall	0	0	1	0	0	0	0	1
Kirk Hammerton	0	0	0	2	0	0	0	2
Masham	0	0	0	1	0	0	0	1
Muscoates	0	0	1	0	0	0	0	1
Richmond	1	0	0	0	0	0	0	1
Ripon	0	0	0	1	2	0	0	3
Sawley	0	0	2	0	0	0	0	2
Selby	0	0	1	0	0	0	0	1
Sowerby (Thirsk)	0	0	0	0	1	0	0	1
Sproxton	0	0	1	0	0	0	0	1

York	3	1	9	9	9	3	0	34
Total	4	2	20	14	13	3	0	56
Yorkshire (West)								
Adel	0	0	0	1	0	0	0	1
Almondbury	0	0	1	0	0	0	0	1
Balidon	0	0	1	0	0	0	0	1
Bawtry	0	0	0	0	1	0	0	1
Becca	0	0	1	0	0	0	0	1
Beeston	0	0	2	1	0	0	0	3
Bingley	0	0	0	0	1	0	0	1
Bradford	0	0	1	1	0	0	3	5
Ferrybridge	0	0	0	0	1	0	0	1
Halifax	0	0	0	0	1	0	1	2
Hemsworth (?)	0	0	1	0	0	0	0	1
Heptonstall	0	0	0	0	2	2	0	4
Leeds	0	0	3	1	0	0	1	5
Lofthouse	0	0	1	0	0	0	0	1
Methley	0	1	0	0	0	0	0	1
Newall	0	0	0	1	0	0	0	1
Northowram	0	0	0	0	0	0	2	2
Otley	0	0	2	0	0	0	0	2
Penistone	0	0	1	0	0	0	0	1
Pontefract	0	1	0	0	1	0	0	2
Rastrick	0	0	2	0	0	0	0	2
Rotherham	0	0	0	0	0	1	0	1
Sowerby	0	0	0	0	2	0	0	2
Stanley juxta Wakefield	0	0	1	1	0	0	0	2
Steeton	0	0	1	0	0	0	0	1
Stockbridge	0	0	1	0	0	0	0	1
Wakefield	1	1	0	0	1	0	0	3
Wallsworth	0	0	0	1	0	0	0	1
White Syke (in Elland)	0	0	0	1	0	0	0	1
Total	1	3	19	8	10	3	7	51
Total number	**17**	**56**	**643**	**453**	**440**	**162**	**555**	**2326**

Source: TNA CP 40B

Notes: This analysis allocates towns and villages to their medieval, not modern counties, but gives their modern names. Bristol is included under Gloucestershire, but became an independent county in 1373. A Farnham clothier is stated to have lived in Hampshire, which is possible, but he has nevertheless been included in this appendix with the other Farnham entries under Surrey. (?) means that the name of the town is unknown or uncertain. Sometimes it is impossible to attribute any county or town to an entry and in this case the entry is only recorded in the total at the bottom. Very occasionally someone is attributed to more than one county in which case they are counted twice.

– Appendix C –

Suffolk's Rich Clothiers

TNA PROB reference	*Name*	*Residence*	*Date will proven*
11/9/202	Archer John	Sudbury	1492
11/9/97	Broke Robert	Sudbury	1492
11/14/196	Caston William	Lavenham	1504
11/16/170	Cowper Robert	Nayland	1510
11/8/158	Elys Thomas	Long Melford	1488
11/6/204	Fen Robert	Stoke-by-Nayland	1474
11/17/598	Foote Geffrey	Long Melford	1514
11/14/325	Forthe William	Hadleigh	1504
11/9/307	Fulby Thomas	Long Melford	1493
11/11/580	Fulsnape Thomas	Hadleigh	1499
11/6/76	Gardiner Stephen	Bury St Edmunds	1472
11/11/213	Golding John	Glemsford	1497
11/8/360	Gosselyn Nicholas	Lavenham	1491
11/8/477	Hach Nicholas the Elder	Nayland	1490
11/11/159	Hache John	Sudbury	1497
11/14/347	Hadley John	Clare	1504
11/6/472	Horrold John	Clare	1478
11/14/319	Horrold Thomas	Clare	1504
11/10/204	Hyde John	Nayland	1494
11/10/435	Hyll John	Long Melford	1495
11/5/126	Kyng John	Hadleigh	1469
11/7/356	Meryell William	Long Melford	1486
11/6/171	Motte John	Bildeston	1473
11/13/349	Newton John	Lavenham	1502
11/18/342	Parker Thomas	Lavenham	1516

11/14/650	Peverell John	East Bergholt	1505
11/16/161	Ponder Roger	Great Waldingfield	1508
11/8/52	Pye John	Long Melford	1487
11/18/55	Raynham Stephen	Nayland	1514
11/11/743	Reynham John	Nayland	1499
11/9/296	Reynham Robert	Stoke-by-Nayland	1493
11/9/295	Ringbell John	Nayland	1493
11/9/365	Risby John	Lavenham	1493
11/15/265	Risby William	Lavenham	1506
11/14/223	Rote John	Nayland	1504
11/14/774	Rysyng Richard	Little Waldingfield	1505
11/12/322	Sergeaunt Robert	Stoke-by-Nayland	1501
11/8/138	Sexteyn Alexander	Lavenham	1488
11/9/310	Smyth John	Long Melford	1493
11/7/352	Spryng Thomas II	Lavenham	1486
11/16/289	Spryng William	Long Melford	1510
11/10/42	Sturmyn Thomas	Lavenham	1488
11/16/82	Tewe Thomas	Nayland	1510
11/14/728	Turnour Hugh	Stoke-by-Clare	1505
11/14/577	Tyll William	Sudbury	1505

– Appendix D –

Probate Inventory of John Reynham

This is the [final] and true inventory of all such/goods, chattels and debts [] of John Reynham/late of Nayland in the diocese of Norwich made and apprised/by Master John [Moroe] appraisor [of the] most reverend father/[] lord cardinal and archbishop of Canterbury [] day/of February in the year of our Lord God 1499

The Hall

Firstly a table and a bench	20d.
Item a board standing on the bench	6d.
Item an old [trammel]	4d.
Item a chair	6d.
Item a back stool	3d.
Item a pair of [stirrups]	2s.
Item 2 pairs of true balance	12d.
Item a 100 lb lead weight	4d.
Item a [small] chest	2s.
Item a pair of iron balance	[]
Total	15s. 8d.

The Parlour

Firstly a spruce table	[]
Item a bench covering and 12 cushions	6s. 8d.
Item a red cloth hanging	2s.
Item a flat table and two back stools	2s.
Item an old chest	6d.
Item a cupboard	[]
Item a candlestick	[]
Total	30s.

The Chamber

Firstly a bedstead	[8]d.
Item a tester and a canopy	12d.
Item a feather bed and a bolster	8s.
Item two pairs of blankets	4s
Item a pair of sheets	3s. 4d.
Item a coverlet of green cloth	20s
Item a russet gown	12s
Item a blue gown lined with fur	16s.
Item a [] gown	12s
Item a [] kirtle	13s. 4d.
[]	10s.
Item a [] girdle decorated with silver	50s.
Item a set of beads	10s.
Total	£8 []

Another Chamber

Firstly a featherbed and a bolster	8s.
Item a coverlet	20s.
Item a pair of blankets	2s.
A pair of sheets	3s. 4d.
[]	3s.
Item a bedstead	4d.
Total	36s. 8d.

Napery

Firstly 3 tablecloths	3s. 4d.
Item 3 towels	2s.
Item 5 pairs of sheets	12s
Total	17s. 4d.

Plate and money

Firstly 2 Christophers weighing 9 ounces, 3s. 2d. per ounce	29s. 6d. (sic)
Item a [plate] of silver weighing 4 ounces, 3s. 2d. per ounce	12s. 8d.
Item 12 silver spoons weighing 6½ ounces, 3s. 2d. per ounce	20s. 7d.
Item 2 mazers	26s. 8d.
Item in money []	£23
Total	£27 9s. 5d.

Land given to the church

Item lands to the value of	£20

The Buttery

Firstly a saltcellar of pewter	[]
Item a latten chafour	2s. 4d.
Item 3 pewter pots	12d.
Item an old bin	7d.
Item a broad [tub]	4d.
Item 30 pieces of pewter	7s. 6d.
Item 4 latten basins	4s.
Item an ewer	10d.
Item 4 candlesticks	4s.
Total	18s. 8d. (sic)

The Kitchen

Firstly 3 brass pans containing 63 lbs, 2½d. per lb	13s. 1½d.
Item a brass pot containing 60 lbs, 2½d. per lb	12s. 9d. (sic)
Item 4 kettles containing 16 lbs, 2d. per lb	2s. 8d.
Item 2 spits	22d.
Item 2 andirons	12d.
Item 2 gridirons	10d.
Item a dripping pan	6d.
Item a frying pan	4d.
Total	33s. ½d.

Chattels

Firstly 8 [horse bits], 6s. per pair	48s.

Corn

Firstly 5 seam of barley, 3s. per seam	15s.
Item 6 seam of rye, 3s. 4d. per seam	20s.
Total	35s.

The Warehouse

Firstly 18 broad cloths, comprising 33 yards a piece, each cloth £3	£54
Item 6 bales of woad, each bale 24s.	£7 4s.
Item 2 bags of alum, each bag 26s. 8d.	53s. 4d.
Item 300 lbs of wool, each hundredweight 40s.	£6
Item 24 handles	2s.
Item an *Îkrath*	8d.
Total	£70
Total of goods apprised	**£137 11s. 9½d.**

Debts owing to the said John Reynham

Firstly William Nytyngall	£100
Item Peter Joye	£60
Item George Lalstrode	£42
Item John Banaster	£40
Item Thomas Taylor	£40
Item John Vygrot	£10
Total	£260
Total of goods apprised and debts	**£397 11s. 9½d.**

Notes: [] means that manuscript is indecipherable or that the author's transcription is his best guess. Words in italics are those in the manuscript that the author cannot transcribe.
Source: TNA PROB 2/87.

Bibliography

MANUSCRIPT SOURCES

British Library London

Ipswich General Court Register	Ms Add. 30,158
Image of woman with spinning wheel from Luttrell Psalter	Add. 41230, f. 193
Image of woman weaver from Egerton Genesis	Egerton 1894, f. 2v
Image of dyers from Bartholomaeus Angelicus, *Des proprietez des choses.* Bruges 1482	Royal MS 15 E III, f. 269

The National Archives (London)

Chancery:	
Chancery pleadings	C 1/70/102, C 1/209/42
Records of the Court of Common Pleas and other courts:	
Plea Rolls, Court of Common Pleas	CP 40/360, 402, 437, 477, 483, 516, 536, 541, 555, 556, 576, 596, 616, 629, 636, 647, 656, 676, 677, 696, 716, 717, 736, 756, 758, 776, 788, 796, 807, 814, 826, 834, 837, 841, 853, 861, 871, 883, 885A, 888, 891, 895, 907, 909, 911, 919, 931, 943, 951, 971, 990, 1013, 1031, 1038, 1042, 1046, 1049, 1055, 1057, 1060, 1064
Exchequer, Records of King's Remembrancer:	
Alnage accounts	E 101/342/8, 10, 21, 25 E 101/343/2, 4, 5, 7, 9–11, 13, 25, 28
Particular customs accounts	E 122/51/39, E 122/52/42, 43, 45, 46, 47, 48 E 122/193/33
Assessment of aliens for tax 1440 and 1483	E 179/180/92, 111
Records of the Prerogative Court of Canterbury	
Inventories compiled before 1661	PROB 2
Will registers	PROB 11
Special Collections	
Ministers' and receivers' accounts	SC 6/1001/3
State papers	SP 1, vol. 4

Cambridge University Library (CUL)

Lakenheath Manor, 1481–89 minister's accounts	EDC 7/16B

Canterbury Cathedral Archive (CCA)

Hadleigh	
Serjeant's accounts	DCc-BR/Hadleigh/16, 21, 22, 25
Farmer's accounts	DCc-MA/8, 168
Court roll	DCc-CR/17
Chartae Antiquae	DCc-ChAnt/C/244

Monks Eleigh

Priory register	DCc-Register/B
Rental 1510	DCc-MA/33

Norfolk Record Office (NRO)

Probate registers	Multon, Popy

Suffolk Record Office, Bury St Edmunds (SROB)

Ordinances of Bury St Edmunds weavers' guild 1477	B9/1/2
Extent of the manor of Melford 1441–2	J/523
Probate registers	Baldwyne, Boner, Hawlee, Hervye, Hoode, Johnson, Mason, Newton, Osbern, Pye

Suffolk Record Office, Ipswich (SROI)

Ipswich

Rental	C/2/1/1/3
Petty court rolls	C/2/3/1
Leet court rolls	C/2/8/1
Composite enrolments (including general court, petty court and leet court)	C/2/10/1
Composite court books	C/2/10/3
Chamberlains' compotus rolls	C/3/3/1
Recognizance rolls	C/2/4/1
Peter Northeast archive	HD 2448/3/351
Probate registers	IC/AA1-8

Thorney Hotot

Rentals	HA/CC2/1/3, 4; HA/CC2/2/1

PRINTED PRIMARY SOURCES

Bacon, Nathaniel *The Annals of Ipswiche . . . by Nathaniel Bacon serving as Recorder and Town Clerk . . . 1654* ed. by W.H. Richardson (Ipswich: Cowell, 1884)

Bailey, Mark (ed.) *The Bailiffs' Minute Book of Dunwich, 1404–1430* (Woodbridge: Boydell for the Suffolk Records Society, 34, 1992)

Calendar of Close Rolls, 1272–1509, 47 volumes (London: HMSO, 1892–1963)

Calendar of Fine Rolls, 1272–1509, 22 volumes (London: HMSO, 1911–62)

Calendar of Patent Rolls, 1232–1509, 52 volumes (London: HMSO, 1891–1916)

Caley, J. (ed.) *Valor Ecclesiasticus*, vol. 3 (London: Records Commission, 1817)

Carus-Wilson, E.M. and Olive Colman (ed.) *England's Export Trade 1275–1547* (Oxford: Clarendon Press, 1963)

Chibnall, Marjorie (ed.) *Select Documents of the English Lands of the Abbey of Bec* (London: Royal Historical Society, 1951)

Crawford, Anne (ed.) *The Household Books of John Howard, Duke of Norfolk* (Stroud: Sutton Publishing, 1992)

Davis, R.H.C. (ed.) *Kalendar of Abbot Samson* (London: Royal Historical Society for the Camden Society, 84, 1954)

Denney, A.H. (ed.) *The Sibton Abbey Estates: Select Documents 1325–1509* (Ipswich: W.S. Cowell for the Suffolk Records Society, 2, 1960)

Dymond, D.P. (ed.) *Register of Thetford Priory*, part 1: *1482–1517* (Oxford: Oxford University Press for Norfolk Records Society, 59, 1994), part 2: *1518–1540* (Oxford: Oxford University Press for Norfolk Records Society, 60, 1995)

Fenwick, C.C. (ed.) *The Poll Taxes of 1377, 1379 and 1381*, part 1: Bedfordshire-Leicestershire (Oxford: Oxford University Press for the British Academy, 1998; repr. 2013), part 2:

Lincolnshire-Westmoreland (Oxford: Oxford University Press for the British Academy, 2001), part 3: Wiltshire-Yorkshire (Oxford: Oxford University Press for the British Academy, 2005)

Fitzherbert, A. *The Boke of Husbandrie* (London: Berthelet edn., 1534; reprinted by English Dialect Society, 1882)

Fortescue, Sir John *De Laudibus Legum Anglia* ed. by S.B. Chrimes (Cambridge: Cambridge University Press, 1942)

Gransden, Antonia (ed.) *William of Hoo's Letter Book 1280–1294* (Ipswich: W.S. Cowell for the Suffolk Records Society, 5, 1963)

Grimwade, M.E. (comp.), *Index of the Probate Records of the Court of the Archdeacon of Suffolk 1444–1700* (London: British Record Soc., 90 and 91, 1979 and 1980)

Grimwade, M.E. (comp.), *Index of the Probate Records of the Court of the Archdeacon of Sudbury 1354–1700* (London: British Record Soc., 95 and 96, 1984)

Hervey, S.H.A. (ed.) *Suffolk in 1327,* Suffolk Green Books, 9 (Woodbridge: Booth, 1906)

Hervey, S.H.A. (ed.) *Suffolk in 1524,* Suffolk Green Books, 10 (Woodbridge: Booth, 1910)

Langland, William *The Vision of Piers Ploughman* translated. by A.V.C. Schmidt (Oxford: Oxford University Press, 1992)

Latham, R.E. (ed.) *Revised Medieval Latin Word-List* (London: Oxford University Press for the British Academy, 1965)

Leland, John *The Itinerary of John Leland in or about the Years 1535–1543, ed. by* Lucy Toulmin Smith, 5 volumes (London: Centaur Press, 1964)

Luders, A. and others (eds) *The Statutes of the Realm (1101–1713),* 11 vols. (London: Rec. Comm., 1810–28)

Middleton-Stewart, Judith (ed.) *Records of the Churchwardens of Mildenhall* (Woodbridge: Boydell for the Suffolk Records Society, 54, 2011)

Millea, Nick (ed.) *The Gough Map: The Earliest Road Map of Great Britain* (Oxford: The Bodleian Library, 2007)

Northeast, Peter (ed.) *Wills of the Archdeaconry of Sudbury,* part 1: *1439–1461,* (Woodbridge: Boydell for the Suffolk Records Society, 44, 2001), part 2: *1462–1474,* (Woodbridge: Boydell for the Suffolk Records Society, 53, 2010)

Pound, John (ed.) *The Military Survey of 1522 for Babergh Hundred* (Woodbridge: Boydell for the Suffolk Records Society, 28, 1986)

Powell, E. (ed.) *A Suffolk Hundred in the Year 1283* (Cambridge: Cambridge University Press, 1910)

Rigby, S.H. (ed.) *The Overseas Trade of Boston in the Reign of Richard II* (Woodbridge: Boydell for the Lincoln Record Society, 93, 2005)

Statham, Margaret (ed.) *Accounts of the Feoffees of the Town Lands of Bury St Edmunds, 1569–1622* (Woodbridge: Boydell for the Suffolk Records Society, 46, 2003)

Strachey, J. and others (eds.) *Rotuli Parliamentorum ut et Petiones et Placita in Parliamento,* 6 vols. (Rec. Com., London, 1767–77)

Tawney, R.H. and Eileen Power (ed.) *Tudor Economic Documents, Vol. I, Agriculture and Industry,* 3 volumes (London: Longmans, Green, 1924)

Twiss, Sir Thomas (ed.) *The Black Book of the Admiralty, Appendix, Part II, Volume II,* 4 volumes (London: Rolls Series, 55, 1873)

Tymms, Samuel (ed.) *Wills and Inventories from the Registers of the Commissary of Bury St Edmunds and the Archdeacon of Sudbury* (London: Camden Society, 46, 1850)

Vanderzee, G. (ed.) *Nonarum Inquisitiones* (London: Records Commission, 1807)

Ward, Jennifer (ed.) *Elizabeth de Burgh: Lady of Clare (1295–1360): Household and Other Records* (Woodbridge: Boydell for the Suffolk Records Society, 57, 2014)

Warner, Sir George (ed.) *The Libelle of Englshe Polycye* (Oxford: Clarendon Press, 1926)

PRINTED SECONDARY SOURCES

Allen, Martin 'The Volume of the English Currency, 1158–1470' *Econ. Hist. Rev.*, 2nd ser. 54 (2001), 595–611

Allen, R.C. and J.L. Weisdorf 'Was there an "industrious revolution" before the industrial revolution? An empirical exercise for England, c. 1300–1830' *Econ. Hist. Rev.*, 2nd ser. 64 (2011), 715–29

Allison, K.J. 'Flock Management in the Sixteenth and Seventeenth Centuries' *Econ. Hist. Rev.*, 2nd ser. 11 (1958), 98–112

Amor, Nicholas R. 'Riding our Recession: Ixworth and Woolpit in the Late Middle Ages' *PSIAH,* 40 (2002), 127–44

Amor, Nicholas R. 'Merchant Adventurer or Jack of All Trades: The Suffolk Clothier in the 1460s' *PSIAH,* 40 (2004), 414–36

Amor, Nicholas R. 'Late Medieval Enclosure – A Study of Thorney, near Stowmarket, Suffolk' *PSIAH,* 41 (2006), 175–97

Amor, Nicholas R. *Late Medieval Ipswich: Trade and Industry* (Woodbridge: Boydell, 2011)

Amor, Nicholas R. 'John Motte – Local Boy Made Good' Bildeston History Quarterly, 2 (Spring 2012), pp. 9–11

Amor, Nicholas R. 'Apprenticeship in Late Medieval Ipswich' *PSIAH,* 42 (2012), 411–23

Andrews, Sybil and Lilian J. Redstone 'Suffolk Courts in English' *PSIAH,* 20 (1929), 199–213

Anon., 'A Descriptive Catalogue of Ancient Deeds in the Public Record Office' *PSIAH,* 10 (1899), 251–344

Arnott, John 'The Church and Priory of S. Mary, Woodbridge' *PSIAH,* 9 (1897), 338–46

Badham, Sally 'Medieval Monuments to the de la Pole and Wingfield Families', in *Wingfield College and its Patrons: Piety and Prestige in Medieval Suffolk* ed. by Peter Bloore and Edward Martin (Woodbridge: Boydell, 2015), pp. 135–76

Bailey, Mark *A Marginal Economy? East-Anglian Breckland in the Later Middle Ages* (Cambridge: Cambridge University Press, 1989)

Bailey, Mark 'Sand into Gold: The Evolution of the Foldcourse System in West Suffolk, 1200–1600' *Ag. Hist. Rev.,* 38 (1990 (A)), 40–57

Bailey, Mark 'Coastal Fishing off South East Suffolk in the Century after the Black Death' *PSIAH,* 37 (1990 (B)), 102–14

Bailey, Mark 'Per impetum maris: Natural Disaster and Economic Decline in Eastern England, 1275–1350', in *Before the Black Death: Studies in the 'crisis' of the early fourteenth century,* ed. by Bruce M.S. Campbell (Manchester: Manchester University Press, 1991 repr. 1992) pp. 184–208

Bailey, Mark 'Rural Society', in *Fifteenth-Century Attitudes: Perceptions of Society in Late Medieval England,* ed. by Rosemary Horrox (Cambridge: Cambridge University Press, 1994) pp. 150–68

Bailey, Mark *Medieval Suffolk: An Economic and Social History* (Woodbridge: Boydell, 2007)

Bailey, Mark 'Technology and the Growth of Textile Manufacture in Medieval Suffolk' *PSIAH,* 42 (2009) 13–20

Bailey, Mark *The Decline of Serfdom in Late Medieval England: From Bondage to Freedom* (Woodbridge: Boydell, 2014)

Baines, Patricia *Spinning Wheels: Spinners and Spinning* (New York: Charles Scribner's Sons, 1977)

Baker, Alan R.H. 'Evidence in the "Nonarum Inquisitiones" of Contracting Arable Lands in England during the early Fourteenth Century' *Econ. Hist. Rev.*, 2nd ser. 19 (1966), 518–532

Barker, Juliet *England, Arise: The People, the King and the Great Revolt of 1381* (London: Little Brown, 2014)

Barron, Caroline M. *London in the Later Middle Ages: Government and People 1200–1500* (Oxford: Oxford University Press, 2004)

Bennett, Judith M. *Ale, Beer and Brewsters in England* (Oxford: Oxford University Press, 1996)
Beresford, M.W. and J.K.S. St Joseph *Medieval England: An Aerial Survey* (Cambridge: Cambridge University Press, 1958; repr. 1979)
Berg, Maxine L. 'Power, Eileen Edna Le Poer (1889–1940)', *DNB*, Oxford University Press, 2004; online edn, Oct 2005 [http://www.oxforddnb.com/view/article/35581, accessed 22 March 2016]
Betterton, Alec 'A Survival from Lavenham's Cloth Industry' *PSIAH,* 35 (1984), pp. 317–9
Betterton, Alec and David Dymond *Lavenham Industrial Town* (Lavenham: Terence Dalton, 1989)
Bettley, James and Nikolaus Pevsner *The Buildings of England, Suffolk: West* (London: Yale University Press, 2015)
Bolton, J.L. *The Medieval English Economy 1150–1500* (London: Dent, 1980)
Bolton, J.L. 'What is Money', in *Medieval Money Matters* ed. by Diana Wood (Oxford: Oxbow Books, 2004), pp. 1–15
Bolton, J.L. *Money in the Medieval English Economy: 973–1489* (Manchester: Manchester University Press, 2012)
Bourke, Helen 'Heaton, Herbert (1890–1973)', Australian Dictionary of Biography, National Centre of Biography, Australian National University, 1983 [http://adb.anu.edu.au/biography/heaton-herbert-6626/text11413, accessed 22 March 2016]
Bowden, Peter J. *The Wool Trade in Tudor and Stuart England* (London: Macmillan, 1962)
Braudel, Fernand *The Wheels of Commerce* (London: Collins, 1982)
Bridbury, A.R. *Economic Growth: England in the Later Middle Ages* (London: Allen and Unwin, 1962; repr. Brighton: Harvester Press, 1975)
Bridbury, A.R. *Medieval Clothmaking: An Economic Survey* (London: Heinemann, 1982)
Britnell, R.H. *Growth and Decline in Colchester, 1300–1525* (Cambridge: Cambridge University Press, 1986)
Britnell, R.H. 'The Occupation of the Land: Eastern England', III, pp. 53–67; 'Farming Practice and Techniques: Eastern England',, III, pp. 194–209; both in *The Agrarian History of England and Wales: 1348–1500* ed. by E. Miller, 8 vols (Cambridge: Cambridge University Press, 1991)
Britnell, R.H. *The Commercialisation of English Society* (Cambridge: Cambridge University Press, 1993; repr. Manchester: Manchester University Press, 1996)
Britnell, R.H. *The Closing of the Middle Ages?: England 1471–1529* (Oxford: Blackwell, 1997)
Britnell, R.H. 'The Economy of British Towns 1300–1540', in *The Cambridge Urban History of Britain: 600–1540* ed. by D.M. Palliser, 3 vols. (Cambridge: Cambridge University Press, 2000), I, pp. 313–33
Britnell, R.H. 'The Woollen Textile Industry of Suffolk in the Later Middle Ages' *The Ricardian* 13 (2003), 86–99
Britnell, R.H. *Britain and Ireland 1050–1530* (Oxford: Oxford University Press, 2004 (A))
Britnell, R.H. 'Uses of Money in Medieval England' in *Medieval Money Matters* ed. by Diana Wood (Oxford: Oxbow Books, 2004 (B)), pp. 16–30
Broadberry, Stephen, Bruce M.S. Campbell, Alexander Klein, Mark Overton, Bas van Leeuwen *British Economic Growth 1270–1870* (Cambridge: Cambridge University Press, 2015)
Burnham, Dorothy K. *A Textile Terminology: Warp and Weft* (London: Routledge & Kegan Paul, 1981)
Burton, Janet and Julie Kerr *The Cistercians in the Middle Ages* (Woodbridge: Boydell, 2011)
Campbell, Bruce M.S. 'Medieval Manorial Structure', in *An Historical Atlas of Norfolk* ed. by Peter Wade-Martins and Jane Everett (Ipswich: Norfolk Museums Service, 1993), pp. 52–3
Campbell, Bruce M.S. *English Seigniorial Agriculture* (Cambridge, Cambridge University Press, 2000)

Campbell, Bruce M.S. 'Nature as Historical Protagonist: Environment and Society in Pre-industrial England' *Econ. Hist. Rev.*, 2nd ser. 63 (2010), 281–314

Campbell, Bruce M.S. 'Grain Yields on English Demesnes after the Black Death' in *Town and Countryside in the Age of the Black Death: Essays in Honour of John Hatcher* ed. by Mark Bailey and Stephen Rigby (Turnout, Belgium: Brepols, 2012), pp. 121–74

Carus-Wilson, Eleanora 'The Aulnage Accounts: a Criticism' *Econ. Hist. Rev.*, A2 (1929), 39–60; reprinted in E.M Carus-Wilson *Medieval Merchant Adventurers* (London: Metheun, 1954; repr. 1967), pp. 279–91

Carus-Wilson, Eleanora 'An Industrial Revolution of the Thirteenth Century' *Econ. Hist. Rev.*, A11 (1941), 114–23; reprinted in E.M Carus-Wilson *Medieval Merchant Adventurers* (London: Metheun, 1954; repr. 1967), pp. 183–210

Carus-Wilson, Eleanora 'The English Cloth Industry in the Late Twelfth and early Thirteenth Centuries' *Econ. Hist. Rev.*, 14 (1944), 32–50; reprinted in E.M Carus-Wilson *Medieval Merchant Adventurers* (London: Metheun, 1954; repr. 1967), pp. 211–38

Carus-Wilson, Eleanora 'The Woollen Industry', in *The Cambridge Economic History of Europe: Trade and Industry in the Middle Ages,* ed. by M.M. Postan and Edward Miller (Cambridge: Cambridge University Press, 1952 repr. 1987), pp. 614–690

Carus-Wilson, Eleanora 'Evidences of Industrial Growth on some Fifteenth-Century Manors' *Econ. Hist. Rev.*, 2nd ser. 12 (1959), 190–205

Catling, H. 'The Evolution of Spinning', in *the Wool Textile Industry in Great Britain* ed. by J. Geraint Jenkins (London: Routledge & Kegan Paul, 1972), pp. 101–16

Childs, Wendy R. *Anglo Castilian Trade in the Later Middle Ages* (Manchester: Manchester University Press, 1978)

Childs, Wendy R. 'The English Export Trade in Cloth in the Fourteenth Century', in *Progress and Problems in Medieval England: Essays in Honour of Edward Miller*, ed. by R.H. Britnell and J. Hatcher (Cambridge: Cambridge University Press, 1996), pp. 121–47

Childs, Wendy R. 'Moving Around', in *A Social History of England 1200–1500* ed. by R. Horrox and W.M. Ormrod (Cambridge: Cambridge University Press, 2006), pp. 260–75

Chorley, Patrick 'The Cloth Exports of Flanders and Northern France during the Thirteenth Century: A Luxury Trade' *Econ. Hist. Rev.*, 2nd ser. 40 (1987), 349–79

Chorley, Patrick 'English cloth exports during the thirteenth and early fourteenth centuries: The continental evidence' *Historical Research* 61 (1988), 1–10

Chorley, Patrick 'The Evolution of the Woollen, 1300–1700', in *The New Draperies in the Low Countries and England* ed. by N.B. Harte (Oxford: Oxford University Press, 1997), pp. 7–33

Clark, C.O. 'The Dyeing of Wool and Woollen Goods', in *The Wool Textile Industry in Great Britain* ed. by J. Geraint Jenkins (London: Routledge & Kegan Paul, 1972), pp. 117–24

Clark, Gregory 'The Long March of History: Farm Wages, Population and Economic Growth, England 1209–1869' *Econ. Hist. Rev.*, 2nd ser. 60 (2007), 97–135

Clarkson, L.A. *Proto-Industrialisation: The First Phase of Industrialisation* (Basingstoke: Macmillan for The Economic History Society, 1985)

Coleman, D.C., 'Proto-Industrialisation: A Concept Too Many' *Econ. Hist. Rev.*, 2nd ser. 36 (1983), 435–448

Corley, T.A.B. 'Unwin, George (1870–1925)', *DNB*, Oxford University Press, 2004 [http://www.oxforddnb.com/view/article/62701, accessed 22 March 2016]

Cornford, Barbara and Andy Reid 'The Uprising of 1381', in *An Historical Atlas of Norfolk* ed. by Peter Wade-Martins and Jane Everett (Ipswich: Norfolk Museums Service, 1993), pp. 86–7

Cornwall, J.C.K *Wealth and Society in Early Sixteenth Century England* (London: Routledge, 1988)

Crawford, Anne 'Howard, John, First Duke of Norfolk (*d.* 1485)', *DNB*, Oxford University Press, 2004; online edn, Jan 2008 [http://www.oxforddnb.com/view/article/13921, accessed 22 March 2016]

Crowfoot, Elisabeth, Frances Pritchard, Kay Staniland *Textiles and Clothing 1150–1450* (London: HMSO, 1992; repr. Woodbridge: Boydell, 2001)
Darby, H.C., R.E. Glasscock, J. Sheail, G.R. Versey 'The Changing Geographical Distribution of Wealth in England: 1086–1334–1525' *Journal of Historical Geography*, 5, 3 (1979), 246–262
Daunton, Martin 'Lipson, Ephraim (1888–1960)', *DNB*, Oxford University Press, 2004 [http://www.oxforddnb.com/view/article/34550, accessed 22 March 2016]
Davis, James *Medieval Market Morality: Life, Law and Ethics in the English Marketplace, 1200–1500* (Cambridge: Cambridge University Press, 2012)
Ditchfield, P.H. and William Page (ed.) *A History of the County of Berkshire: Volume 3* (London: Victoria County History, 1923)
Duffy, Eamon 'Wingfield College and the Cult of Purgatory', in *Wingfield College and its Patrons: Piety and Prestige in Medieval Suffolk* ed. by Peter Bloore and Edward Martin (Woodbridge: Boydell, 2015), pp. 49–59
Dunn, Penelope 'Trade', in *Medieval Norwich* ed. by Carole Rawcliffe and Richard Wilson (London: Hambledon, 2004), pp. 213–34
Dyer, Alan 'Appendix: Ranking Lists of English Medieval Towns', in *The Cambridge Urban History of Britain: 600–1540* ed. by D.M. Palliser, 3 vols. (Cambridge: Cambridge University Press, 2000), I, pp. 747–70
Dyer, Christopher, *Standards of Living in the Later Middle Ages: Social Change in England c. 1200–1520* (Cambridge: Cambridge University Press, 1989)
Dyer, Christopher, *Everyday Life in Medieval England* (London: Hambledon and London, 2000)
Dyer, Christopher *Making a Living in the Middle Ages* (London: Yale University Press, 2002)
Dyer, Christopher *An Age of Transition?: Economy and Society in England in the Later Middle Ages* (Oxford: Oxford University Press, 2005)
Dyer, Christopher *A Country Merchant 1495–1520: Trading and Farming at the End of the Middle Ages* (Oxford: Oxford University Press, 2012 (A))
Dyer, Christopher 'Poverty and its Relief in Late Medieval England' Past and Present, 216 (2012 (B)), 41–78
Dymond, David 'Terriers, Tithes and Farming' *PSIAH*, 42 (2012), 434–54
Dymond, David 'Suffolk's Cloth Industry: A History' in *Suffolk Threads Heritage Trail*
Dymond, David 'Camping Closes', in *An Historical Atlas of Suffolk* ed. by David Dymond and Edward Martin (Ipswich: Suffolk County Council, 1988; repr. 1989 and 1999), pp. 154–5
Dymond, David 'The Woollen Cloth Industry', in *An Historical Atlas of Suffolk* ed. by David Dymond and Edward Martin (Ipswich: Suffolk County Council, 1988; repr. 1989 and 1999), pp. 140–1
Dymond, David 'Medieval and Later Markets', in *An Historical Atlas of Norfolk* ed. by Peter Wade-Martins and Jane Everett (Ipswich: Norfolk Museums Service, 1993), pp. 76–7
Dymond, David and Peter Northeast *A History of Suffolk* (Chichester: Phillimore, 1985; repr. 1995)
Dymond, David and Clive Paine *Five Centuries of an English Parish Church: 'The State of Melford Church', Suffolk* (Cambridge: EAH Press, 2012)
Dymond, David and Roger Virgoe 'The Reduced Population and Wealth of Early-Fifteenth Century Suffolk' *PSIAH*, 36 (1986), 72–100
Edmonds, John *The History of Woad and the Medieval Woad Vat* (Little Chalfont: John Edmonds, 1998)
Edmonds, John *Medieval Textile Dyeing* (Little Chalfont: John Edmonds, 2003)
Edwards, James Frederick and Brian Paul Hindle 'The transportation system of medieval England and Wales' *Journal of Historical Geography*, 17, 2 (1991), 123–34
Egan, Geoffrey *Lead Cloth Seals and Related Items in the British Museum* (London: The British Museum, 1994)

Ellis, P. 'The techniques of weaving', in *the Wool Textile Industry in Great Britain* ed. by J. Geraint Jenkins (London: Routledge & Kegan Paul, 1972), pp. 125–56

Epstein, S.R., 2008. 'Craft Guilds in the Pre-modern Economy; A Discussion', *Econ. Hist. Rev.*, 2nd ser., 61 (2008), 155–74

Evans, Nesta *The East Anglian Linen Industry: Rural Industry and Local Economy, 1500–1850* (Aldershot: Gower, 1985)

Farmer, David L. 'Marketing the Produce of the Countryside, 1200–1500', in *The Agrarian History of England and Wales: 1348–1500* ed. by E. Miller, 8 vols. (Cambridge: Cambridge University Press, 1991), III, pp. 324–430

Farmer, David L. 'Prices and Wages, 1350–1500', in *The Agrarian History of England and Wales: 1348–1500* ed. by E. Miller, 8 vols. (Cambridge: Cambridge University Press, 1991), III, pp. 431–525

Fenwick, Valerie and Vic Harrup *Untold Tales from the Suffolk Sandlings* (Woodbridge: Butley Research Group, 2009)

Fradley, Michael *Water Street, Lavenham, Suffolk: A Desk-based Assessment of a Brick Culvert,* Research Department Report Series 7/2207 (Cambridge: English Heritage, 2007)

Fryde, E.B. *Peasants and Landlords in Later Medieval England* (Stroud: Alan Sutton, 1996; repr. 1999)

Galloway, James A. 'Urban Hinterlands in Later Medieval England', in *Town and Country in the Middle Ages: Contrasts, Contacts and Interconnections 1100–1500* ed. by Kate Giles and Christopher Dyer (Leeds: Maney, 2003), pp. 111–30

Gascoyne, Adrian and Maria Medlycott *Essex Historic Grazing March Project* (Chelmsford: Essex County Council, 2014) [http://content.historicengland.org.uk/images-books/publications/essex-historic-grazing-marshes-project/Final_Grazing_Marshes.pdf/, accessed 18 April 2016]

Goddard, Richard 'Surviving Recession' in *Survival and Discord in Medieval Society: Essays in Honour of Christopher Dyer* ed. by Richard Goddard, John Langdon and Miriam Muller (Turnout, Belgium: Brepols, 2010), pp. 69–87

Gransden, Antonia *A History of the Abbey of Bury St Edmunds 1182–1256: Samson of Tottington to Edmund of Walpole* (Woodbridge: Boydell, 2007)

Gransden, Antonia *A History of the Abbey of Bury St Edmunds 1257–1301: Simon of Luton and John of Northwold* (Woodbridge: Boydell, 2015)

Gray, H.L. 'The Production and Exportation of English Woollens in the Fourteenth Century' *The English Historical Review*, 39, 153 (1924), 13–35

Grimwood, C.G and S.A. Kay *History of Sudbury Suffolk* (Sudbury: Grimwood and Kay, 1952)

Harari, Yuval Noah *Sapiens: a Brief History of Humankind* (London; Harvill Secker, 2014)

Hare, J.N. 'Growth and Recession in the Fifteenth-century Economy: The Wiltshire Textile Industry and the Countryside' *Econ. Hist. Rev.*, 2nd ser. 52 (1999), 1–26

Hare, John *A Prospering Society: Wiltshire in the later Middle Ages* (Hatfield: University of Hertfordshire, 2011)

Hare, John 'The Economic Context of the Brokerage Books', pp. 25–34; 'Southampton's Trading Partners: Salisbury', pp. 53–63; 'Southampton's Trading Partners: Beyond Hampshire and Wiltshire', pp. 105–10; 'Commodities: The Cloth Industry', pp. 153–60, all in *English Inland Trade 1430–1540: Southampton and its Region* ed. by Michael Hicks (Oxford: Oxbow Books, 2015)

Harriss, G. L. 'Fastolf, Sir John (1380–1459)', *DNB*, Oxford University Press, 2004 [http://www.oxforddnb.com/view/article/9199, accessed 22 March 2016]

Hastings, Margaret *The Court of Common Pleas in Fifteenth Century England: A Study of Legal Administration and Procedure* (New York: Cornell University Press, 1947; repr. 1971)

Hatcher, John *Plague, Population and the English Economy, 1348–1530* (London: Macmillan, 1977)

Hatcher, John 'The Great Slump of the Mid-fifteenth Century', in *Progress and Problems in Medieval England: Essays in Honour of Edward Miller* ed. by Richard Britnell and John Hatcher (Cambridge: Cambridge University Press, 1996), pp. 237–72

Hatcher, John 'Unreal Wages', in *Commercial Activity, Markets and Entrepreneurs in the Middle Ages* ed. by Ben Dodds and Christian D. Liddy (Woodbridge: Boydell, 2011), pp. 1–24

Hatcher, John and T.C. Barker *A History of British Pewter* (London: Longman for the Master Wardens and Commonalty of the Mystery of Pewterers of the City of London, 1974)

Heaton, Herbert *The Yorkshire Woollen and Worsted Industries, From the Earliest Times up to the Industrial Revolution* (Oxford: Oxford University Press, 1920)

Hervey, John 'Extent of Hadleigh Manor, 1305' *PSIAH,* 11 (1902), 152–72

Hicks, Michael 'The Freight Transport of Southampton', in *English Inland Trade 1430–1540: Southampton and its Region* ed. by Michael Hicks (Oxford: Oxbow Books, 2015), pp. 43–51

Hilton, Rodney *Bondmen Made Free: Medieval Peasant Movements and the English Rising of 1381* (London: Routledge, 1973)

Hilton, Rodney 'Capitalism – What's in a Name?', in *The Transition from Feudalism to Capitalism* ed. by Rodney Hilton (London: Verso, 1978), pp. 145–58

Hinde, Andrew *England's Population: A History Since the Domesday Survey* (London: Hodder Arnold, 2003)

Holt, Richard *The Mills of Medieval England* (Oxford: Basil Blackwell, 1988)

Hudson, Rev. W. 'The Assessment of the Townships of the County of Norfolk for the King's Tenths and Fifteenths, as settled in 1334' Norfolk Archaeology, 12 (1895) 243–97

Hunt, Robert *Mineral Statistics of the United Kingdom of Great Britain and Ireland for 1853 and 1854* (London: HMSO, 1855)

Hurry, Jamieson B. *The Woad Plant and its Dye* (Oxford: Oxford University Press, 1930)

Hyde, Nina 'Wool: Fabric of History' National Geographic, Vol. 173, No. 5 (May 1988), 552–591

Iredale, John A. 'Preparation of Wools' in *the Wool Textile Industry in Great Britain* ed. by J. Geraint Jenkins (London: Routledge & Kegan Paul, 1972), pp. 65–70

James, Margery Kirkbride *Studies in the Medieval Wine Trade,* ed. by Elspeth M. Elspeth (Oxford: Clarendon Press, 1971)

Keene, Derek *A Survey of Medieval Winchester* 2 vols. (Oxford: Oxford University Press, 1985)

Kerling, Nelly Johanna Martina *Commercial Relations of Holland and Zeeland with England from the Late 13th Century to the Close of the Middle Ages* (Leiden: E.J. Brill, 1954)

Kermode, Jennifer *Medieval Merchants: York, Beverley and Hull in the Later Middle Ages* (Cambridge: Cambridge University Press, 1998; repr. 1999)

Kermode, Jennifer 'The Great Towns', in *The Cambridge Urban History of Britain: 600–1540* ed. by D.M. Palliser, 3 vols. (Cambridge: Cambridge University Press, 2000), I, pp. 441–65

Kerridge, Eric 'Wool Growing and Wool Textiles in Medieval and Early Modern Times', in *the Wool Textile Industry in Great Britain* ed. by J. Geraint Jenkins (London: Routledge & Kegan Paul, 1972), pp. 19–33

Kerridge, Eric *Textile Manufactures in Early Modern England* (Manchester: Manchester University Press, 1985)

Kightly, Charles 'Hungerford, Walter, First Baron Hungerford (1378–1449)', *DNB*, Oxford University Press, 2004; online edn, Jan 2008 [http://www.oxforddnb.com/view/article/14181, accessed 22 March 2016]

Knowles, David and R. Neville Hadcock *Medieval Religious Houses* (London: Longman, Green and Co, 1953)

Kok, Annette 'A Short History of the Orchil Dyes' *The Lichenologist,* 3 (1966), 248–72

Kowaleski, Maryanne *Local Markets and Regional Trade in Medieval Exeter* (Cambridge: Cambridge University Press, 1995)

Kowaleski, Maryanne 'Port Towns: England and Wales 1300–1540', in *The Cambridge Urban History of Britain: 600–1540* ed. by D.M. Palliser, 3 vols. (Cambridge: Cambridge University Press, 2000), I, pp. 467–494

Kowaleski, Maryanne 'The Shipmaster as Entrepreneur in Medieval England', in *Commercial Activity, Markets and Entrepreneurs in the Middle Ages* ed. by Ben Dodds and Christian D. Liddy (Woodbridge: Boydell, 2011), pp. 165–82

Kowaleski, Maryanne 'The Demography of Maritime Communities in Late Medieval England', in *Town and Countryside in the Age of the Black Death: Essays in Honour of John Hatcher* ed. by Mark Bailey and Stephen Rigby (Turnout, Belgium: Brepols, 2012), pp. 87–118

Lambert, Bart and Milan Pajic, 'Drapery in Exile: Edward III, Colchester and the Flemings, 1351–1366' History: *The Journal of the Historical Association*, 99: 338, 2014, 733–753

Langdon, John 'Horse Hauling: A Revolution in Vehicle Transport in Twelfth- and Thirteenth- Century England?' *Past and Present*, 103 (1984), 37–66

Langdon, John *Mills in the Medieval Economy* (Oxford: Oxford University Press, 2004)

Langdon, John 'The Efficiency of Inland Water Transport in Medieval England' in *Waterways and Canal Building in Medieval England* ed. by John Blair (Oxford: Oxford University Press, 2007)

Lee, John S. *Cambridge and its Economic Region 1450–1560* (Hatfield: University of Hertfordshire, 2005)

Lee, John S. 'The Role of Fairs in the Later Middle Ages' in *Town and Countryside in the Age of the Black Death: Essays in Honour of John Hatcher* ed. by Mark Bailey and Stephen Rigby (Turnout, Belgium: Brepols, 2012), pp. 407–37

Lee, John S 'Crises in the Late Medieval English Cloth Trade', in *Crises in Economic and Social History* ed. by A.T. Brown, Andy Burn and Rob Doherty (Woodbridge: Boydell, 2015)

Lemon, Hugo 'The Development of Hand Spinning Wheels' *Textile History*, 1 (1968), 83–91

Lemon, Hugo 'The Evolution of Combing', in *the Wool Textile Industry in Great Britain* ed. by J. Geraint Jenkins (London: Routledge & Kegan Paul, 1972), pp. 85–100

Lipson, E *The History of the Woollen and Worsted Industries* (London, A. & C. Black, 1921)

Livingstone, Marilyn 'A Snapshot in Time', in *Peasants and Lords in the Medieval English Economy: Essays in Honour of Bruce M.S. Campbell* ed. by Maryanne Kowaleski, John Langdon and Phillipp R. Schofield (Turnout, Belgium: Brepols, 2015), pp. 325–51

Lloyd, T.H 'Some Costs of Cloth Manufacturing in Thirteenth-Century England' *Textile History*, 1 (1970), 332–36

Lloyd, T.H *The Movement of Wool Prices in Medieval England* (Cambridge: Cambridge University Press for the Economic History Society, 1973)

Lloyd, T.H (review) 'A.R. Bridbury. Medieval English Clothmaking: An Economic Survey' *Econ. Hist. Rev.*, 2nd ser. 36 (1983), 449–50

Lobel, M.D. 'A list of the Aldermen and Bailiffs of Bury St Edmunds from the Twelfth to the Sixteenth Century' *PSIAH*, 22 (1934), 17–28

Lucas, Adam *Ecclesiastical Lordship, Seigneurial Power and the Commercialisation of Milling in Medieval England* (Farnham: Ashgate 2014)

Maddison, Angus *The World Economy: Historical Statistics* (Paris: OECD, 2003)

Martin, Edward 'Hundreds and Liberties', in *An Historical Atlas of Suffolk* ed. by David Dymond and Edward Martin (Ipswich: Suffolk County Council, 1988; repr. 1989 and 1999), pp. 26–7

Martin, Edward and Max Satchell *Wheare most Inclosures be. East Anglian Fields: History, Morphology and Management* (Ipswich: Suffolk County Council for East Anglian Archaeology, 124, 2008)

Masschaele, James *Peasants, Merchants, and Markets* (Basingstoke: Macmillan, 1997)

Mate, Mavis 'Agrarian Economy After the Black Death: The Manors of Canterbury Cathedral Priory, 1348–91' *Econ. Hist. Rev.*, 2nd ser., 37 (1984), 341–54

Mate, Mavis E. *Trade and Economic Developments, 1450–1550; The Experience of Kent, Surrey and Sussex* (Woodbridge: Boydell, 2006)

Mathias, Peter 'Wilson, Eleanora Mary Carus- (1897–1977)', rev. *DNB*, Oxford University Press, 2004 [http://www.oxforddnb.com/view/article/30905, accessed 22 March 2016]

McClenaghan, Barbara *The Springs of Lavenham and the Suffolk Cloth Trade in the XV and XVI Centuries* (Ipswich: W.E. Harrison, 1924)

McIntosh, Marjorie Keniston *Poor Relief and Community in Hadleigh, Suffolk 1547–1600* (Hatfield: University of Hertfordshire Press, 2013)

Millea, Nick (ed.) *The Gough Map: The Earliest Road Map of Great Britain* (Oxford: The Bodleian Library, 2007)

Miller, Edward 'The Fortunes of the English Textile Industry during the Thirteenth Century' *Econ. Hist. Rev.*, 2nd ser. 18 (1965), 64–82

Miller, Edward 'The Occupation of the Land: the Southern Counties', in *The Agrarian History of England and Wales: 1348–1500* ed. by E. Miller, 8 vols. (Cambridge: Cambridge University Press, 1991), III, pp. 136–151

Miller, Edward and John Hatcher *Medieval England: Towns, Commerce and Crafts* (Harlow: Longman, 1995)

Muldrew, Craig " 'Th'ancient Distaff' and 'Whirling Spindle': Measuring the Contribution of Spinning to Household Earnings and the National Economy in England, 1550–1770" *Econ. Hist. Rev.*, 2nd ser. 65 (2012), 498–526

Munro, John 'An Economic Aspect of the Collapse of the Anglo-Burgundian Alliance, 1428–1442' *The English Historical Review*, 85 (1970), 225–44

Munro, John 'Wool-Price Schedules and the Qualities of English Wools in the Later Middle Ages c. 1270–1499' *Textile History*, 9 (1978), 118–69

Munro, John 'The 1357 Wool-Price Schedule and the Decline of Yorkshire Wool Values' *Textile History*, 10 (1979), 211–19

Munro, John H. 'Industrial Transformation in the North-west European Textile Trades: Economic Progress or Economic Crisis?', in *Before the Black Death: Studies in the 'Crisis' of the Early Fourteenth Century*, ed. by Bruce M.S. Campbell (Manchester: Manchester University Press, 1991 repr. 1992) pp. 110–48

Munro, John H. 'The Origins of the English "New Draperies": The Resurrection of an Old Flemish Industry 1270–1570', in *The New Draperies in the Low Countries and England* ed. by N.B. Harte (Oxford: Oxford University Press, 1997), pp. 35–127

Munro, John H. 'The "Industrial Crisis" of the English Textile Towns, c. 1290 – c. 1330', in *Thirteenth-century England VII: Proceedings of the Durham conference*, ed. by M. Prestwick, R.H. Britnell and R. Frame (Woodbridge: Boydell, 1999)

Munro, John H. 'Medieval Woollens: Textiles, Textile Technology and Industrial Organisation' and 'Medieval Woollens: The Western European Woollen Industries and their Struggle for International Markets', in *The Cambridge History of Western Textiles*, ed. by D. Jenkins, 2 vols. (Cambridge: Cambridge University Press, 2003), I, pp. 181–324

Munro, John 'Three Centuries of Luxury Textile Consumption in the Low Countries and England, 1330–1570: Trends and Comparisons of Real Values of Woollen Broadcloths (Then and Now)', in *The Medieval Broadcloth: Changing Trends in Fashions, Manufacturing and Consumption*, ed. by Katherine Vestergard Pedersen and Marie-Louise B. Nosch (Oxford: Oxbow Books, 2009), pp. 1–73

Munro, John 'The Late Medieval Decline of English Demesne Agriculture: Demographic, Monetary, and Political-Fiscal Factors', in *Town and Countryside in the Age of the Black Death: Essays in Honour of John Hatcher* ed. by Mark Bailey and Stephen Rigby (Turnout, Belgium: Brepols, 2012), pp. 299–348

Nightingale, Pamela *A Medieval Mercantile Community: The Grocer's Company & the Politics & Trade of London* 1000–1485 (London: Yale University Press, 1995)

Northeast, Peter 'Medieval Religious Houses', in *An Historical Atlas of Suffolk* ed. by David Dymond and Edward Martin (Ipswich: Suffolk County Council, 1988; repr. 1989 and 1999), pp. 70–1

Ogilvie, S.C. 'Guilds, Efficiency and Social Capital: Evidence from German Proto-industry' *Econ. Hist. Rev.*, 2nd ser., 57 (2004), 286–333
Oldland, John 'The Finishing of English Woollens', in *Medieval Clothing and Textiles, Vol. 3,* ed. by Robin Netherton and Gale R. Owen-Crocker (Woodbridge: Boydell, 2007), pp. 97–118
Oldland, John 'The London Fullers and Shearmen, and their Merger to Become the Clothworkers' Company' *Textile History,* 39 (2008), 172–92
Oldland, John "Fyne Worsted which is Almost like Silke': Norwich's Double Worsted' *Textile History,* 42(2) (2011), 181–99
Oldland, John 'Wool and Cloth Production in Late Medieval and Early Tudor England' *Econ. Hist. Rev.*, 2nd ser. 67 (2014), 25–47
Penn, Thomas *Winter King: the Dawn of Tudor England* (London: Penguin, 2012)
Phillipp R. Schofield, 'Spring Family (*per. c.*1400–*c.*1550)', *DNB*, Oxford University Press, 2004; online edn, Jan 2008 [http://www.oxforddnb.com/view/article/55402, accessed 22 March 2016]
Phythian-Adams, Charles *Desolation of a City: Coventry and the Urban Crisis of the Late Middle Ages* (Cambridge: Cambridge University Press, 1979)
Pilgrim, J.E. 'The Cloth Industry in East Anglia', in *The Wool Textile Industry in Great Britain* ed. by J. Geraint Jenkins (London: Routledge & Kegan Paul, 1972), pp. 252–68
Platt, Colin *Medieval Southampton: The Port and Trading Community, AD 1000–1600* (London: Routledge and Kegan Paul, 1973)
Ponting, Kenneth G. *The Woollen Industry of South-West England* (Bath: Adams & Dart, 1971)
Ponting, K.G. 'Cloth Finishing', pp. 170–84; 'The West of England Cloth Industry', pp. 235–41; both in *the Wool Textile Industry in Great Britain* ed. by J. Geraint Jenkins (London: Routledge & Kegan Paul, 1972),
Poos, L.R. *A Rural Society after the Black Death: Essex 1350–1525* (Cambridge: Cambridge University Press, 1991)
Postan, M. 'Some Economic Evidence of Declining Population in the Later Middle Ages' *Econ. Hist. Rev.*, 2nd ser., II (1950), 221–46
Postan, M. 'Economic Foundations of Medieval Society' in *Essays on Medieval Agriculture and General Problems of the Medieval Economy* (Cambridge: Cambridge University Press, 1973 (A))
Postan, M. *Medieval Trade and Finance* (Cambridge; Cambridge University Press, 1973 (B))
Pound, J.F. 'Rebellion and Poverty in early Sixteenth-Century Suffolk: The 1525 Uprising against the Amicable Grant' *PSIAH,* 39 (1999), 317–30
Power, Eileen *Medieval People* (London: Metheun, 1924; repr. 1925)
Power, Eileen 'The Wool Trade in the Fifteenth Century' in *Studies in English Trade in the Fifteenth Century* (ed.) by Eileen Power and M.M. Postan (London: Routledge, 1933), pp. 39–90
Power, Eileen *The Wool Trade in English Medieval History* (Oxford: Oxford University Press, 1941; repr. 1942, 1949, 1955, 1965)
Pugh, Ralph B. 'Salzman, Louis Francis (1878–1971)', rev. *DNB*, Oxford University Press, 2004 [http://www.oxforddnb.com/view/article/31652, accessed 22 March 2016]
Rackham, Oliver *The History of the Countryside* (London: Dent, 1986)
Ramsey, G.D. *The English Woollen Industry, 1500–1750* (Basingstoke: Macmillan for The Economic History Society, 1982)
Rawcliffe, C. '"That Kindliness Should be Cherished More, and Discord Driven Out": the Settlement of Commercial Disputes by Arbitration in Later Medieval England', in *Enterprise and Individuals in Fifteenth-Century England* ed. by Jennifer Kermode (Stroud: Alan Sutton, 1991), pp. 99–117
Rawcliffe, Carole *Urban Bodies: Communal Health in Late Medieval English Towns and Cities* (Woodbridge: Boydell, 2013)

Redstone, V.B. 'St Edmund's Bury and Town Rental for 1295' *PSIAH,* 13 (1908), 191–222
Redstone, V.B. 'Early Sudbury Clothiers' *PSIAH,* 14 (1910), 99–104
Richmond, C. *John Hopton: A Fifteenth-Century Suffolk Gentleman* (Cambridge: Cambridge University Press, 1981)
Rigby, Stephen H. 'Urban Population in Late Medieval England' *Econ. Hist. Rev.*, 2nd ser., 63 (2010), 391–417
Rogers, Alan *Noble Merchant: William Browne (c1410–1489) and Stamford in the Fifteenth Century* (Bury St Edmunds: Abramis Academic, 2012)
Rutledge, Elizabeth 'Medieval and Later Ports, Trade and Fishing up to 1600', in *An Historical Atlas of Norfolk* ed. by Peter Wade-Martins and Jane Everett (Ipswich: Norfolk Museums Service, 1993), pp. 78–9
Ryder M.L. 'The Origin of Spinning' *Textile History,* 1 (1968), 73–82
Ryder, M.L., 'The Wools of Britain', in *the Wool Textile Industry in Great Britain* ed. by J. Geraint Jenkins (London: Routledge & Kegan Paul, 1972), pp. 51–64
Ryder, M.L 'Medieval Sheep and Wool Types' *Ag. Hist. Rev.*, 32 (1984), 14–28
Ryder, M.L. 'Fleece Grading and Wool Sorting' *Textile History,* 26 (1995), 3–22
Salzmann, L.F. *English Industries of the Middle Ages* (London: Constable, 1913)
Scarfe, Norman 'Medieval and Later Markets', in *An Historical Atlas of Suffolk* ed. by David Dymond and Edward Martin (Ipswich: Suffolk County Council, 1988; repr. 1989 and 1999), pp. 76–7
Scarfe, Norman 'Medieval and Later Fairs', in *An Historical Atlas of Suffolk* ed. by David Dymond and Edward Martin (Ipswich: Suffolk County Council, 1988; repr. 1989 and 1999), pp. 78–9
Schofield, R.S. 'The Geographical Distribution of Wealth in England, 1334–1649' *Econ. Hist. Rev.*, 2nd ser. 18 (1965), 483–510
Scott Thomas, Gladys 'Three Suffolk Figures: Thomas Wolsey: Stephen Gardiner: Nicholas Bacon: a Study in Social History' *PSIAH,* 25 (1950), 149–63
Slade, Mrs Justice *Tolley's Employment Handbook 28th Edition 2014 (London: LexisNexis, 2014)*
Smedley, Norman 'Iron Age Weaving Combs from Cambridgeshire and Suffolk' *Proceedings of the Cambridge Antiquarian Society,* 54 (1960), 47–9
Smith, Adam *An Enquiry into the Nature and Causes of the Wealth of Nations* (Radford, Virginia: Wilder, 2000)
Smith, Richard M. 'Measuring Adult Mortality in an Age of Plague', in *Town and Countryside in the Age of the Black Death: Essays in Honour of John Hatcher* ed. by Mark Bailey and Stephen Rigby (Turnout, Belgium: Brepols, 2012), pp. 43–85
Spufford, Peter *Power and Profit: The Merchant in Medieval Europe* (London: Thames and Hudson, 2002)
Stenton, F.M. 'The Road System of Medieval England' *Econ. Hist. Rev.*, 1st ser. 7 (1936), 1–21
Stephens, W.B. (ed.) *A History of the County of Warwick: Volume 8, the City of Coventry and Borough of Warwick* (Oxford: Oxford University Press for the Victoria County History, 1969)
Stephenson, M.J. 'Wool Yields in the Medieval Economy' *Econ. Hist. Rev.*, 2nd ser. 41 (1988), 368–91
Stevens, Matthew Frank 'Londoners and the Court of Common Pleas in the Fifteenth Century', in *London and Beyond: Essays in Honour of Derek Keene* ed. by Matthew Davies and James A. Galloway (London: University of London, 2012), pp. 225–45
Stone, David 'The Productivity and Management of Sheep in Late Medieval England' *Ag. Hist. Rev.*, 32 (2003), 1–22
Stone, David *Decision-Making in Medieval Agriculture* (Oxford: Oxford University Press, 2005)
Sutton, Anne F. 'The Early Linen and Worsted Industry of Norfolk and the Evolution of the London Mercers Company' *Norfolk Archaeology*, 40 (1989) 201–25

Sutton, Anne F. *The Mercery of London: Trade, Goods and People, 1130–1578* (Aldershot: Ashgate Publishing, 2005)
Sutton, Anne F. 'London Mercers from Suffolk c. 1200–1570: Benefactors, Pirates and Merchant Adventurers' *PSIAH,* Part I, 42 (2009) 1–12; Part II, 42 (2010) 162–85
Swanson, Heather *Medieval Artisans* (Oxford: Basil Blackwell, 1989)
Tawney, R.H. *Religion and the Rise of Capitalism* (Middlesex: Penguin Books, 1938)
Thirsk, Joan *The Rural Economy of England* (London: Hambledon, 1984)
Thornton, Gladys A. *A History of Clare Suffolk* (Cambridge, W. Heffer, 1928)
Thrupp, Sylvia L. *The Merchant Class of Medieval London* (Chicago, 1948; repr. Ann Arbor: University of Michigan Press, 1962)
Todd, Hilary and David Dymond 'Population Densities, 1327 and 1524', in *An Historical Atlas of Suffolk* ed. by David Dymond and Edward Martin (Ipswich: Suffolk County Council, 1988; repr. 1989 and 1999), pp. 80–3
Unger, Richard W. 'Thresholds for Market Integration in the Low Countries and England in the Fifteenth Century', in *Money, Markets and Trade in Late Medieval Europe: Essays in Honour of John H. A. Munro,* ed. by Lawrin Armstrong, Ivana Elbl and Martin M. Elbl (Leiden: Brill, 2007), pp. 349–80
Unwin, George 'Woollen Cloth – The Old Draperies' in *Victoria County History, Suffolk,* ed. by William Page, 2 volumes (London: Archibald Constable, 1907), II, pp. 254–66
Van Bavel, Bas *Manors and Markets: Economy and Society in the Low Countries, 500–1600* (Oxford: Oxford University Press, 2010)
Van Bavel, Bas and Auke Rijpma 'How Important were Formalized Charity and Social Spending before the Rise of the Welfare State? A Long-run Analysis of Selected Western European Cases, 1480–1500' *Econ. Hist. Rev.*, 2nd ser. 69 (2016), 159–87
Veale, Elspeth M. *The English Fur Trade in the Later Middle Ages* (Oxford: Oxford University Press, 1966; repr. London: London Record Society, 38, 2003)
Whewell, C. S. 'Milling and Milling Machinery', in *the Wool Textile Industry in Great Britain* ed. by J. Geraint Jenkins (London: Routledge & Kegan Paul, 1972), pp. 157–69
Whittle, Jane *The Development of Agrarian Capitalism: Land and Labour in Norfolk 1440–1580* (Oxford: Oxford University Press, 2000)
Wild, M.T. 'The Yorkshire Wool Textile Industry', in *The Wool Textile Industry in Great Britain* ed. by J. Geraint Jenkins (London: Routledge & Kegan Paul, 1972), pp. 185–234
Wordie, J.R. 'The Chronology of English Enclosure, 1500–1914' *Econ. Hist. Rev.*, 2nd ser., 36 (1983), 483–505
Wymer, John 'Solid Geology', in *An Historical Atlas of Suffolk* ed. by David Dymond and Edward Martin (Ipswich: Suffolk County Council, 1988; repr. 1989 and 1999), pp. 16–7
Yates, Margaret 'Change and Continuities in Rural Society from the Later Middle Ages to the Sixteenth Century: The Contribution of West Berkshire' *Econ. Hist. Rev.*, 2nd ser. 52 (1999), 617–37
Yates, Margaret *Town and Countryside in Western Berkshire, c. 1327–c. 1600: Social and Economic Change* (Woodbridge: Boydell, 2007)
Zell, Michael *Industry in the countryside: Wealden society in the sixteenth century* (Cambridge: Cambridge University Press, 1994)

Unpublished Papers, Dissertations and Presentations

Munro, J. 'Distaff I – Focus on Fibre' presentation to International Medieval Congress: University of Leeds, 4 July 2013
Quinton, Eleanor Jane Powys 'The Drapers and the Drapery Trade of Late Medieval London, c.1300–c.1500' (PhD thesis, University of London, 2001)
Sear, Jo 'Late Medieval Consumption in East Anglian Market Towns', in The Economic History Society Annual Conference 2013 programme, pp. 133–7

Sear, Jo 'Consumption and Trade in East Anglian Market Towns and their Hinterlands in the Late Middle Ages' (PhD thesis, University of Cambridge, 2015)
Slavin, Philip 'The Beginning of the End: Sheep Panzootics and the Fortunes of the Wool Industry in England 1250–1330', presentation to Economic History Conference, 30 March 2014
Thirsk, J. 'Women's Initiatives in Early Modern England', Presentation to 'Women's Initiatives in early Modern England, 1500–1750' Conference, London, 4 June 1994
Wigmore, E. 'Long Melford in 1441/2' (unpublished paper)

On-line Sources

http://aalt.law.uh.edu [accessed 19 April 2016]
http://aalt.law.uh.edu/Indices/CP40Indices/CP40_Indices.html [accessed 19 April 2016]
http://adb.anu.edu.au [accessed 19 April 2016]
http://en.wikipedia.org [accessed 19 April 2016]
http://list.historicengland.org.uk [accessed 19 April 2016]
http://www.access.arch.cam.ac.uk [accessed 19 April 2016]
http://www.bartleby.com [accessed 19 April 2016]
http://www.bbc.co.uk/news/science-environment-24332239 [accessed 19 April 2016]
http://www.british-history.ac.uk [accessed 19 April 2016]
http://www.campop.geog.cam.ac.uk/research/projects/occupations/transport[accessed 19 April 2016]
http://www.econsoc.hist.cam.ac.uk/podcast-campbell.html [accessed 19 April 2016]
http://www.freemaptools.com/how-far-is-it-between.htm [accessed 19 April 2016]
http://www.history.ac.uk/cmh/gaz/gazweb2.html [accessed 19 April 2016]
http://www.historyofparliamentonline.org [accessed 19 April 2016]
http://www.londonroll.org [accessed 19 April 2016]
http://www.mapdevelopers.com/elevation_calculator.php [accessed 19 April 2016]
http://www.medieval-graffiti-suffolk.co.uk [accessed 19 April 2016]
http://www.nature.com [accessed 19 April 2016]
http://www.oed.com [accessed 22 March 2016]
http://www.oxforddnb.com [accessed 22 March 2016]
https://jacobweisdorf.files.wordpress.com/2012/04/unreal-wages_humphries-and-weisdorf_jan-2016.pdf [accessed 19 April 2016]
https://www.englandsimmigrants.com [accessed 19 April 2016]

Index

(Unless otherwise stated rivers, towns and villages are in Suffolk)